Smart Innovation, Systems and Technologies

Volume 70

Series editors

Robert James Howlett, Bournemouth University and KES International,
Shoreham-by-sea, UK
e-mail: rjhowlett@kesinternational.org

Lakhmi C. Jain, University of Canberra, Canberra, Australia;
Bournemouth University, UK;
KES International, UK
e-mail: jainlc2002@yahoo.co.uk; Lakhmi.Jain@canberra.edu.au

About this Series

The Smart Innovation, Systems and Technologies book series encompasses the topics of knowledge, intelligence, innovation and sustainability. The aim of the series is to make available a platform for the publication of books on all aspects of single and multi-disciplinary research on these themes in order to make the latest results available in a readily-accessible form. Volumes on interdisciplinary research combining two or more of these areas is particularly sought.

The series covers systems and paradigms that employ knowledge and intelligence in a broad sense. Its scope is systems having embedded knowledge and intelligence, which may be applied to the solution of world problems in industry, the environment and the community. It also focusses on the knowledge-transfer methodologies and innovation strategies employed to make this happen effectively. The combination of intelligent systems tools and a broad range of applications introduces a need for a synergy of disciplines from science, technology, business and the humanities. The series will include conference proceedings, edited collections, monographs, handbooks, reference books, and other relevant types of book in areas of science and technology where smart systems and technologies can offer innovative solutions.

High quality content is an essential feature for all book proposals accepted for the series. It is expected that editors of all accepted volumes will ensure that contributions are subjected to an appropriate level of reviewing process and adhere to KES quality principles.

More information about this series at http://www.springer.com/series/8767

Vladimir L. Uskov · Jeffrey P. Bakken
Robert J. Howlett · Lakhmi C. Jain
Editors

Smart Universities

Concepts, Systems, and Technologies

 Springer

Editors
Vladimir L. Uskov
Department of Computer Science
   and Information Systems
Bradley University
Peoria, IL
USA

Jeffrey P. Bakken
The Graduate School, Bradley University
Peoria, IL
USA

Robert J. Howlett
KES International
Shoreham-by-sea, West Sussex
UK

Lakhmi C. Jain
University of Canberra
Canberra, ACT
Australia

ISSN 2190-3018   ISSN 2190-3026   (electronic)
Smart Innovation, Systems and Technologies
ISBN 978-3-319-86628-4   ISBN 978-3-319-59454-5   (eBook)
DOI 10.1007/978-3-319-59454-5

Printed on acid-free paper

This Springer imprint is published by Springer Nature
The registered company is Springer International Publishing AG
The registered company address is: Gewerbestrasse 11, 6330 Cham, Switzerland

Foreword

For over a decade, the mission of KES International has been to provide a professional community, networking and publication opportunities for all those who work in knowledge-intensive subjects. At KES, we are passionate about the dissemination, transfer, sharing and brokerage of knowledge. The KES community consists of several thousand experts, scientists, academics, engineers, students and practitioners who participate in KES activities.

KES Conferences

For nearly 20 years, KES has run conferences in different countries of the world on leading edge topics:

- Intelligent Systems: including Intelligent Decision Technologies, Intelligent Interactive Multimedia Systems and Services, Agent and Multi-Agent Systems and Smart Education and e-Learning.
- Sustainable Technology: including Sustainability in Energy and Buildings, Smart Energy and Sustainable Design and Manufacturing.
- Innovation, Knowledge Transfer, Enterprise and Entrepreneurship: including Innovation and Knowledge Transfer and Innovation in Medicine and Healthcare.
- Digital Media: including Archiving Tomorrow and Innovation in Music.

KES Journals

KES edits a range of journals and serials on knowledge-intensive subjects:

- International Journal of Knowledge Based and Intelligent Engineering Systems.
- Intelligent Decision Technologies: an International Journal.

- InImpact: the Journal of Innovation Impact.
- Sustainability in Energy and Buildings: Research Advances.
- Advances in Smart Systems Research.

Book Series

KES edits the Springer book series on Smart Innovation, Systems and Technologies. The series accepts conference proceedings, edited books and research monographs. KES Transactions (published by Future Technology Press) is a book series containing the results of applied and theoretical research on a range of leading edge topics. Papers contained in KES Transactions may also appear in the KES Open Access Library (KOALA), our own online gold standard open access publishing platform.

Training and Short Courses

KES can provide live and online training courses on all the topics in its portfolio. KES has good relationships with leading universities and academics around the world and can harness these to provide excellent personal development and training courses.

Dissemination of Research Project Results

It is essential for research groups to communicate the outcomes of their research to those that can make use of them. But academics do not want to run their own conferences. KES has specialists and knowledge of how to run a conference to disseminate research results. Or a research project workshop can be run alongside a conference to increase dissemination to an even wider audience.

The KES-IKT Knowledge Alliance

KES works in partnership with the Institute of Knowledge Transfer (IKT), the sole accredited body dedicated to supporting and promoting the knowledge professional: those individuals involved in innovation, enterprise, and the transfer, sharing and exchange of knowledge. The IKT accredits the quality of innovation and knowledge transfer processes, practices activities, and training providers, and the professional status of its members.

About KES International

Formed in 2001, KES is an independent worldwide knowledge academy involving about 5,000 professionals, engineers, academics, students and managers, operated on a not-for-profit basis, from a base in the UK. A number of universities around the world contribute to its organisation, operation and academic activities. KES International Operations Ltd is a company limited by guarantee that services the KES International organisation.

Robert J. Howlett
Lakhmi C. Jain

Preface

Smart University is an emerging and rapidly growing area that represents a creative integration of smart technologies, smart features, smart software and hardware systems, smart pedagogy, smart curricula, smart learning and academic analytics, and various branches of computer science and computer engineering.

This is the main reason that in June of 2013, a group of enthusiastic and visionary scholars from all over the world arrived with the idea to organise a new professional event and community that would provide an excellent opportunity for faculty, scholars, Ph.D. students, administrators and practitioners to propose and evaluate innovative ideas and approaches, collaborate on design and development of new systems and technologies, and discuss findings and outcomes of research, development, case studies and best practices in Smart Education, Smart e-Learning, Smart University and related areas.

The research, design and development topics of our interest in those areas include but are not limited to (1) conceptual modelling of Smart Education, Smart e-Learning, Smart Universities, Smart Campuses, Smart Analytics; (2) infrastructure, main characteristics, functions and features of Smart Universities, Smart Campuses and Smart Classrooms; (3) Smart University-wide software, hardware, security, safety, communication, collaboration and management systems; (4) Smart Education and Smart e-Learning strategies, approaches and environments; (5) Smart Pedagogy; (6) Smart Learning Analytics and Smart Academic Analytics; (7) modelling of Smart Student/Learner and Smart Faculty; (8) assessment and quality assurance in Smart Education and Smart e-Learning; (9) professional development in Smart Education and Smart e-Learning; (10) social, cultural and ethical dimensions and challenges of Smart Education; and (11) educational applications of various innovative smart technologies such as Internet of Things, cloud computing, ambient intelligence, smart agents, sensors, wireless sensor networks and context awareness technology, and numerous other topics.

Since that initial meeting in 2013, the following books were published as the logical outcomes of a mutually beneficial collaboration between experts in Smart Education, Smart e-Learning, Smart University areas and the KES International professional association, including (1) Neves-Silva, R., Tshirintzis, G., Uskov, V.,

Howlett, R., Lakhmi, J. (Eds). Smart Digital Futures, IOS Press, Amsterdam, The Netherlands (2014); (2) Uskov, V.L., Howlett, R.J., Jain, L.C. (Eds). Smart Education and Smart e-Learning, Springer, Heidelberg, Germany (2015); (3) Uskov, V.L., Howlett, R.J., Jain, L.C. (Eds). Smart Education and e-Learning 2016, Springer, Heidelberg, Germany (2016); and (4) Uskov, V.L., Howlett, R.J., Jain, L.C. (Eds). Smart Education and e-Learning 2017, Springer, Heidelberg, Germany (in print). Those publications contain top quality contributions by individual researchers and research teams in Smart Education, Smart e-Learning, and Smart University areas.

This book serves as one more excellent example of fruitful collaboration within organised professional community of specialists in the Smart University area. It contains peer-reviewed contributions from researchers from nine countries including USA, UK, Australia, Japan, Russia, China, Norway, Romania and Denmark. Those contributions cover various topics such as conceptual modelling of Smart Universities, software systems and technologies for Smart Universities, software systems for students with disabilities at Smart Universities, best practices of building smarter college/university and Smart Campus, innovative approaches to and best practices of Smart Education, Smart Learning, long-life learning and professional development.

It is our sincere hope that this book will serve as a useful source of valuable data and information about current research and developments, and provide a solid foundation for further progress in Smart University, Smart Education, Smart e-Learning and related areas.

Peoria, USA	Prof. Vladimir L. Uskov, Ph.D.
Peoria, USA	Prof. Jeffrey P. Bakken, Ph.D.
Shoreham-by-sea, UK	Prof. Robert J. Howlett, Ph.D.
Canberra, Australia	Prof. Lakhmi C. Jain, Ph.D.
May 2017	

Contents

Part II   Smart Universities: Concepts, Systems and Technologies

About the Editors

Prof. Vladimir L. Uskov, Ph.D., is a Professor of Computer Science and Information Systems and Director of the InterLabs Research Institute, Bradley University (USA). Dr. Uskov obtained his Ph.D. degree from Moscow Aviation Institute—Technical University in 1986. His areas of expertise include (1) Software Engineering (1980–present), (2) Web-Based Education, Online Education and e-Learning (1995–present), (3) Applications of Information Technologies in Education (1985–present), (4) Software/IT Project Management (2004–present), and (5) Smart Universities, Smart Education, Smart Classrooms, and Smart e-Learning (2012 - present). Dr. Uskov has more than 325 professional publications, including 6 textbooks, 5 chapters and 51 published articles in referred journals in 3 different languages. Dr. Uskov received numerous grants in his areas of expertise from the National Science Foundation (NSF), Microsoft Corporation and various governmental institutions, including NSF grants ## 0420506, 0196015, 9950029 and 0002219. Dr. Uskov was a chair of the annual international conferences on Web-Based Education (WBE-2002–WBE-2013) and Computers and Advanced Technology in Education (CATE-2002–CATE-2013). From 2014, Dr. Uskov is a chair of the KES International's annual international conference on Smart Education and e-Learning (SEEL). In 1995–2016, Dr. Uskov worked as a visiting scholar and/or visiting professor in various universities of Japan, Germany, France, Italy, Holland and Norway. Dr. Uskov is a Senior Member of the IEEE (2004).

Dr. Jeffrey P. Bakken, Ph.D., is Professor and Associate Provost for Research and Dean of the Graduate School at Bradley University (USA) where he has held that position since June of 2012. He has a Bachelor's degree in Elementary Education from the University of Wisconsin-LaCrosse and graduate degrees in the area of Special Education-Learning Disabilities from Purdue University. Dr. Bakken has received the College of Education and the University Research Initiative Award, the College of Education Outstanding College Researcher Award, the College of Education Outstanding College Teacher Award and the Outstanding University Teacher Award from Illinois State University. His specific areas of interest include: response to intervention, collaboration, transition, teacher effectiveness, assessment, learning strategies, technology, Smart Classrooms and Smart Universities. He has written more than 155 academic publications that include books, chapters, journal articles, proceedings at international conferences, audio tapes, encyclopaedia articles, newsletter articles, book reviews, a monograph, a manual and one publisher Website. In reference to presentations made to others at conferences, he has made 230 presentations at International/National and Regional/State conferences, and received over one million dollars in external funding.

Prof. Robert J. Howlett, Ph.D., M.Phil., B.Sc. (Hons) is a member of the Institution of Engineering and Technology and the British Computer Society. He is a Chartered Engineer and a Chartered Information Technology Practitioner. Bob is the Executive Chair of the KES (Knowledge-Based and Intelligent Engineering Systems) International, an organisation dedicated to supporting and facilitating research, in three areas: intelligent systems; sustainability in energy and buildings; and innovation and knowledge transfer. Bob is a Director of the Institute of Knowledge Transfer. He is a nationally known figure in knowledge transfer and the UK Government Knowledge Transfer Partnerships (KTP) programme. His work on knowledge transfer and innovation has been recognised by his appointment as Visiting Professor, Enterprise, at Bournemouth University. He has supervised about 20 projects transferring university expertise to companies, mainly small to medium enterprises (SMEs), and has set up many more. He has a number of years experience of applying neural networks, expert systems, fuzzy paradigms and other intelligent techniques to industrial problem domains, e.g. sustainability; control, modelling and simulation of renewable energy systems; monitoring and control of internal combustion engines, particularly small engines for off-road and power generation applications; and a range of condition monitoring and fault diagnosis problems. He led a research team, funded by grants and industrial contracts. He has published widely on the subject and has presented invited talks, keynote addresses, etc. He is Editor-in-Chief of the International Journal of Knowledge-Based Intelligent Engineering Systems and Honorary Editor of Intelligent Decision Technologies: an International Journal. He supports a number of other journals through regional editorships and membership of advisory and review boards. He is a past and current member of the scientific committees of a number of conferences. He is on the editorial board of various book series. He has authored over 50 publications in refereed journals and conferences, and edited over 20 books. He has reviewed research project applications for the EPSRC and internationally as an Expert Evaluator under Framework and for other EU programmes.

Prof. Lakhmi C. Jain, Ph.D., B.E.(Hons), M.E., Fellow (Engineers Australia), serves as a Visiting Professor in Bournemouth University, United Kingdom and Adjunct Professor in the Faculty of Education, Science, Technology and Mathematics in the University of Canberra, Australia. KES International was initiated for providing a professional community the opportunities for publications, knowledge exchange, cooperation and teaming. Involving around 5,000 researchers drawn from universities and companies worldwide, KES facilitates international cooperation and generate synergy in teaching and research. KES regularly provides networking opportunities for professional community through one of the largest conferences of its kind in the area of KES. His interests focus on the artificial intelligence paradigms and their applications in complex systems, security, e-education, e-healthcare, unmanned air vehicles and intelligent systems.

Chapter 1
Innovations in Smart Universities

Vladimir L. Uskov, Jeffrey P. Bakken, Robert J. Howlett
and Lakhmi C. Jain

Abstract This chapter provides a brief overview of current innovative research in Smart Universities area, including projects on smart technologies, software/hardware systems for Smart Classrooms, and Smart Pedagogy.

Keywords Smart university · Smart technology · Smart systems · Smart pedagogy

1.1 Introduction

Fast proliferation of smart phones, smart devices, smart systems, and smart technologies provides academic institutions, students, faculty, professional staff, and administration with enormous opportunities in terms of new highly technological approaches to increase the quality of teaching strategies and learning outcomes. In addition, these technological advances provide remarkably effective management and administration of the main functions and services of colleges/universities.

V.L. Uskov (✉)
Department of Computer Science and Information Systems, and InterLabs Research Institute, Bradley University, Peoria, USA
e-mail: uskov@fsmail.bradley.edu; uskov@bradley.edu

J.P. Bakken
The Graduate School, Bradley University, Peoria, USA
e-mail: jbakken@fsmail.bradley.edu

R.J. Howlett
KES International, Leeds, UK
e-mail: rjhowlett@kesinternational.org

L.C. Jain
Bournemouth University, Poole, UK;
University of Canberra, Canberra, Australia
e-mail: Lakhmi.jain@canberra.edu.au; jainlakhmi@gmail.com

© Springer International Publishing AG 2018
V.L. Uskov et al. (eds.), *Smart Universities*, Smart Innovation,
Systems and Technologies 70, DOI 10.1007/978-3-319-59454-5_1

The concept of a 'Smart University' is an emerging and fast evolving area that represents the creative integration of innovative concepts, smart software and hardware systems, Smart Classrooms with state-of-the-art technologies and technical platforms, Smart Pedagogy based on modern teaching and learning strategies, Smart Learning Analytics and academic analytics, and various branches of computer science and computer engineering. The papers presented at the KES annual international conferences [1–4] on Smart Education and e-Learning (http://www.kesinternational.org/) clearly show that, in the near future, Smart University concepts, Smart Classrooms, and Smart Pedagogy will be actively deployed and used by leading academic institutions and training organizations around the world.

As Albert Einstein said, "We cannot solve our problems with the same thinking we used when we created them". We need to develop new conceptual models and identify unique features, systems and technologies for the next level of university's evolution—*a Smart University*. This is probably the main reason that multiple researchers and research teams around the world are actively working on the design, development, testing, and implementation of various innovative smart technologies, software and hardware systems, and smart devices on their university campuses, providing the ability for integration into the educational processes utilized. For example, research projects on applications of smart technology in Smart Classrooms and on Smart Universities campuses mainly focus on (1) Internet-of-Things technology, (2) cloud computing technology, (3) Radio Frequency Identification (RFID) technology, (4) ambient intelligence technology, (5) smart agents technology, (6) augmented and virtual reality technology, (7) remote (virtual) labs, (8) location and situation awareness technologies (indoor and outdoor), (9) Wireless Sensor Networking (WSN) technology, (10) sensor technology (motion, temperature, light, humidity, etc.), as well as many other types of emerging and advanced technologies.

The current design and development projects aimed to provide advanced software/hardware systems for Smart Classrooms usually are focused on systems of the following types: (1) Smart Learning/Teaching Analytics (big data analytics) systems, (2) serious games and systems for gamification of learning/training process, (3) context (situation) awareness systems (including geographic location awareness, learning context awareness, security/safety awareness systems), (4) automatic translation systems (from/to English language), (5) speaker/instructor tracking—smart cameraman—systems (in the Smart Classroom), (6) conferencing systems for smooth one-to-many and many-to-many audio/video interaction/communication/collaboration between local/in-classroom and remote/online students/learners, (7) gesture/emotion/activity recognition systems, (8) face recognition systems, (9) intelligent cyber-physical worlds' systems (for safety and security in buildings and on campus), (10) smart software agents and smart robotics, as well as other advanced software systems [5].

Most of the current research projects being done in the area of Smart Pedagogy are focused on various innovative technology-based student-centered learning and teaching approaches such as (1) learning-by-doing (including active use of virtual labs), (2) collaborative learning, (3) adaptive teaching, (4) serious games- and gamification-based learning, (5) flipped classrooms, (6) learning analytics and academic analytics, (7) context-based learning, (8) e-books, (9) personal enquiry based learning, (10) crossover learning, and other innovative strategies.

Despite the fact that multiple publications are available on the above-mentioned topics (most of those publications are listed in the References sections of chapters below, including references [6–8] of this chapter), we were not able to locate any published books on the topic of Smart Universities. As a result, we arrived with the idea that there should be a book with a well-thought collection of contributions—a book that covers concepts, models, smartness levels, features, components, systems, and technologies to be used by smart academic institutions. This is the main reason that we initiated the development of this pioneering book in the area of Smart Universities.

1.2   Chapters of the Book

This book includes 13 chapters grouped into 4 separate parts: Part 1—Smart Universities: Literature Review and Creative Analysis, Part 2—Smart Universities: Concepts, Systems and Technologies, Part 3—Smart Education: Approaches and Best Practices, and Part 4—Smart Universities: Smart Long Life Learning. A brief description of those chapters, each of which is a peer-reviewed contribution presented by various international research, design, and development teams, is given below.

Chapter 2 presents the outcomes of a systematic literature review and creative analysis of professional publications available in Smart Universities and related areas. The performed systematic creative analysis proved to be a useful approach to identify and briefly compare various professional publications with a wide range of proposed ideas and approaches, developed technical platforms, software and hardware systems, introduced smartness levels, best practices, etc. for Smart Universities. This could potentially help administrators, faculty, and professional staff at traditional universities to understand, identify, and evaluate various potential paths for a transformation of their university into a Smart University. This chapter describes the proposed and developed Smart Maturity Model for Smart Universities —a model that can be viewed broadly as Smart University level-by-level "smartness" evolution and improvement of its main functions.

Chapter 3 presents a comprehensive approach to conceptual modeling of Smart Universities. It is based on the modeling of smartness levels, smart software and hardware systems, smart technology, Smart Pedagogy, and several other distinctive features and components of a smart university. The obtained research findings and outcomes clearly show that main distinctive features, components, technologies and systems of a Smart University go well beyond those in a traditional university with predominantly face-to-face learning activities. This can be seen through examples such as the recommendation of multiple types of software systems for deployment at Smart Universities and in Smart Classrooms. Particularly, this chapter presents information about 100 + analyzed software systems that can support various activities at a smart university; it also provides recommendations for specific software systems that could benefit both faculty and students at a smart university.

Chapter 4 discusses one of the most distinctive features of a smart university. This is its ability of adaptation to and smooth accommodation of special students, i.e. students with various types of disabilities including physical, visual, hearing, speech, cognitive, and other types of impairments. This chapter presents the outcomes of systematic identification, analysis, and testing of available open source and commercial text-to-voice, voice-to-text, and gesture recognition software systems. These systems could significantly benefit students with disabilities attending a smart university. Particularly, this chapter presents information about 70 analyzed software systems and provides recommendations for text-to-voice, voice-to-text, and gesture recognition software systems for possible implementation at a smart university.

Chapter 5 describes an excellent practice to build a smarter college. This is a practical approach used by National Institute of Technology, Gifu College (NIT, Gifu College). The main idea is that, in a smarter college, various components should be interconnected to achieve the college's most effective functioning. These include hardware, software, systems, faculty development, information sharing, and institutional policies such as admission policy, curriculum policy, and diploma policy. The presented Active Learning practice at NIT, Gifu College is characterized by intensive use of educational systems, information and communication technologies, and equipment. From this point of view, Active Learning, as developed at NIT, Gifu College, has almost the same meaning as smart education.

Chapter 6 provides a brief overview of several modern technologies that emerged in the last few years, including Big Data, Internet of Things (IoT), Future Internet, and Crowdsourcing. A smart university could potentially be a proper place where all of these technologies could be examined and applied continuously as a sustainable evolution. Based on the performed analysis done of software systems and related aspects of this evolution, this chapter presents an open architecture for easily extensible services. These services are responsible for providing the users with value-added information; as a result, it is expected that they will increase the smartness level of university campus. Particularly, three directions were investigated for this reason: IoT to involve sensing, cloud computing and ubiquitous computing to make services available everywhere, and the transformation of service consumers to content generators or data producers and/or developers for new data source connections. Additionally, Big Data is considered as a forthcoming addition to these designated technologies to enhance the analytical capabilities of smart communities.

Chapter 7 introduces a novel framework for a smart virtual university hospital. The experiences to develop and test the solutions for training interprofessional team communication and collaboration are also described. The main premise behind this approach is that establishing a smart virtual university hospital mirroring a real life hospital can prepare students for direct patient contact such as practice placement and clinical rotation. This could optimize and sometimes also increase students' time on task. Additionally, such a virtual arena will support student learning by providing adaptive and flexible solutions for practicing a variety of clinical situations at the students' own pace. Based on multiple performed experiments and observed practices of student groups—medicine and nursing students, who worked in groups with the clinical scenarios in a virtual hospital using desktop PCs alone and with virtual reality goggles—it may be concluded that a smart virtual university hospital could be a feasible alternative for collaborative interprofessional learning.

Chapter 8 introduces an advanced approach to make mathematical education smarter on the basis of applying various information and communications technologies to mathematical education. It presents a developed innovative educational system—EdLeTS system—to support practical components of mathematical classes in various aspects. The distinctive features of this systems, such as support of personalization of education, the system's availability "at any time and from anywhere", and self-learning mode, are aimed at resolving some of the identified problems in math education.

Chapter 9 presents another progressive approach to make education smarter. In this case, it is based on a developed modular framework and an accompanying spiral design process that facilitates the design, development, and continued improvement of smarter serious games. The implementation of this framework has been systematically explored through an evolving serious game developed using the framework, which is a game focused on teaching precalculus at the college level.

Chapter 10 studies a challenging topic—the individual completion of programming exercises at higher education institutions. The problem is that, while some students can easily solve the programming problems independently, there are still many students who require a lot of additional time and/or efforts to solve the assigned programming problems. This chapter, in general, is focused on making initial computer science education smarter by developing a smart educational environment and by supporting instructors and teaching assistants in their provision of smart pedagogy for students. Particularly, this chapter presents the proposed novel support functions utilized to assess the learning conditions of a programming practicum. The goal of these functions is to reduce the burden on instructors by supporting the assessment of learning conditions in order to improve the quality of instruction.

Chapter 11 discusses the aspects of Smart Life-Long Education and continuous professional development. The developed approach is based on the idea that a smart university may facilitate self-regulated learning of learners through the introduction of a personal development e-Portfolio. This assists learners in planning their professional development path and reflecting upon their own learning outcomes. An implementation of the developed approach in the City University of Hong Kong shows a model combining training content in massive online open courses with clearly specified intended learning outcomes. This, combined with competence-based definitions and personalized training portfolios, may help professionals fulfill their continuous professional development and training requirements.

Chapter 12 describes an interesting case study of introducing smart technologies and approaches into the educational process at a traditional polytechnic university.

Particularly, it examines several techniques aimed at enhancing student motivation in terms of independent work and in relation to independent and lifelong learning. The following techniques are considered in this chapter: a "role reversal" methodology of education, e-learning and blended learning, involving the professional community into the educational process, and professional training in English. The applications of several innovative and smart techniques, such as training sessions, group teaching methods, role-playgames, the use of smart components, etc., are analyzed and tested with students majoring in Power Engineering. The chapter provides a lot of statistical data and findings relevant to this interesting case study.

Chapter 13 focuses on the meeting of Constructivism (as a learning theory) and Smart Learning. This, in turn, theorizes the idea of Smart Constructivist Learning with applications in smart learning environments. Relying on the phenomena of "meaning construction" and "meaningful understanding production" in the framework of smart constructivism, this chapter is focused on the analysis of Smart Constructivist Knowledge Building and analysis of Learning-and-Constructing-Together as a smart constructivist model. It is expected that the presented theory and models could support the development of innovative smart learning strategies.

1.3   Concluding Remarks

We believe this book will serve as a useful source of research data and findings, design and development outcomes, best practices, and case studies for those interested in the rapidly growing area of Smart Universities, including faculty, scholars, Ph.D. students, administrators, and practitioners. It is our sincere hope that this book will provide a foundation of further progress and inspiration for research projects and advanced developments in the area of Smart Universities.

References

1. Neves-Silva, R., Tshirintzis, G., Uskov, V., Howlett, R., Lakhmi, J. (eds.): Smart Digital Futures. IOS Press, Amsterdam, The Netherlands (2014)
2. Uskov, V.L., Howlett, R.J., Jain, L.C. (eds.): Smart Education and Smart e-Learning. Springer, pp. 3–14, June 2015, 514 p., ISBN: 978-3-319-19874-3 (2015)
3. Uskov, V.L., Howlet, R.J., Jain, L.C. (eds.): Smart Education and e-Learning 2016. Springer, 643 p., ISBN: 978–3-319-39689-7 (2016)
4. Uskov, V., Howlet, R. Jain, L. (eds.): Smart Education and e-Learning 2017. Springer (in print)
5. Uskov, V.L., Bakken, J.P., Pandey, A.: The ontology of next generation smart classrooms. In: Uskov, V.L., Howlett, R.J., Jain, L.C. (eds.) Smart Education and Smart e-Learning. Springer, pp. 3–14, June 2015, 514 p. ISBN: 978-3-319-19874-3 (2015)
6. Hwang, G.J.: Definition, framework and research issues of smart learning environments—a context-aware ubiquitous learning perspective. Smart Learn. Environ.: A Springer Open J. 1, 4 Springer (2014)

7. Tikhomirov, V., Dneprovskaya, N.: Development of strategy for smart University, 2015 Open Education Global international conference, Banff, Canada, April 22–24 (2015)
8. Uskov, V.L, Bakken, J.P., Pandey, A., Singh, U., Yalamanchili, M., Penumatsa, A.: Smart university taxonomy: features, components, systems. In: Uskov, V.L., Howlett, R.J., Jain, L.C. (eds.) Smart Education and e-Learning 2016. Springer, pp. 3–14, June 2016, 643 p. ISBN: 978-3-319-39689-7 (2016)

Part I
Smart Universities: Literature Review and Creative Analysis

Chapter 2
Smart University: Literature Review and Creative Analysis

Colleen Heinemann and Vladimir L. Uskov

Abstract Research, design, and development of smart universities, smart education, smart classrooms, smart learning environments, smart pedagogy, smart learning and academic analytics, and related topics became the main themes of various pioneering international and national events and projects, governmental and corporate initiatives, institutional agendas, and strategic plans. This paper presents the outcomes of an ongoing project aimed at a systematic literature review and creative analysis of professional publications available in those areas. The premise is that the outcomes of the performed systematic creative analysis will enable researchers to identify the most effective and well-thought ideas, approaches, developed software and hardware systems, technical platforms, smart features and smartness levels, and best practices for the next evolutionary generation of a university—a smart university. The presented Smart Maturity Model can be viewed as an evolutionary approach for a traditional university to progress to various levels of maturity of smart university.

Keywords Smart university · Literature review · Systematic approach · Creative analysis

2.1 Introduction

In several recent years, the ideas of smart education (SmE), smart university (SmU), smart classroom (SmC), smart learning environments (SLE), and related topics became the main themes of various pioneering international and national events and projects, governmental and corporate initiatives, institutional agendas, and strategic plans.

C. Heinemann · V.L. Uskov (✉)
Department of Computer Science and Information Systems, InterLabs Research Institute, Bradley University, Peoria, IL, USA
e-mail: uskov@bradley.edu; uskov@fsmail.bradley.edu

C. Heinemann
e-mail: cheinemann@mail.bradley.edu

V.L. Uskov et al. (eds.), *Smart Universities*, Smart Innovation,
Systems and Technologies 70, DOI 10.1007/978-3-319-59454-5_2

The KES International professional organization organizes annual conferences on Smart Education and e-Learning, beginning in 2014 [1]. The first annual international conference on Smart Learning Environments (ICSLE) was also held in 2014 [2]. Additionally, the Springer open international journal on Smart Learning Environments was launched in 2014 [3].

One of the most well-known national initiatives in SmE area is the Smart Education Initiative (SEI) in South Korea. "Smart education has been a topic of conversation in all Korean schools since June 29, 2011 when the Korean government announced its smart education initiative. … First, smart education calls for a new pedagogy. That is, the new pedagogy should not just deal with letters and numbers but also address sounds and images together with all other types of multimedia. Second, teachers and students as workforce have the same importance in classrooms. Third, a resource-enriched learning environment will be implemented in the form of content cloud where teachers and students can freely and safely upload and download open educational resources and content together" [4]. In general, it presents a paradigm shift from traditional education to SMART Education, that is an abbreviation of Self-directed, Motivated, Adaptive, Resource enriched, Technology embedded education (Fig. 2.1) [5].

"The IBM company has created the Smarter Educational Framework to demonstrate how it works in partnership with clients to support and enable—through the use of technology—smarter education establishments, making them

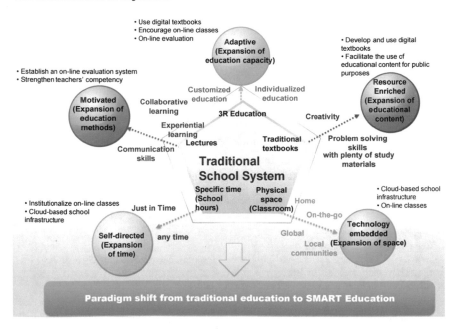

Fig. 2.1 Paradigm shift from traditional education to SMART education [5]

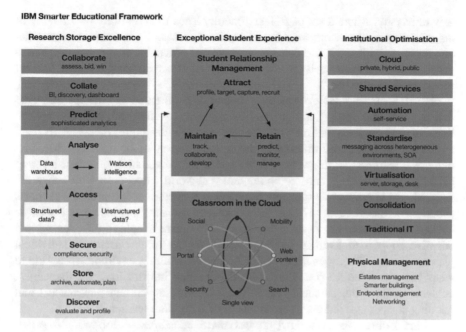

Fig. 2.2 IBM Smarter Educational Framework [6]

more efficient, productive and competitive. … Smarter education drives sustainable performance improvement by leveraging information to make better decisions, anticipate problems and resolve them proactively, and coordinate resources to operate effectively" (Fig. 2.2) [6].

The Smart School solution by the Samsung company has three core components: (1) the interactive management solution, (2) the learning management system, and (3) the student information system. Its multiple unique features and functions are targeted at smart school impact on education and benefits, including (1) increased interactivity, (2) personalized learning, (3) efficient classroom management, and (4) better student monitoring [7].

2.2   Literature Review: a Classic Approach

Recently, various creative researchers and developers began presenting their visions of SmU, SmC, SmE, SLE, smart campus, smart teacher, smart pedagogy, etc.; a brief summary of several remarkable publications on those topics—a classic literature review—is given below.

Smart University. Tikhomirov in [8] presented his vision of smart education as follows: "*Smart University* is a concept that involves a comprehensive modernization of all educational processes. … The *smart education* is able to provide a new university, where a set of ICT and faculty leads to an entirely new quality of the processes and outcomes of the educational, research, commercial and other university activities. ... The concept of *Smart* in education area entails the emergence of technologies such as smart boards, smart screens and wireless Internet access from everywhere".

Smart Learning Environments. Hwang [9] presented a concept of *smart learning environments* "... that can be regarded as the technology-supported learning environments that make adaptations and provide appropriate support (e.g., guidance, feedback, hints or tools) in the right places and at the right time based on individual learners' needs, which might be determined via analyzing their learning behaviors, performance and the online and real-world contexts in which they are situated. ... (1) A smart learning environment is context-aware; that is, the learner's situation or the contexts of the real-world environment in which the learner is located are sensed... (2) A smart learning environment is able to offer instant and adaptive support to learners by immediate analyses of the needs of individual learners from different perspectives... (3) A smart learning environment is able to adapt the user interface (i.e., the ways of presenting information) and the subject contents to meet the personal factors (e.g., learning styles and preferences) and learning status (e.g., learning performance) of individual learners".

Smart Education. IBM [10] defines *smart education* as follows: "A smart, multi-disciplinary student-centric education system—linked across schools, tertiary institutions and workforce training, using: (1) adaptive learning programs and learning portfolios for students, (2) collaborative technologies and digital learning resources for teachers and students, (3) computerized administration, monitoring and reporting to keep teachers in the classroom, (4) better information on our learners, (5) online learning resources for students everywhere".

Cocoli et al. [11] describes *smart education* as follows: "Education in a smart environment supported by smart technologies, making use of smart tools and smart devices, can be considered smart education..... In this respect, we observe that novel technologies have been widely adopted in schools and especially in universities, which, in many cases, exploit cloud and grid computing, Next Generation Network (NGN) services and portable devices, with advanced applications in highly interactive frameworks ... smart education is just the upper layer, though the most visible one, and other aspects must be considered such as: (1) communication; (2) social interaction; (3) transport; (4) management (administration and courses); (5) wellness (safety and health); (6) governance; (7) energy management; (8) data storage and delivery; (9) knowledge sharing; (10) IT infrastructure".

Smart Campus. Kwok [12] defines *intelligent campus* (*i-campus*) as "... a new paradigm of thinking pertaining to a holistic intelligent campus environment which encompasses at least, but not limited to, several themes of campus intelligence, such as holistic e-learning, social networking and communications for work collaboration, green and ICT sustainability with intelligent sensor management systems, protective and preventative health care, smart building management with automated security control and surveillance, and visible campus governance and reporting".

Xiao [13] envisions *smart campus* as follows: "*Smart campus* is the outcome of the application of integrating the cloud computing and the internet of things. ...The application framework of smart campus is a combination of IoT and cloud computing based on the high performance computing and internet".

Smart Teachers. Abueyalaman [14] argues that "A smart campus depends on an overarching strategy involving people, facilities, and ongoing faculty support as well as effective use of technology.... A smart campus deploys *smart teachers* and gives them smart tools and ongoing support to do their jobs while assessing their pedagogical effectiveness using smart evaluation forms".

Smart Learning Communities. Adamko et al. [15] describe features of smart learning community applications as follows: "... the requirements of the smart community applications are the following: (1) sensible—the environment is sensed by sensors; (2) connectable—networking devices bring the sensing information to the web; (3) accessible—the information is published on the web, and accessible to the users; (4) ubiquitous—the users can get access to the information through the web, but more importantly in mobile any time and any place; (5) sociable—a user can publish the information through his social network; (6) sharable—not just the data, but the object itself must be accessible and addressable; (7) visible/augmented —make the hidden information seen by retrofitting the physical environment".

Smart Classrooms. A detailed overview of first generation smart classrooms and requirements for second-generation smart classrooms is available in [16]. For example, Huang et al. in [17] proposed "... a SMART model of smart classroom which characterized by showing, manageable, accessible, interactive and testing. ... A smart classroom relates to the optimization of teaching content presentation, convenient access of learning resources, deeply interactivity of teaching and learning, contextual awareness and detection, classroom layout and management, etc." Pishva and Nishantha in [18] define a smart classroom as an intelligent classroom for teachers involved in distant education that enables teachers to use a real classroom type teaching approach to instruct distant students. "Smart classrooms integrate voice-recognition, computer-vision, and other technologies, collectively referred to as intelligent agents, to provide a tele-education experience similar to a traditional classroom experience" [18]. Glogoric et al. in [19] addressed the potential of using Internet-of-Things (IoT) technology to build a smart classroom. "Combining the IoT technology with social and behavioral analysis, an ordinary classroom can be transformed into a smart classroom that actively listens and analyzes voices, conversations, movements, behavior, etc., in order to reach a conclusion about the lecturers' presentation and listeners' satisfaction" [19].

Our vision of the next generation SmC is based on the idea that SmC—as a smart system—should implement and demonstrate significant maturity at various "smartness" levels such as (1) adaptation, (2) sensing (awareness), (3) inferring (logical reasoning), (4) self-learning, (5) anticipation, and (6) self-organization and re-structuring. The components of an SmC include, but are not limited to,

(a) hardware, smart devices, or equipment, (b) software systems, applications, and smart technologies, (c) various activities related to smart learning and teaching or Smart Pedagogy, and (d) systems for learning, teaching, and performing academic analytics in SmC—Smart Analytics systems [16, 20].

The performed classic review of the above-mentioned, as well as 100 + additional publications and reports relevant to SmU, SmE, SmC, and SLE areas, unfortunately, does not provide readers with a clear understanding of focus, scope, and important details of analyzed publications. Moreover, the classic literature review usually does not help readers to compare the proposed approaches, features, smartness levels, and details in those publications. This was the main reason that we proposed and used a systematic approach to literature review in SmU, SmE, SmC, and SLE areas, using a developed framework as described below.

2.3 Literature Review: Creative Analysis Based on Systematic Approach

In order to (a) overcome problems of a classic literature review approach and (b) be able to compare various publications in SmU, SmE, SmC, SLE and related areas, we developed a framework for systematic creative analysis of those publications. In the general case, this framework may contain multiple sections (rubrics) that may include, but are not limited to,

(1) general information about a publication (i.e. main topic or title of a publication, list of authors, publisher, year of publication, reference number, etc.);
(2) main topic(s);
(3) proposed idea or approach and main details (as described by authors);
(4) identified list of main features (or, functions) that are relevant to SmU or SmE concepts;
(5) identified smartness levels of SmU addressed by proposed approach, developed framework or system, etc.

The outcomes of performed systematic creative analysis for 8 selected publications are presented in Tables 2.1, 2.2, 2.3, 2.4, 2.5, 2.6, 2.7 and 2.8 below; however, the outcomes for additional 13 examples are available in Tables 2.11, 2.12, 2.13, 2.14, 2.15, 2.16, 2.17, 2.18, 2.19, 2.20, 2.20, 2.21, 2.22 and 2.23 in Appendix.

2.3.1   Approaches to Develop Smart Universities

Table 2.1 A smart University using RFID technology [27]

Item	Details
Main topic	Smart university—technology
Approach used	Smart university development approach is based on implementation and active utilization of RFID technology to enhance varying aspects such as employee/equipment tracking, security, etc.
Details	"…present how emerging technology of RFID can be used for building a smart university. … Several actors and assets should be tagged: • Each employee will be tagged using smart employee card • Each student will also be tagged having their unique ID i.e. roll numbers • Different office items will be tagged using RFID based labels • Each office, classroom, lab will be assigned a unique ID that will be stored in RFID reader unit RFID reader units are placed at strategic places as follows: • RFID reader unit will be placed next to the door of each room • Reader unit will also be placed at the University entrance and exit • University cafeteria and common rooms will be equipped with reader node • Labs and classrooms will also be having at-least one reader at the entrance and exit" [27]
Identified features	• Facial recognition system • Intelligent cyber-physical systems (for safety and security) • Location awareness technologies (indoor and outdoor) • Gesture (activity) recognition systems • Context (situation) awareness systems
Smartness levels addressed	• Sensing—by technology specifically used for identification and human sensing • Anticipation—keeps track of location awareness and intelligent cyber-physical systems for safety and security

Table 2.2 Smart University taxonomy: features, components, systems [20]

Item	Details
Main topic	Smart university—smartness levels, systems and technology, teaching strategies
Approach used	SmU development approach is based on the idea of next generation of Smart Class-room systems that should significantly emphasize not only software/hardware features but also "smart" features and functionality of smart systems. Therefore, next generation of smart classrooms should pay more attention to implementation of "smartness" maturity levels or "intelligence" levels, and abilities of various smart technologies

(continued)

Table 2.2 (continued)

Item	Details
Details	"The objectives of this project were to create a taxonomy of a smart university … and identify SmU's main (1) features, (2) components, and (3) systems that go well beyond those in a traditional university with predominantly face-to-face classes and learning activities. … The premise is that to-be-developed SmU taxonomy will (1) enable us to identify and predict most effective software, hardware, pedagogy, teaching/learning activities, services, etc. for the next generation of a university—smart university, and (2) help traditional universities to understand, identify and evaluate paths for a transformation into a smart university" [20]
Identified features	*Software systems to be deployed by SmU* • Web-lecturing systems (with video capturing and computer screen capturing functions) for learning content development pre-class activities • Smart classroom in-class activities recording systems • Smart cameraman software systems • Systems for seamless collaborative learning (of both local and remote students) in smart classroom and sharing learning content/documents • Collaborative Web-based audio/video one-to-one and many-to-many communication systems • Systems to host, join, form and evaluate group discussions (including both local and remote students) • Systems to replay automatically recorded class activities and lectures for post-class review and activities (by both local and remote students) • Repositories of digital learning content and online (Web) resources, learning portals • Smart learning analytics and smart teaching analytics systems • Speaker/instructor motion tracking systems • Speech/voice recognition systems • Speech-to-text systems • Text-to-voice synthesis systems • Face recognition systems • Emotion recognition systems • Gesture (activity) recognition systems • Context (situation) awareness systems • Automatic translation systems (from/to English language) • Intelligent cyber-physical systems (for safety and security) • Various smart software agents • Power/light/HVAC consumption monitoring system(s) *Technologies to be deployed by SmU* • Internet-of-Things technology • Cloud computing technology • Web-lecturing technology • Collaborative and communication technologies • Ambient intelligence technology • Smart agents technology • Smart data visualization technology • Augmented and virtual reality technology • Computer gaming (serious gaming) technology • Remote (virtual) labs • 3D visualization technology

(continued)

Table 2.2 (continued)

Item	Details
	• Wireless sensor networking technology • RFID (radio frequency identification) technology • Location awareness technologies (indoor and outdoor) • Sensor technology (motion, temperature, light, humidity, etc.)
	Hardware systems to deployed by SmU • Panoramic video cameras • Ceiling-mounted projectors (in some cases, 3D projectors) • SMART boards and/or interactive white boards • Smart pointing devices • Controlled and self-activated microphones and speakers • Interconnected big screen monitors or TVs ("smart learning cave") • Interconnected laptops or desktop computers • Smart card readers • Biometric-based access control devices • Robotic controllers and actuators
	Smart curricula to be deployed by SmU • Adaptive programs of study—major and minor programs, concentration and certificate programs—with variable structures adaptable to types of students/learners, smart pedagogy, etc. • Adaptive courses, lessons and learning modules with variable components and structure suitable for various types of teaching—face-to-face, blended, online, types of students/learners, smart pedagogy, etc.
	Smart pedagogy to be deployed by SmU • Learning-by-doing (including active use of virtual labs) • Collaborative learning • e-books • Learning analytics • Adaptive teaching • Student-generated learning content • Serious games- and gamification-based learning • Flipped classroom • Project-based learning • Bring-Your-Own-Device • Smart robots (robotics) based learning
Smartness levels addressed	• Adaptation • Sensing (awareness) • Inferring (logical reasoning) • Self-learning • Anticipation • Self-organization and restructuring

The outcomes of creative analysis of several other publications relevant to current developments in smart universities are available in Tables 2.11, 2.12, 2.13 and 2.14 in Appendix.

2.3.2   Approaches to Develop Smart Campuses

Table 2.3 An ontology-based framework for model movement on a smart campus [26]

Item	Details
Main topic	Smart campus—concepts
Approach used	Smart campus development approach is based on ontology-based framework as a way to represent, analyze, and visualize human mobility and movement, specifically when it relates to scheduling activities
Details	"This work introduces an ontology-based framework for modeling, analyzing, and visualizing human movement associated with scheduled activities through integrates semantic web technologies" [26]
Identified features	• Smart learning/teaching analytics (big data analytics) systems • 3D visualization technology • Smart data visualization technology • Speaker/instructor motion tracking systems
Smartness levels addressed	• Sensing—by technology specifically used for modeling, analyzing, and visualizing human movement • Inferring—analyzing the information gathered through the sensors and sensing systems

Table 2.4 Constructing smart campus based on cloud computing platform and internet of things [31]

Item	Details
Main topic	Smart campus—technology
Approach used	Smart campus development approach is based on a higher stage of education information systems that connect everything through RFID technology, sensors, and the Internet of Things technology
Details	"…smart campus includes portal architecture, management and service, smart management, infrastructure, etc. Smart campus system integrates hardware device of digital school, and cloud storage as the means of data storage is applied…" [31]
Identified features	• Internet-of-Things technology • Systems for seamless collaborative learning (of both local and remote students) in smart classroom and shared learning content/documents • Repositories of digital learning content and online (web) resources, learning portals
Smartness levels addressed	• Adaptation—of classroom model • Sensing—by technology specifically used for identification

The outcomes of creative analysis of several other publications relevant to current developments in smart campuses are available in Tables 2.15, 2.16, 2.17 and 2.18 in Appendix.

2.3.3 Approaches to Develop Smart Learning Environments

Table 2.5 Conditions for effective smart learning environments [21]

Item	Details
Main topic	Smart learning environment—technology
Approach used	SLE development approach is based on an idea of setting of a set of varying levels of physical and digital locations through which a student can learn through context awareness
Details	"… Human Learning Interfaces (HLIs) that can facilitate the research and development of SLEs…" [21]
Identified features	• Augmented and virtual reality technology • Various smart software agents • Collaborative Web-based audio/visual one-to-one and many-to-many communication systems • Systems to host, join, form and evaluate group discussions (including both local and remote students) • Learning analytics
Smartness levels addressed	• Adaptation —of learning methods and classroom models • Self-learning—active use of innovative hardware such as Web-lecturing systems

The outcomes of creative analysis of an additional publication relevant to current developments in smart learning environments are available in Table 2.19 in Appendix.

2.3.4 Approaches to Develop Smart Classrooms

Table 2.6 Incorporating a smart classroom 2.0 Speech-Driven PowerPoint System (SDPPT) into University teaching [34]

Item	Details
Main topic	Smart classroom—systems and technology
Approach used	SMC development approach is based on idea of having a classroom equipped with networked computers and audiovisual devices to allow instructors to teach students in remote locations as well as physically in the classroom
Details	"The newly developed SDPPT system utilized voice recognition technology to identify certain keywords as they were spoke and the system then automatically responded by presenting the corresponding PowerPoint slides on the overhead screen" [34]
Identified features	• Face recognition system • Repositories of digital learning content and online (web) resources, learning portals • Speech/voice recognition system • Augmented and virtual reality technology • SMART boards and/or interactive white boards • Adaptive programs of study—major and minor programs, concentration and certificate programs—with variable structures adaptable to types of students/learners, smart pedagogy, etc. • Systems to host, join, form and evaluate group discussions (including both local and remote students)
Smartness levels addressed	• Adaptation—of teaching style • Sensing—by technology specifically used for identification and sensing ability • Inferring—based on Power Point systems information gathered

Table 2.7 The ontology of next generation smart classrooms [16]

Item	Details
Main topic	Smart classroom—smartness levels, systems and technology, teaching strategies
Approach used	SmC development approach is based on the idea of next generation of Smart Class-room systems that should significantly emphasize not only software/hardware features but also "smart" features and functionality of smart systems. Therefore, next generation of smart classrooms should pay more attention to implementation of "smartness" maturity levels or "intelligence" levels, and abilities of various smart technologies

(continued)

Table 2.7 (continued)

Item	Details
Details	"… the next generation of smart classroom systems should significantly emphasize not only software/hardware features but also "smart" features and functionality of smart systems … Therefore, next generation of smart classrooms should pay more attention to implementation of "smartness" maturity levels and abilities of various smart technologies. … The main goal of next generation smart classroom systems is to demonstrate significant maturity at various "smartness" levels, including (1) adaptation, (2) sensing (awareness), (3) inferring (logical reasoning), (4) self-learning, (5) anticipation, and (6) self-organizations and restructuring…" [16]
Identified features	*SmC hardware/equipment (with relevance to various types of users)* **Common for entire SmC:** • Array of video cameras installed to capture main classroom activities, movements, discussions, expressions, gestures, etc. • Ceiling-mounted projector(s) with 1 or 2 big size screen to display main activities in actual classroom; in some cases—3D projectors • Student boards (big screen displays or TV) to display images of remote/online students from different locations • One or many (depending on class size, number of remote students, learning needs and workload) hidden computer systems to actually run the software and components of the Smart Classroom system • Bluetooth and Internet enabled devices like cell phones, smart phones, PDAs and laptops to facilitate communication and information/data/notes exchange • Network equipment (for example, Wi-Fi routers, zig bee transceivers, infrared, RFID readers and tags) to facilitate authorization and other forms of inter- device secure and reliable communication • Access to the Internet (mobile Web) • Wireless sensor network • Sensors (location detection, voice detection, motion sensors, thermal sensors, humidity, sensors for facial and voice recognition, etc.) • Robotic controllers and actuators to perform functions like intensity control, temperature control, movement, etc. • Devices: context aware devices, virtual mouse, biometric based login devices, automated zoom-in devices • Controlled and self-activated microphones(s) for instructor and students • Various type of speakers • Various types of lights **Specific for instructor:** • Instructor's tablet PC (to write formulas, equations, run PPT presentations, • video and audio clips, etc. in real time) • Big size smart board (to write formulas, equations, etc. in real time) • Document camera (connected to projector) **Specific for local (in-classroom) student/learner:** • Array of mobile devices: smart interconnected mobile devices—smart phones, PDAs, laptops, smart headphones, etc. • (In some cases only): 3D goggles **Specific for remote/online student/learner:** • Desktop or tablet PC or laptop with connected or built-in microphone, speakers • Access to the broadband Internet

(continued)

Table 2.7 (continued)

Item	Details
	SmC software systems (with relevance to various types of users) **Common for entire SmC:** • Agent-based systems to enable various types of communication between devices in the Smart Classroom system • Learning management system (LMS) or access to university wide LMS • Advanced software for rich multimedia streaming, control and processing • Software systems to address needs of special students, for example, visually impaired students (speech and gesture based writing/editing/navigation and accessibility tools to facilitate reading and understanding) • Smart cameraman software (for panoramic cameras) • Recognition software: face, voice, gesture • Motion or hand motion stabilizing software • Noise cancellation software • Security system for a secure log-in and log-out of registered student • Implementation of Internet-of-Things technology • Implementation of elements of various emerging technologies (for example, Smart Environments, Ambient Intelligence, Smart Agents) **Specific to instructor:** • Smart drawing tools (for example, Laser2cursor) for drawing on smart boards, navigating and giving remote students floor to speak • Situation and/or context aware analytical system (that may generate hints and/or recommendations to instructor) • Analytical systems to analyze and rank class performance and outcomes • Systems to analyze presence, attendance, etc. **Specific for local (in-classroom) student/learner:** • Smart notebook/laptop/tablet PC software • Main office software applications • Same view and smart view software Systems to analyze presence, attendance, etc. **Specific for remote/online student/learner:** • Remote client programs to facilitate remote learning • Main office software applications • Software systems to analyze presence, attendance, etc. *SmC types of teaching/learning strategies to be deployed* **Common for entire SmC:** • Smart classroom pedagogy (or, smart technology based teaching) • Learning-by-doing • Collaborative learning • Project-based learning • Advanced technology-based learning • e-Learning pedagogy • Games-based learning and pedagogy • Flipped classroom pedagogy
Smartness levels addressed	• Adaptation • Sensing (awareness) • Inferring (logical reasoning) • Self-learning • Anticipation • Self-organizations and restructuring

The outcomes of creative analysis of several other publications relevant to current developments in smart classrooms are available in Tables 2.20, 2.21 and 2.22 in Appendix.

2.3.5   Approaches to Develop Smart Education or Smart Learning

Table 2.8 Smart approach to innovated education for 21st century [22]

Item	Details
Main topic	Smart education—concepts
Approach used	SmE development is based on an educational concept that does not simply rely on technological education, but is also self-directed, motivated, adaptive, resource-enriched, as well as technology-embedded education
Details	"Why School Education Should be Innovated? • Address desynchronization issues between students and education systems • Quality of education outcomes emerges a key issue • New quality frameworks pay more attention to evidence-based planning rather than examination-dominated assessment • Education systems should be reformed to accommodate the behavior and characteristics of digital native students: what, and how to educate students • Leveraging technology is vital factor to reform education system: mobile network, Learning Analytics, OER, OCW, MOOCs, open platforms. • Schools and classrooms should be reformed to accommodate changes in education environment: smart school, future schools, restructuring classroom settings" [22]
Identified features	• Flipped classroom teaching/learning strategy • Repositories of digital learning content and online (web) resources, learning portals • Internet of Things technology • RFID (radio frequency identification) technology • System for seamless collaborative learning (of both local and remote students) in smart classroom and shared learning content/documents • E-Books • Interconnected laptops or desktop computers
Smartness levels addressed	• Adaptation—of classroom models • Sensing—by technology specifically used for identification and sensing ability • Self-learning—active use of innovative hardware such as Web-lecturing system

2.4 Smartness Levels of Smart Universities

We believe that SmU as a smart system should significantly emphasize, not only state-of-the-art software/hardware features, emerging technology and innovative technical platforms, pioneering teaching and learning strategies, but also "smart" features of smart systems [20]. Therefore, the designers of SmU should pay more attention to the maturity of smart features of SmU that may occur on various levels of SmU's smartness, or smartness levels, such as adaptation, sensing, inferring, self-learning, anticipation, and self-organization.

The performed creative analysis of designated publications enabled us to identify sets of smartness levels addressed in those publications. These sets are presented in Tables 2.1, 2.2, 2.3, 2.4, 2.5, 2.6, 2.7 and 2.8 in Sect. 2.3 and also in Tables 2.11, 2.12, 2.13, 2.14, 2.15, 2.16, 2.17, 2.18, 2.19, 2.20, 2.20, 2.21, 2.22 and 2.23 in Appendix. Additionally, we present a brief summary of examples of addressing SmU smartness levels in various publications (Table 2.9).

Table 2.9 Context awareness systems for smart university: A list of most important features in existing systems

Smartness level	Target audience	Details	Ref. # or our analysis
Adaptation	Students	SmC should be able to adapt to various types of students, for example, students with disabilities	[20]
	Faculty and students	The course content used by the faculty can be posted on the website and content could be available to different users in different formats, making it adaptable to every user	[15]
	Faculty	Smart Pedagogy should help faculty with adaptation to new style of learning and/or teaching (learning-by-doing, flipped classrooms, etc.) and/or courses (MOOCs, SPOCs, open education and/or life-long learning for retirees, etc.)	[20]
	Professional staff	University network should be able to adapt to new technical platforms and mobile devices (mobile networking, tablets, mobile devices with iOS and Android operating systems, etc.)	Our analysis

(continued)

Table 2.9 (continued)

Smartness level	Target audience	Details	Ref. # or our analysis
Sensing	Students and faculty	One of the most sensible options is to integrate the attendance monitoring and recording processes into an existing access control system, thereby optimizing value and minimizing cost, disruption, and system duplication	[23]
	Faculty	Local information services connect to the sensors and provide their measurements to the instructor	[15]
	Professional staff	A single smart card solution can work both with wired and wireless locks installed on secure storage receptacles such as cabinets. This locks can be connected to the building's online access control system for near online and near real-time control. Cards also can be used to protect and monitor higher-value resources	[23]
	University administrators	The smart card can be used as a credit/debit card. This allows the card to be used effectively in e-commerce applications. Smart cards greatly improve the convenience and security of any transaction	[27]
Inferring	Students	When a student enters a room, his unique ID is sent to the control circuit and the central attendance server. The person is identified and a log is made according to the time and location. If the person identified has a valid ID, the control circuit performs the room automation function as described in its profile	[27]
		Contactless smart card technology provides the necessary security, not only to get through entrances, etc., but also to make meal purchases, pay for laundry, printing, copying, and additional campus services. Students can check out books automatically, with all of their information recorded and loaded into the student database. This provides real-time information on the status of overdue books	[23]
	Faculty	Faculty can use data pulled from sensors and devices to sense that a student is not grasping a concept, a student is bored, etc.	[23]
		RFID technology provides faculty/instructors with access to the labs and classrooms	[27]
	Technical staff	All lighting and parking systems at the university become a part of the Internet of Things technology. The smart campus system controls manage the lights, intelligent parking assistance, etc.	[23]
	University administrators	The smart classroom's ability to sense a student's gestures or facial expressions helps bolster the university's reputation because students can get help from instructors on a more personal level to ensure that everyone understands the material presented	[23]
		University administrators can gather information regarding where students spend their time, what they purchase, etc. based on information gathered from the student's smart card in order to optimize the utilization of the university	[26]

(continued)

Table 2.9 (continued)

Smartness level	Target audience	Details	Ref. # or our analysis
Anticipation	Professional staff	Issuing a smart ID card to all faculty, students, staff, and visitors allows security departments and technical staff to keep better tabs on what resources will be necessary and how much of each will be necessary based on how many people there are	[23]
		Data regarding how many people are in a classroom, how many parking spaces are available, etc. can be gathered in order to plan ahead to let everyone know that the parking lot is full, the classroom is full, etc. RFID-enabled cards present the optimal solution for planning for visitors, having resources ready, etc.	[27]
	University administrators	Access to data stored on the card allows the university administrators to analyze the data and anticipate necessary resources, funding, etc. for projects or classes	[27]
		Seeing the data that tells where students spend their time, what resources they use most, etc. allows university administrators to anticipate what resources need to be repurchased or reallocated	[23]
		Enrollment Management System to predict, anticipate, and control variations on student enrollment	[20]
		Campus-wide Safety System (CSS) to anticipate, recognize and act accordingly in case of various events on campus	[20]
Learning	Students	Gesture recognition systems and facial recognition systems can gather data regarding if a student understands a concept or is having trouble with a concept. This data can be interpreted by the system to show what percentage of the time, for example, the student is understanding the content	Our analysis
	Faculty	The online repositories of resources can determine what resources the students spend the most time with and request that the instructor provide additional resources of a similar nature. The collected data gives the ability to do statistical investigation from a pedagogical point of view	[15]
	Professional staff	Based on how sensors are set up in the classroom or building, it may take the system some time to learn where other sensors are in order to develop the optimal path for sending or receiving information	[27]
		The databases are not only collecting large volumes of data but are also learning to understand the relationships between entities by mapping their connection to the system	[15]
	University administrators	Learning from anonymous Opinion Mining System (OMS)	[20]
		Learning from different types of classes—MOOCs, blended, online, SPOCs, etc.	[20]

(continued)

Table 2.9 (continued)

Smartness level	Target audience	Details	Ref. # or our analysis
Self-organization	Students	Sensors determine whether or not a student is in a classroom, the lights are on, etc. and readjust accordingly. This may need to be changed numerous times depending on where students are or how long they are there	Our analysis
	Faculty	Automatic re-configuration of SmC systems, performance parameters, sensors, actuators and features in a smart classroom in accordance with instructor's profile	[20]
		Faculty can walk around classrooms freely and allow gesture recognition systems or motion detection systems to reorganize and ensure that the instructor is always in the camera view; it is done automatically	Our analysis
	Professional staff	Automatic re-configuration of wireless sensor network (WSN) because nodes may join or leave spontaneously (i.e. evolving network typology), university-wide cloud computing (with multiple clients and services), etc.	[20]
		Sensors determine whether or not someone is in a classroom, the lights are on, etc. and readjust accordingly	Our analysis

2.5   Smart University: Smartness Maturity Model

The performed systematic creative analysis of publications in SmU, SmC, SmE, and SLE areas clearly shows that various research, design, and development teams, as well as colleges and universities, addressed different levels of college/university maturity as a smart college/university. For example, one extreme case (SmU's initial level of maturity in terms of SmE) may deal with the following situation: a college/university purchased a SMART board for a classroom and, therefore, may say that they already developed or implemented smart education. The other case may be as follows: a university already has multiple smart classrooms that are equipped with various software and hardware systems to provide smart education and smart pedagogy; however, that university may not have university-wide policies and strategies regarding SmE yet.

In order to recognize and classify the current status of SmE education development in a college/university or status of evolution of a traditional university towards SmU, we proposed and developed the Smart Maturity Model (SMM) for SmU. This model was inspired by the CMMI (Capability Maturity Model Integrated) model [43, 44] that is actively used in the software development industry for improving the quality of software engineering process and final product —software systems.

Definition. Smart Maturity Model is a methodology used to design, develop, and continuously improve a smart university's main business functions such as

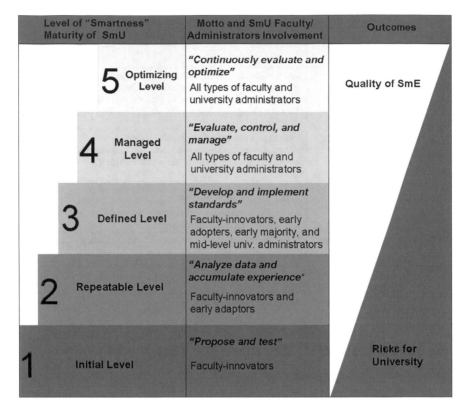

Level of "Smartness" Maturity of SmU	Motto and SmU Faculty/ Administrators Involvement	Outcomes
5 Optimizing Level	"Continuously evaluate and optimize" All types of faculty and university administrators	Quality of SmE
4 Managed Level	"Evaluate, control, and manage" All types of faculty and university administrators	
3 Defined Level	"Develop and implement standards" Faculty-innovators, early adopters, early majority, and mid-level univ. administrators	
2 Repeatable Level	"Analyze data and accumulate experience" Faculty-innovators and early adaptors	
1 Initial Level	"Propose and test" Faculty-innovators	Risks for University

Fig. 2.3 The proposed smartness maturity model for SmU: level-based hierarchical structure (inspired by CMM Irepresentation in [45])

education, teaching, learning, research, services, enrollment, management, administration, control, security, safety, etc.

SMM broadly can be viewed as SmU evolution and/or improvement approach to progress level-by-level from traditional university status to SmU status.

Levels. SMM defines 5 levels of university "smartness" maturity in terms of a university's readiness to implement SmE, create and actively use multiple SmC, involve various types of faculty to learn and use innovative technology in teaching, implement and use smart pedagogy, etc.

In terms of a university's maturity of implementation and use of smart education on campus, creation of smart classrooms, use of smart pedagogy, deployment and use of smart technology, smart analytics on campus, etc., the SMM for SmU (Fig. 2.3 and Table 2.10) uses the following levels of "smartness" maturity: (1) initial level (lowest level of maturity), (2) repeatable level, (3) defined level, (4) managed level, and (5) optimizing level (highest level of maturity). If a university does not use smart devices or SMART board(s) in classrooms or smart technologies and systems on campus at all, then this model assumes that the "smartness" maturity level of that university in terms of SmE is equal to zero.

Table 2.10 The proposed Smartness Maturity Model for SmU: levels of "smartness" maturity and examples

Level # and name	*Motto* and types of faculty/admins involved	Examples of main activities relevant to research, design and development of SmU, SmE, SmC, SLE, Smart Pedagogy, Smart Analytics, etc.	Examples (Ref. #)
1 Initial level (lowest level of maturity)	*"Propose-and-test"* Faculty-innovators (2–3% of faculty)	• Propose innovative ideas/approaches and test them (for example, use of interconnected mobile devices in classrooms, or use of flipped classroom learning strategy); • Propose new type of learning activity and test it in SmC (for example, perform experiments with a joint work and collaboration of in-classroom and remote/online students when they work on joint course project—collaborative learning) • Perform stand-alone experiments and testing with smart devices in teaching/learning (for example, use of a single SMART Board in a classroom or use of just one classroom on campus) • Process testing data and get information • Compare obtained outcomes with current practices (for example, compare *learning-by-doing* approach with *learning-by-listening* approach in education)	[13, 14, 18, 19, 23, 25, 26, 30–33]
2 Repeatable level	*"Accumulate and analyze"* Faculty-innovators and early adopters (13–15% of faculty)	• Repeat proposed and best practices for different types if students/learners, in different locations and setups, for different majors (for example, use of the same smart classroom by Computer Science majors and Communication majors or creation of multiple smart classrooms on campus) • Measure and creatively analyze obtained outcomes • Generalize accumulated experience/findings/outcomes/best practices (internal and external), get information, and make conclusions • Identify user requirements for software, hardware, technology, teaching and learning styles, etc. (for example, faculty requirements for teaching Programming classes in a smart classroom for in-classroom and remote/online student)	[2, 12, 15–17]

(continued)

Table 2.10 (continued)

Level # and name	*Motto* and types of faculty/admins involved	Examples of main activities relevant to research, design and development of SmU, SmE, SmC, SLE, Smart Pedagogy, Smart Analytics, etc.	Examples (Ref. #)
3 Defined level	"*Develop and implement standards*" Faculty-innovators, early adopters and early majority (about 30-35% of all faculty) and mid-level university administrators	• Develop standards at SmU in terms of smart education, smart teaching, smart learning, smart pedagogy • Identify standard sets of required software and hardware systems, technology for "standard' smart classroom on campus • Develop standards for smart education, software and hardware systems and smart technologies to be used by various types of students (including students with disabilities) and various types of faculty (for example, faculty with different background in SmE and experience) • Create multiple SmC on campus, create smart campus • Create and implement faculty development programs in SmE and SmC	[5, 22, 36]
4 Managed level	"*Evaluate, control, and manage*" All types of faculty, including late majority (30-35%) and laggards (about 15%) and upper-level university administrators	• Develop university policies on SmE, smart teaching, smart leaning, active use of SmC, etc. for all types of faculty and students • Identify well-defined quantitative indicators of SmE effectiveness (including tangible and intangible benefits) • Active (if necessary—mandatory) faculty development of all faculty groups in SmE, smart pedagogy, SmC software and hardware systems areas	
5 Optimizing level (highest level of maturity)	"*Continuously evaluate and optimize*" All types of faculty and university administrators	• Continuous evaluation of current SmE outcomes, smart teaching, smart learning, smart pedagogy, etc. and continuous comparative "expected vs actual outcomes" analysis • Causal analysis and resolution, correction and/or optimization of identified drawbacks or weaknesses • Continuous implementations of new well-tested systems, hardware, technologies, smart pedagogy for SmU, SmE and SmC • Continuous improvement of SmU main business functions • Continuous improvement of SmU management and administration	

Based on our obtained experience in SmC and SmE, therelations between (a) highly-technological SmE, SmC, and SLE (asinnovative technology-based approaches to education) and (b) various types of college/university faculty may be described by Roger's "Diffusion of Innovation" typology (Fig. 2.4) [35]. As a result, specific groups of faculty should be involved in SmE and SmC only on designated levels of SMM to optimize the "cost-benefits" ratio.

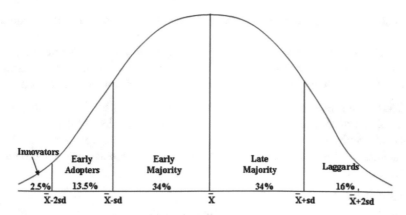

Fig. 2.4 Diffusion of innovation [35] or Roger's typology

The performed literature review and creative analysis enabled us (to the level of information presented in a publication) to classify multiple related projects in accordance with introduced SMM levels—some analysis outcomes are available in the most right column in Table 2.10.

2.6 Strengths of Smart Universities

Our vision of SmU is based on the idea that SmU—as a smart system—should implement and demonstrate significant maturity at various "smartness" levels such as (1) adaptation, (2) sensing (awareness), (3) inferring (logical reasoning), (4) self-learning, (5) anticipation, and (6) self-organization and re-structuring—see details and examples in Table 2.9 above. We believe these are the most important advantages of SmU in comparison with traditional universities. However, the performed literature review and creative analysis enabled us to arrive with an additional list of reported advantages of SmU and opportunities for traditional universities if they decide to evolve to certain level of SmU.

SmU strengths provide advantages in numerous aspects of education, such as staying abreast with advanced technology, providing exposure to this advanced technology, flexible learning styles and modes of learning content delivery, ubiquitous access to educational materials, etc.

Technological advances. An advance provided by SmU is the implementation and active utilization of advanced technology; some examples of this advanced technology implemented in current variations of Smart Universities are as follows:

- RFID tags allow for easier logistics and product-related information storage and retrieval [23, 27];
- contactless ID cards for identification and financial transaction performance [31, 32];

- cloud computing technologies allow for file sharing and access for all students [25];
- RFID tags provide security enhancements as they provide identification;
- cloud computing technology allows easier access to information for students so that they can work at their own pace and access information as needed.

Educational advances. SmU are advantageous due to the fact that they utilize more advanced technology than traditional universities. SmU provides both students and faculty with exposure to technological advances as well as their active use in teaching and learning. This provides the ability for students, learners, and faculty to use, for example:

- state of the art facilities [37];
- technology in combination with human interaction [38];
- well-equipped laboratories, libraries, and IT facilities [39].

Each of these provides users at SmU with unique hands-on experience with advanced technology—smart technology, smart systems, smart devices, smart classrooms, etc. — that perhaps would not be used in a traditional university setting. Having exposure to this advanced technology provides a holistic education for students that also contains customization, personalization, and subjection to unique teaching and learning styles [21].

Additionally, SmU provides the advantage of having flexibility when it comes to learning. This can be done because of SmU's ability to tailor its educational style to in-classroom learning, e-learning, or a combination of the two as blended learning (b-learning). This is a major strength of SmU over traditional universities due to the fact that these days (and, as expected, in the future) so many students have unique situations that perhaps require being out of the classroom for a majority of the time. The advantages of this concept provide SmU with the following strengths:

- learning at one's convenience (any place, any time, any device) [40];
- potential for independent learning as well as structured learning [41];
- curriculum designed to meet both local needs and international standards [39];
- resources available anytime and anywhere [42].

The listed strengths above provide students, who normally might not have access to higher education, with the ability to learn in any location and at any time.

These reported advantages of SmU provide the foundation to continually expand on the concepts discussed above. This can provide opportunities for a wider variety of students to take advantage of the educational resources available to them, providing them with a richer educational experience at SmU.

2.7 Conclusions

The performed literature review, creative analysis, proposed SMM model, and obtained research findings and outcomes enabled us to make the following conclusions:

(1) In several recent years, the ideas of smart education, smart university, smart classroom, smart learning environments, and related topics became the main topics of various pioneering international and national events and projects, governmental and corporate initiatives, institutional agendas, and strategic plans.

(2) Our vision of SmU is based on the idea that SmU—as a smart system—should implement and demonstrate significant maturity at various "smartness" levels such as (1) adaptation, (2) sensing (awareness), (3) inferring (logical reasoning), (4) self-learning, (5) anticipation, and (6) self-organization and re-structuring.

(3) The performed classic literature review of 100 + publications and reports relevant to SmU, SmE, SmC, and SLE areas usually does not provide readers with a clear understanding of focus, scope, and main details of analyzed publications; moreover, it usually does not help to compare the proposed approaches, features, smartness levels, and details in those publications.

(4) In order to overcome problems of the classic literature review approach, and in order to be able to compare various publications in SmU, SmE, SmC, SLE and related areas, we developed a framework for systematic creative analysis of those publications. In the general case, this framework may contain multiple sections (rubrics) that may include, but are not limited to, a) general information about a publication (i.e. main topic or title of a publication, list of authors, publisher, year of publication, reference number, etc.); b) main topic; c) proposed idea or approach and main details (as described by authors); d) identified list of main features (or functions) that are relevant to SmU concepts; e) identified smartness levels of SmU addressed by proposed approach, developed system, and other rubrics. Examples of applications of the developed framework to creatively analyze various related publications are presented in Tables 2.1, 2.2, 2.3, 2.4, 2.5, 2.6, 2.7 and 2.8 in Sect. 2.3 and Tables 2.11, 2.12, 2.13, 2.14, 2.15, 2.16, 2.17, 2.18, 2.19, 2.20, 2.20, 2.21, 2.22 and 2.23 in Appendix.

(5) Examples of utilization of proposed SmU smartness levels in several analyzed publications are described in Table 2.9 (Sect. 2.4).

(6) In order to recognize and classify the current status of smart education development in a particular college/university or status of evolution of a traditional university towards smart university, we proposed and developed the Smart Maturity Model (SMM) for SmU (Table 2.10 in Sect. 2.5).

(7) The relations between SmE, SmC, and SLE (as innovative approaches to education) on one side and types of faculty on the other side are well described by Roger's "Diffusion of Innovation" Typology (Fig. 2.3 in Sect. 2.5). As a result, we conclude that specific types of faculty should be involved in SmE and SmC initiates only on designated levels of SmU "smartness" maturity to optimize "cost-benefits" ratio.

(8) Finally, the performed literature review and creative analysis enabled us (to the level of information presented in each publication) to classify multiple related projects in accordance with introduced SMM model levels (some of examples are available in the most right column in Table 2.10).

Acknowledgements The authors would like to thank Dr. Jeffrey P. Bakken, Associate Provost for Research and Dean of the Graduate School, Dr. Cristopher Jones, Dean of the LAS College, and Sandra Shumaker, Executive Director, Office of Sponsored Programs at Bradley University for their strong support of our research, design and development activities in Smart University and Smart Education areas.

This project is partially supported by grant REC # 1326809 from Bradley University.

Appendix

1. **Approaches to develop smart universities: the outcomes of systematic creative analysis** (Tables 2.11, 2.12, 2.13 and 2.14)
2. **Approaches to develop smart campuses: the outcomes of systematic creative analysis** (Tables 2.15, 2.16, 2.17 and 2.18)
3. **Approaches to develop smart learning environments: the outcomes of systematic creative analysis** (Table 2.19)
4. **Approaches to develop smart classrooms: the outcomes of systematic creative analysis** (Tables 2.20, 2.21 and 2.22)
5. **Approaches to develop smart learning: the outcomes of systematic creative analysis** (Table 2.23)

Table 2.11 Smarter Universities: a vision for the fast changing digital era [11]

Item	Details
Main topic	Smart university—concepts
Approach used	SmU is considered as a place where knowledge is shared in a seamless way and is a system that is green, robust, personalized, responsible, interactive, and adaptive, as well as accessible anywhere, anytime, and from any device
Details	"…We analyze the current situation of education in universities, with particular reference to the European scenario. Specifically, we observe that recent evolutions, such as pervasive networking and other enabling technologies, have been dramatically changing human life, knowledge acquisition, and the way works are performance and people learn" [11] • "Opinion mining—The first step of the process is collecting different option, which will be later organized and structured. • Needs collection—The second phase of the proposed model corresponds to an in-depth analysis of the needs emerging from the area, the communities and the organizations • Vision—The presence of multiple variables and constraints encourages the creation of a "strategic" vision that must be translated into clear objectives, ambitious yet realistic • Priorities—The above mentioned objectives are then ordered as a two-dimensional array according to their priority and their measure of urgency. • Common contents—The model extracts common contents, knowledge and skills that an individual must have in multiple scientific areas • Domain specific contents—The vertical part of the T is represented by the knowledge and the skills that individuals must possess in a specific domain.

(continued)

Table 2.11 (continued)

Item	Details
	• Competencies, standards and policies—The competencies are described in the ECF and teachers' skills are taken into account • Matching—One of the most challenging parts of the model is the task of matching the choices with the needs • Monitoring and analytics—The proposed model provides an abstract representation of a vision"
Identified features	• Collaborative and communication technologies • Systems for seamless collaborative learning (of both local and remote students) in smart classroom and shared learning content/documents • Elements of Internet-of-Things technology • Cloud computing technology in place • Elements of Smart Agents technology
Smartness levels addressed	• Adaptation—various classroom models can be used • Sensing—by observing the recent evolutions • Inferring—based on opinion mining and needs that were collected previously

Table 2.12 Designing smart University using RFID and WSN [23]

Item	Details
Main topic	Smart university—technology
Approach used	Smart university design approach is based on implementation and active use of Radio Frequency Identification (RFID) technology and wireless sensor network (WSN) technology to improve student's experience whether they experience it remotely or physically in the classroom
Details	"There is a necessity to build a smart system that decrease load in managing the attendance and improves the performance of colleges, universities and any education institute. ... The developed prototype shows how evolving technologies of RFID and WSN can add in improving student's attendance method and power conservation. RFID technology provides the means of persons identification and forms the basis of: • Student attendance record • Employee attendance record • Authentication of attendance • Automation of electrical appliances."[23]
Identified features	• RFID technology • Smart card readers • Biometric-based access control devices • Robotic controllers and actuators • Intelligent cyber-physical systems • Flipped classrooms • Remote (virtual) labs
Smartness levels addressed	• Sensing—by technology specifically used for identification and sensing ability • Anticipation (of student behavior)—by keeping track of attendance records, authentication of attendance, etc.

Table 2.13 System architecture for a smart university building [25]

Item	Details
Main topic	Smart university—buildings
Approach used	Smart university development approach is based on idea of having smart buildings that are able to remotely monitor and manage processes while being energy efficient
Details	"…An intelligent platform is proposed, that integrates sensors within a university building and campus based on Web Services middleware. The aim is to provide automation of common processes, reduce the energy footprint and provide control of devices in a remote manner" [25]
Identified features	• Sensor technology (motion, temperature, light, humidity, etc.) • RFID technology • WSN technology • Power/light/HVAC consumption monitoring system(s)
Smartness levels addressed	• Sensing—by technology specifically used for identification and sensing ability • Inferring—make inferences based on information gathered by sensors to enhance ability of smart classroom • Self-organization—use systems and sensors that are automatically configured to run smart classroom

Table 2.14 Smart education environment system [30]

Item	Details
Main topic	Smart university—library
Approach used	Smart university development approach is based on a framework that upgrades the traditional, book-based library to a digital resource library to meet the needs of an emerging IT-aware generation
Details	"…a framework for an Smart Education Environment System (SEES), which will provide a library with an integrated database incorporating three core sub-systems: 'Electronic Bookshelves', for automating access to the bookshelves; 'Virtual White Space', for the discussion of information found in the library; and 'Innovation and Social Network Database (ISND)', for disseminating and storing new ideas and concepts…" [30]
Identified features	• E-books • SMART boards and/or interactive white boards • RFID technology • Repositories of digital learning content and online (Web) resources, learning portals • Systems for seamless collaborative learning (of both local and remote students) in smart classroom and shared learning content/documents
Smartness levels addressed	• Adaptation—of classroom model • Sensing—by technology specifically used for identification

Table 2.15 Exploration on security system structure of smart campus based on cloud computing [32]

Item	Details
Main topic	Smart campus—technology
Approach used	Smart campus development approach is based on a cloud computing technology and the Internet of Things technology to ensure everything is completely digital and all information, class interaction, and lectures are available from remote locations
Details	"This paper studies such problems as smart campus, smart campus system structure and campus cloud structure from the perspective of information security, designs security system structure of smart campus on this basis and describes the security system structure at six levels" [32]
Identified features	• Intelligent cyber-physical system (for safety and security) • Various smart software agents • Internet-of-Things technology • Repositories of digital learning content and online (web) resources, learning portals
Smartness levels addressed	• Sensing—by technology specifically used for identification

Table 2.16 New challenges in smart campus applications [15]

Item	Details
Main topic	Smart campus—concepts
Approach used	Smart campus development approach is based on a framework that can be used by both local and remote participants. This framework contains everything to improve the usability of the system and provide users with a better and more effective learning experience
Details	"…create an architecture framework which allows various members of the community to create and use services based on the data that is collected in a university environment. … The overall goal is to improve the usability of the system and provide better user experience by applying personalization" [15]
Identified features	• Sensor technology (motion, temperature, light, humidity, etc.) • Repositories of digital learning content and online (Web) resources, learning portals • Collaborative Web-based audio/visual one-to-one and many-to-many communication systems • E-books • Adaptive courses, lessons and learning modules with variable components and structure suitable for various types of teaching—face-to-face, blended, online, types of students/learners, smart pedagogy, etc. • Systems to host, join, form and evaluate group discussions (including both local and remote students)
Smartness levels addressed	• Adaptation—of teaching style and classroom model • Sensing—by technology specifically used for identification and sensing ability

Table 2.17 A vision for the development of i-campus [12]

Item	Details
Main topic	Smart campus—concepts
Approach used	Smart campus development approach is based on holistic e-learning, social networking and communications for work collaboration, green and CIT sustainability with intelligent sensor management systems and smart building management with security control and surveillance
Details	"…envision the possible ways of intelligent campus development. … Personalized learning is a board term which includes personalized study plan and learning path, personalized information gathering and learning resource retrieval, and personalized learning activities. … Adaptive learning refers to the fulfillment of dynamic needs of learners and provision of feedbacks to the current state of learning in a smart learning environment. … There are some possible research and development opportunities relating to the development of intelligent campus which may include learning management systems and knowledge management, personalized learning, adaptive learning and e-portfolio, immersive educational space, and safe learning environment" [12]
Identified features	• Smart data visualization technology • Repositories of digital learning content and online (Web) resources, learning portals • Web-lecturing technology • Interconnected laptops or desktop computers • Elements of Internet-of-Things technology
Smartness levels addressed	• Adaptation—of teaching style and classroom model

Table 2.18 Future content-aware pervasive learning environment: smart campus [24]

Item	Details
Main topic	Smart campus—technology
Approach used	Smart campus development approach is based on technology to allow remote learning, interaction between local and remote students, and access for all students to all digital resources
Details	"…we proposed a novel design of interactive pervasive learning environment. … The proposed solution is actually a service oriented framework in which data mining is used for knowledge discovery, context is identified by using more precise approach, and then automatic service discovery, activation and execution takes place to facilitate the user in pervasive learning environment" [24]
Identified features	• Repositories of digital learning content and online (Web) resources, learning portals • Systems for seamless collaborative learning (of both local and remote students) in smart classroom and shared learning content/documents • Collaborative Web-based audio/visual one-to-one and many-to-many communication systems • Smart learning/teaching analytics (big data analytics) systems
Smartness levels addressed	• Adaptation—of classroom model • Inferring—based on big data analytics systems information gathered

Table 2.19 Definition, framework and research issues of smart learning environments-a context-aware ubiquitous learning perspective [9]

Item	Details
Main topic	Smart learning environment—concepts
Approach used	SLE development approach is based on an idea of student's ability to access digital resources and interact with learning systems from any pack at any time. It also actively provides guidance through context awareness
Details	"The definition and criteria of smart learning environments are presented from the perspective of context-aware ubiquitous learning. A framework is also presented to address the design and development considerations of smart learning environments to support both online and real-world learning activities. … An expert model or expert knowledge model that contains the teaching materials, a student model or learner model that evaluates students' learning status and performance, an instructional model or pedagogical knowledge model that determines teaching content, educational tools and presentation methods based on the outcomes of the student model, and a user interface for interacting with students" [9]
Identified features	• Ubiquitous computing technology • Context (situation) awareness systems • Adaptive teaching • Systems for seamless collaborative learning (of both local and remote students) in smart classroom and shared learning content/documents
Smartness levels addressed	• Adaptation—of teaching style and classroom model • Sensing—done by ubiquitous computing technology used in a context aware system

Table 2.20 Smart classrooms for distance education and their adoption in multiple classroom architecture [18]

Item	Details
Main topic	Smart classroom—systems and technology
Approach used	SMC development approach is based on implementation and active use of integrated voice-recognition, computer-vision, as well as other varying technologies, to provide a tele-education experience for remote learners that is similar to a traditional classroom experience
Details	"…an overview of the technologies used in smart classrooms for distance education by classifying smart classrooms into four categories and discussing the type of technologies used in their implementation…" [18]
Identified features	• Web-lecturing systems (with video capturing and computer capturing functions) for learning content development pre-class activities • Systems for seamless collaborative learning (of both local and remote students) in smart classroom and shared learning content/documents • Collaborative Web-based audio/visual one-to-one and many-to-many communication systems • Intelligent cyber-physical systems (for safety and security) • Repositories of digital learning content and online (Web) resources, learning portals

(continued)

Table 2.20 (continued)

Item	Details
	• Panoramic cameras • Ceiling-mounted projectors (in some cases, 3D projects) • Interconnected big screen monitors or TVs ("smart learning cave") • Interconnected laptops or desktop computers • SMART boards and/or interactive white boards
Smartness levels addressed	• Adaptation—of teaching style and classroom model • Anticipation—keeps track of location awareness and intelligent cyber-physical systems for safety and security

Table 2.21 Classroom in the era of ubiquitous computing smart classroom [28]

Item	Details
Main topic	Smart classroom—systems and technology
Approach used	SMC development approach is based on context-awareness, such as understanding a user's intentions based on audio/visual inputs or situational information
Details	"…four essential characteristics of futuristic classroom in the upcoming era of ubiquitous computing: natural user interface, automatic capture of class events and experience, context-awareness and proactive service, collaborative work support. This elaborates the details in the details in the design and implementation of the ongoing Smart Classroom project…" [28]
Identified features	• Collaborative Web-based audio/visual one-to-one and many-to-many communication systems • Gesture (activity) recognition and speaker/instructor motion tracking systems • Speech/voice recognition systems • Controlled and self-activated microphones and speakers
Smartness levels addressed	• Adaptation—of classroom model • Sensing—by technology specifically used for identification and human sensing • Self-organization—through classroom's controlled and self-activated systems

Table 2.22 Smart classroom: enhancing collaborative learning using pervasive computing technology [29]

Item	Details
Main topic	Smart classroom—hardware and technology
Approach used	SMC development approach is based on an implementation and active use of a collection of embedded, wearable, and handheld devices that are wirelessly connected to enhance the idea of collaborative learning
Details	"..Smart Classroom that uses pervasive computing technology to enhance collaborative learning among college students… integrate mobile and handheld devices, such as Personal Digital Assistants (PDAs), with fixed computing infrastructures, such as PCs, sensors, etc. in a wireless network environment inside a classroom…" [29]

(continued)

Table 2.22 (continued)

Item	Details
Identified features	• Interconnected laptops or desktop computers • Systems to relay automatically recorded class activities and lectures for post-class review and activities (by both local and remote students) • Systems to host, join, form and evaluate group discussions (including both local and remote students) • Collaborative and communication technologies • Web-lecturing systems (with video capturing and computer screen capturing functions) for learning content development pre-class activities
Smartness. levels addressed	• Adaptation—of teaching style and classroom model

Table 2.23 Competence analytics [33]

Item	Details
Main topic	Smart learning—concepts
Approach used	Smart learning development is based on idea of personalized learning experiences, taking place through learning analytics, analysis, discovery, etc.
Details	"This article introduces a framework called Smart Competence Analytics in Learning (SCALE) that tracks finer-level learning experiences and translates them into opportunities for customized feedback, reflection, and regulation. The SCALE framework is implemented in four layers: the sensing layer, the analysis layer, the competence layer, and the visualization layer. ... The SCALE framework has been designed for seamless integration with different types of learning-related artifacts such as the Moodle learning management system...The current SCALE system does not focus as much on the physical context of a student (e.g., classroom, instructor availability) as it does with the student's learning context (e.g., learners' background knowledge, learner motivation)..." [33]
Identified features	• Sensor technology • Smart learning/teaching analytics (big data analytics) systems • Smart data visualization technology • Adaptive programs of study—major and minor programs, concentration and certificate programs—with variable structures adaptable to types of students/learners, smart pedagogy, etc.
Smartness levels addressed	• Adaptation—of classroom model • Sensing—by technology specifically used for identification

References

1. KES International, Annual KES international conference on Smart Education and e-Learning (SEEL). http://seel-17.kesinternational.org/
2. International conference on Smart Learning Environments (ICSLE). http://www.iasle.net/index.php/icsle2017
3. Springer Open International Journal on Smart Learning Environments. https://slejournal.springeropen.com/
4. Chun, S.: Korea's smart education initiative and its pedagogical implications. CNU J. Educ. Stud. 34(2), 1–18 (2013)
5. Chun, S. Lee, O.: Smart education society in South Korea. In: GELP Workshop on Building Future Learning Systems, Durban, South Africa (19–21, April 2015). http://gelpbrasil.com/wp-content/uploads/2015/05/smart_education_workshops_slides.pdf
6. Smarter education with IBM. https://www-935.ibm.com/services/multimedia/Framework_-_Smarter_Education_With_IBM.pdf
7. The Next-Generation Classroom: Smart, Interactive and Connected Learning (2012). http://www.samsung.com/es/business-images/resource/white-paper/2012/11/EBT15_1210_Samsung_Smart_School_WP-0.pdf
8. Tikhomirov, V., Dneprovskaya, N.: Development of strategy for smart University. In: 2015 Open Education Global International Conference, Banff, Canada (2015). 22–24, April
9. Hwang, G.J.: Definition, framework and research issues of smart learning environments-a context-aware ubiquitous learning perspective. Smart Learning Environments—a Springer Open Journal, 1:4, Springer (2014)
10. IBM, Smart Education. https://www.ibm.com/smarterplanet/global/files/au__en_uk__cities__ibm_smarter_education now.pdf
11. Coccoli, M., Guercio, A., Maresca, P., Stanganelli, L.: Smarter Universities: a vision for the fast changing digital era, J. Vis. Lang Comput 25, 1003–1011, Elsevier (2014)
12. Kwok, L.F.: A vision for the development of i-campus, Smart Learning Environments—a Springer Open Journal, 2:2, Springer (2015)
13. Xiao, N.: Constructing smart campus based on the cloud computing platform and the internet of things. In: Proceedings of 2nd International Conference on Computer Science and Electronics Engineering (ICCSEE 2013), Atlantis Press, Paris, France, pp. 1576–1578 (2013)
14. Abueyalaman, E.S. et al.: Making a smart campus in Saudi Arabia. EDUCAUSE Q. 2, 1012 (2008)
15. Adamko, A., Kadek, T., Kosa, M.: Intelligent and adaptive services for a smart campus visions, concepts and applications. In: Proceedings of 5th IEEE International Conference on Cognitive Infocommunications, 5–7 Nov. 2014. Vietri sul Mare, Italy, IEEE (2014)
16. Uskov, V.L., Bakken, J.P., Pandey, A.: The ontology of next generation smart classrooms. In: Proceedings of the 2nd International Conference on Smart Education and e-Learning SEEL-2016, 17–19 June 2015. Sorrento, Italy, Springer, pp. 1–11 (2015)
17. Huang, R., Hub, Y.,Yang, J., Xiao, G.: The functions of smart classroom in smart learning age. In: Proceedings of 20th International Conference on Computers in Education ICCE. Nanyang Technological University, Singapore (2012)
18. Pishva, D., Nishantha, G.G.D.: Smart classrooms for distance education and their adoption to multiple classroom architecture. J. Netw. 3(5) (2008)
19. Gligorić, N., Uzelac, A., Krco, S.: Smart classroom: real-time feedback on lecture quality. In: Proceedings of 2012 IEEE International Conference on Pervasive Computing and Communications Workshops (PERCOM Workshops), pp. 391–394, 19–23 Mar. 2012. Lugano, Switzerland, IEEE (2012). doi:10.1109/PerComW.2012.6197517
20. Uskov, V.L, Bakken, J.P., Pandey, A., Singh, U., Yalamanchili, M., Penumatsa, A.: Smart University taxonomy: features, components, systems. In: Uskov, V.L., Howlett, R.J., Jain, L.C. (eds.) Smart Education and e-Learning 2016, Springer, pp. 3–14, June 2016, 643 p., ISBN: 978-3-319-39689-7 (2016)

21. Koper, R.: Conditions for effective smart learning environments, in Smart Learning Environments: a Springer Open Journal (2014)
22. Hwang, D. Smart approach to innovative education for 21st century. In: International Conference IITE-2014 "New Challenges for Pedagogy and Quality Education: MOOCs, Clouds and Mobiles" (2014)
23. Al Shimmary, M.K., Al Nayar, M.M., Kubba, A.R.: Designing smart University using RFID and WSN. In: Int. J. Comput. Appl. 34–39 (2015)
24. Khan, M.T., Zia, K.: Future context-aware pervasive learning environment: smart campus. In: Proceedings of International Conference on Information Technology in Science Education, Gazimagusa, Turkish Republic of Northern Cyprus (2007)
25. Stavropoulos, T., Tsioliaridou, A., Koutitas, G., Vrakas, D., Vlahavas, I.: System architecture for a smart University building. In: Diamantaras, K., Duch, W., Iliadis, L.S. (eds.) ICANN 2010, LNCS, vol. 6354, pp. 477–482. Springer, Heidelberg (2010)
26. Fan, J., Stewart, K.: An ontology-based framework for modeling movement on a smart campus. In: Analysis of Movement Data, GIScience Workshop, Vienna, Austria (2014)
27. Rehman, A., Abbasi, A.Z., Shaikh, Z.A.: Building a smart University using RFID Technology. In: Proceedings of IEEE International Conference on Computer Science and Software Engineering, vol. 5, pp. 641–644. IEEE (2008)
28. Jiang, C., Shi, Y., Xu, G., Xie, W.: Classroom in the era of ubiquitous computing smart classroom. In: Proceedings of IEEE International Conference on Wireles LANs and Home Networks, pp. 14–26. IEEE (2001)
29. Yau, S.S., Gupta, S.K.S., Karim, F., Ahamed, S.I., Wang, Y., Wang, B.: Smart classroom: enhancing collaborative learning using pervasive computer technology. In: Proceedings of 2003 Annual Conference of American Society for Engineering Education, pp. 1–9 (2003)
30. Salah, A.M., Lela, M., Al-Zubaidy, S.: Smart education environment system. In: GESJ: Computer Science and Telecommunications, pp. 21–26 (2014)
31. Nie, X.: Constructing smart campus based on the cloud computing platform and the internet of things. In: Proceedings of 2nd International Conference on Computer Science and Electronics Engineering (2013)
32. Zhou, W.: Exploration on security system structure of smart campus based on cloud computing. In: Proceedings of 3rd International Conference on Science and Social Research (2014)
33. Kumar, V., Boulanger, D., Seanosky, J., Kinshuk, Panneerselvam, K., Somasundaram, T.S. Competence analytics. J. Comput. Educ. 251–270 (2014)
34. Chen, C.L., Chang, Y.H., Chien, Y.T., Tijus, C., Chang, C.Y.: Incorporating a smart classroom 2.0 Speech-Driven PowerPoint System (SDPPT) into University Teaching, in Smart Learning Environments: a SpringerOpen Journal (2015)
35. Rogers, E.M.: Diffusion of Innovations, 5th edn, p. 221. The Free Press, New York, USA (2003)
36. Uskov, V.L., Bakken J.P., Karri, S., Uskov, A.V., Heinemann, C., Rachakonda, R.: Smart University: conceptual modeling and system design. In: Uskov, V., Bakken, J.P., Howlett, R., Jain, L. (eds.) Smart Universities. Springer (accepted for publication in 2017)
37. Murray, B.: A SWOT analysis of the teaching process of math and science at Worcester's North High School. https://web.wpi.edu/Pubs/E-project/Available/E-project-052711-122422/unrestricted/BMurray_North_SWOT_v2.pdf
38. Jha-Thakur, U., Fischer, T.B.: 25 Years of the UK EIA system: strengths, weaknesses, opportunities and threats, in environmental impact assessment review, pp. 19–26 (2016)
39. Yigit, A.S., Al-Ansary, M.D., Al-Najem, N.M.: SWOT analysis and strategic planning as an effective tool for improving engineering education at Kuwait University. https://pdfs.semanticscholar.org/275c/3b04f153059daeca4b01cae700bb23cd969e.pdf
40. Odeh, M., Warwick, K., Garcia-Perez, A.: The impacts of cloud computing adoption at Higher Education Institutions: a SWOT analysis. Int. J. Comput. Appl. 127(4), 15–21 (2015)

41. Grcic, M., Picek, R.: Using SWOT analysis for defining strategies of mobile learning. In: Proceedings of 2013 International Conference on Systems, Control, Signal Processing and Informatics, pp. 307–312 (2013)
42. Kim, T., Cho, J.Y., Lee, B.G.: Evolution to smart learning in public education: a case study of Korean public education. In: Ley, T., Ruohonen, M., Laanpere, M., Tatnall, A. (eds.) Open and Social Technologies for Networked Learning, IFIP Advances in Information and Communication Technology, vol 395. Springer (2013)
43. Software Engineering Institute, Carnegie Mellon University. https://www.sei.cmu.edu/cmmi/
44. CMMI Institute. http://cmmiinstitute.com/
45. CMMI (Capability Maturity Model Integration). https://sultanalmasoud.files.wordpress.com/2014/05/cmmi-pyramid.gif

Part II
Smart Universities: Concepts, Systems and Technologies

Chapter 3
Smart University: Conceptual Modeling and Systems' Design

Vladimir L. Uskov, Jeffrey P. Bakken, Srinivas Karri,
Alexander V. Uskov, Colleen Heinemann and Rama Rachakonda

Abstract The development of Smart University concepts started just several years ago. Despite obvious progress in this area, the concepts and principles of this new trend are not clarified in full yet. This can be attributed to the obvious novelty of the concept and numerous types of smart systems, technologies and devices available to students, learners, faculty and academic institutions. This paper presents the outcomes of a research project aimed at conceptual modeling of smart universities as a system based on smartness levels of a smart system, smart classrooms, smart faculty, smart pedagogy, smart software and hardware systems, smart technology, smart curriculum, smart campus technologies and services, and other distinctive components. The ultimate goal of this ongoing research project is to develop smart university concepts and models, and identify the main distinctive features, components, technologies and systems of a smart university—those that go well beyond features, components and systems used in a traditional university with predominantly face-to-face classes and learning activities. This paper presents the up-to-date outcomes and findings of conceptual modeling of smart university.

Keywords Smart university · Smartness levels · Smart university components · Conceptual modeling · Software systems for a smart university

3.1 Introduction

The "smart university" (SmU) concept and several related concepts, such as smart classrooms [1–11], smart learning environments [12, 13], smart campus [14–17], smart education [18–23], and smart e-learning [22–24], were introduced just several

V.L. Uskov (✉) · S. Karri · A.V. Uskov · C. Heinemann · R. Rachakonda
Department of Computer Science and Information Systems and InterLabs Research Institute,
Bradley University, Peoria, IL, USA
e-mail: uskov@fsmail.bradley.edu; uskov@bradley.edu

J.P. Bakken
The Graduate School, Bradley University, Peoria, IL, USA
e-mail: jbakken@fsmail.bradley.edu

© Springer International Publishing AG 2018   49
V.L. Uskov et al. (eds.), *Smart Universities*, Smart Innovation,
Systems and Technologies 70, DOI 10.1007/978-3-319-59454-5_3

years ago; they are in permanent evolution and improvement since that time. The introduced ideas and approaches to build SmU, as well as smart education (SmE), are rapidly gaining popularity among the leading universities in the world because modern, sophisticated high-tech-based smart technologies, systems, and devices create unique and unprecedented opportunities for academic and training organizations in terms of higher teaching standards and expected learning/training outcomes. "The analysts forecast the global smart education market to grow at a CAGR of 15.45% during the period 2016–2020" [25]. "Markets and Markets forecasts the global smart education and learning market to grow from $193.24 Billion in 2016 to $586.04 Billion in 2021, at a Compound Annual Growth Rate (CAGR) of 24.84%" [26].

3.1.1 Literature Review

To-date, most researchers are focused on perspectives of contemporary higher education and tendencies that correspond to the concepts of SmU and SmE. For example, the authors of [27–29] discuss various aspects of contemporary universities and their future perspectives in the context of applications of smart information and communication technologies (ICT) in education. On the other hand, the authors of [30] presented one of the first research studies on new opportunities for universities in the context of smart education.

The other significant part of research in SmE is focused on the problem of educational outcomes in the contemporary educational systems—smart learning environments (SLE). For example, the concept of outcomes-based education has been proposed and this approach can be regarded as a part of the SmE concept. The main part of obtained outcomes includes the studies of cognitive abilities, needs, skills, and their training through e-learning or, in general, to 21st century skills [31]. There are also resources available on instructional design and cognitive science, which consider the problem of learning, structuring material, communication, and forming cognitive competence in this group, for example [32].

Different attempts to make conceptual models of ICT infrastructure of SmU and smart e-learning systems are currently being actively researched; for example, the topics concerning smart e-learning standards, smart gadgets, and learning equipment are discussed [33]. Several research projects were aimed at organizational aspects of SmU, smart education such as organizational structure, educational trajectories, learning strategies, etc. These projects usually emphasize the fact that many aspects of contemporary education need new flexible organizational structures, which can be referred to as "smart" [30, 34].

The detailed literature review on existing approaches to build SmU is available in another chapter of this book, specifically in [35].

Despite the obvious progress in SmU area, the main concepts and conceptual models of smart universities are not clarified in full yet due to obvious uniqueness, innovativeness, and complexity of this research area.

3.1.2 Research Project Goal and Objectives

The performed analysis of above-mentioned and multiple additional publications and reports relevant to (1) smart universities, (2) university-wide smart software and hardware systems and technologies, (3) smart classrooms, 94) smart learning environments, and (5) smart educational systems, undoubtedly shows that SmU-related topics will be in the focus of multiple research, design, and development projects in the upcoming 5–10 years. It is expected that, in the near future, SmU concepts and hardware/software/technological solutions will start to play a significant role and be actively deployed and used by leading academic intuitions in the world.

Project Goal. The overall goal of this ongoing multi-aspect research project is to develop conceptual modeling of a smart university, i.e. to identify and classify SmU's main smart features, components, relations (links) between components, interfaces, inputs, outputs, and limits/constraints. The premise is that SmU conceptual modeling will (1) enable us to identify and predict the most effective software, hardware, pedagogy, teaching/learning activities, services, etc., for the next generation of a university—smart university—and (2) help traditional universities understand the strengths, weaknesses, opportunities, and threats of becoming a smart university and also identify and evaluate paths for a possible transformation from traditional university into a smart one.

Project Objectives. The objectives of this project were to identify SmU's main (1) smartness levels (or smart features), (2) components, and (3) specific software and hardware systems that go well beyond those in a traditional university with predominantly face-to-face classes and learning activities. Due to limited space, we present a summary of up-to-date research outcomes below.

3.2 Smart University: Conceptual Modeling

3.2.1 Smartness University: Modeling of Smartness Levels

Based on our vision of SmU and up-to-date obtained research outcomes, we believe that SmU as a smart system should significantly emphasize, not only pioneering software/hardware features and innovative modern teaching/learning strategies, but also "smart" features of smart systems (Table 3.1) [36, 37]. Therefore, the designers of SmU should pay more attention to a maturity of smart features of SmU that may occur on various levels of SmU's smartness—smartness levels.

Several examples of SmU possible distinctive functions for every proposed SmU smartness level are presented in Table 3.2.

Table 3.1 Classification of smartness levels of a smart system [36, 37]

Smartness levels	Details
Adaptation	Ability to modify physical or behavioral characteristics to fit the environment or better survive in it
Sensing	Ability to identify, recognize, understand, and/or become aware of phenomenon, event, object, impact, etc.
Inferring	Ability to make logical conclusion(s) on the basis of raw data, processed information, observations, evidence, assumptions, rules, and logic reasoning
Learning	Ability to acquire new or modify existing knowledge, experience, behavior to improve performance, effectiveness, skills, etc.
Anticipation	Ability of thinking or reasoning to predict what is going to happen or what to do next
Self-organization and re-structuring (optimization)	Ability of a system to change its internal structure (components), self-regenerate, and self-sustain in purposeful (non-random) manner under appropriate conditions but without an external agent/entity

Table 3.2 SmU distinctive features (that go well beyond features of a traditional university) [38]

SmU smartness levels	Details	Possible examples (limited to 3)
Adaptation	SmU ability to automatically modify its business functions, teaching/learning strategies, administrative, safety, physical, behavioral and other characteristics, etc. to better operate and perform its main business functions (teaching, learning, safety, management, maintenance, control, etc.)	• SmU easy adaptation to new style of learning and/or teaching (learning-by-doing, flipped classrooms, etc.) and/or courses (MOOCs, SPOCs, open education and/or life-long learning for retirees, etc.)
		• SmU easy adaptation to needs of students with disabilities (text-to-voice or voice-to-text systems, etc.)
		• SmU easy network adaptation to new technical platforms (mobile networking, tablets, mobile devices with iOS and Android operating systems, etc.)
Sensing (awareness)	SmU ability to automatically use various sensors and identify, recognize, understand and/or become aware of various events, processes, objects, phenomenon, etc. that may have impact (positive or negative) on SmU's operation, infrastructure, or well-being of its components—students, faculty, staff, resources, properties, etc.	• Various sensors of a Local Action Services (LAS) system to get data regarding power use, lights, temperature, humidity, safety, security, etc.
		• Smart card (or biometrics) readers to open doors to mediated lecture halls, computer labs, smart classrooms and activate

(continued)

Table 3.2 (continued)

SmU smartness levels	Details	Possible examples (limited to 3)
		features/software/hardware that are listed in user's profile
		• Face, voice, gesture recognition systems and corresponding devices to retrieve and process data about students' class attendance, class activities, etc.
Inferring (logical reasoning)	SmU ability to automatically make logical conclusion(s) on the basis of raw data, processed information, observations, evidence, assumptions, rules, and logic reasoning.	• Student Analytics System (SAS) to create (update) a profile of each local or remote student based on his/her interaction, activities, technical skills, etc.
		• Local Action Services (LAS) campus-wide system to analyze data from multiple sensors and make conclusions (for ex: activate actuators and close/lock doors in all campus buildings and/or labs, turn off lights, etc.)
		• SAS can recommend administrators take certain pro-active measures regarding a student
Self-learning	SmU ability to automatically obtain, acquire or formulate new or modify existing knowledge, experience, or behavior to improve its operation, business functions, performance, effectiveness, etc. (A note: Self description, self-discovery and self-optimization features are a part of self-learning)	• Learning from active use of innovative software/hardware systems—Web-lecturing systems, class recording systems, flipped class systems, etc.
		• Learning from anonymous Opinion Mining System (OMS)
		• Learning from different types of classes—MOOCs, blended, online, SPOCs, etc.
Anticipation	SmU ability to automatically think or reason to predict what is going to happen, how to address that event, or what to do next	• Campus-wide Safety System (CSS) to anticipate, recognize, and act accordingly in case of various events on campus
		• Enrollment Management System to predict, anticipate, and control variations of student enrollment
		• University-wide Risk Management System (snow days, tornado, electricity outage, etc.)

(continued)

Table 3.2 (continued)

SmU smartness levels	Details	Possible examples (limited to 3)
Self-organization and configuration, re-structuring, and recovery	SmU ability automatically to change its internal structure (components), self-regenerate, and self-sustain in purposeful (non-random) manner under appropriate conditions but without an external agent/entity. (A note: Self-protection, self-matchmaking, and self-healing are a part of self-organization)	• Automatic configuration of systems, performance parameters, sensors, actuators and features in a smart classroom in accordance with instructor's profile
		• Streaming server automatic closedown and recovery in case of temp electrical outage
		• Automatic re-configuration of wireless sensor network (WSN) because nodes may join or leave spontaneously (i.e. evolving network typology), university-wide cloud computing (with multiple clients and services), etc.

3.2.2 Smartness University: Conceptual Model

The proposed conceptual model of a SmU, labeled as *CM-SmU*, can be described as follows [40].

Definition 3.1 Smart University is described as *n*-tuple of n elements that can be chosen from the following main sets:

$$
\begin{aligned}
CM - SmU = \; <&\{SmU\_FEATURES\}, \{SmU\_STAKEHOLDERS\}, \\
&\{SmU\_CURRICULA\}, \{SmU\_PEDAGOGY\}, \\
&\{SmU\_CLASSROOMS\}, \\
&\{SmU\_SOFTWARE\}, \\
&\{SmU\_HARDWARE\}, \{SmU\_TECHNOLOGY\}, \\
&\{SmU\_RESOURCES\} >
\end{aligned}
\tag{3.1}
$$

where:

SmU_FEATURES   a set of most important smart features of SmU, including adaptation, sensing, inferring, self-learning, anticipation, self-optimization or re-structuring;

SmU_STAKEHOLDERS   a set of SmU stakeholders; for example, it includes a subset of SmU faculty (instructors) at SmU, i.e. those who are trained and predominantly teach classes in smart classrooms and actively use smart boards, smart systems, smart technology, etc.,

SmU_CURRICULA	a set of smart programs of study and smart courses at SmU—those that can, for example, change (or optimize) its structure or mode of learning content delivery in accordance with given or identified requirements (due to various types of students or learners);
SmU_PEDAGOGY	a set of modern pedagogical styles (strategies) to be used at SmU;
SmU_CLASSROOMS	a set of smart classrooms, smart labs, smart departments and smart offices at SmU;
SmU_SOFTWARE	a set of university-wide distinctive smart software systems at SmU (i.e. those that go well beyond those used at a traditional university);
SmU_HARDWARE	a set of university-wide smart hardware systems, devices, equipment and smart technologies used at SmU (i.e. those that go well beyond those used at a traditional university);
SmU_TECHNOLOGY	a set of university-wide smart technologies to facilitate main functions and features of SmU;
SmU_RESOURCES	a set of various resources of SmU (financial, technological, human, etc.)

In general, SmUs may have multiple additional minor sets; however, for the purpose of this research project, we will limit a number of SmU most important sets as presented in (3.1).

3.2.3 Smart University: Modeling Distinctive Components and Elements

SmU may have numerous components of a traditional university; however, it must have multiple additional components to implement actively use and maintain SmU distinctive features that are described in Table 3.2. Based on our vision of SmU and outcomes of our research, the SmU main sets should include at least those set elements (or SmU distinctive sub-components) that are described in Table 3.3.

3.2.4 Smart University: "Components and Features" Matrix

The performed research enabled us to arrive with a very important outcome —"Smart University: Components and Smartness Levels" matrix (Fig. 3.1). This matrix clearly shows that there should be a one-to-one correspondence between a

Table 3.3 SmU main sets (components) and main distinctive sub-sets (sub-components) that go well beyond components of a traditional university

SmU sets (components)	Examples of SmU distinctive sub-components (that go well beyond those in a traditional university)
SmU_SOFTWARE	• Pre-class learning content development systems • In-class activities' recording systems • Post-class activities' supporting systems (for example, systems to replay automatically recorded lectures, in-class discussions, and activities; learning content management systems, post-class review, and discussion systems for both local and remote students, etc.) • Smart cameraman software systems to automatically record and synchronize various in-class activities • Systems for seamless collaborative learning (of both local and remote students) in smart classroom and sharing of learning content/notes/documents • Web-based audio/video conferencing systems for smooth one-to-one and many-to-many communication/collaboration/interaction between local/in-class and remote/online students/learners • Systems to host, join, form, and evaluate group discussions (including both local and remote students) • Repositories of digital learning content and online (Web) resources, learning portals, digital libraries on appropriate learning modules • Smart learning/teaching analytics (big data analytics) systems • Speech/voice recognition systems • Speech-to-text systems • Text-to-voice synthesis systems • Automatic translation systems (from/to English language) • Speaker/instructor tracking systems (in the classroom) • Gesture (activity) recognition systems • Face recognition systems • Emotion recognition systems • Context (situation) awareness systems (including location awareness, learning context awareness, security/safety awareness systems) • Intelligent cyber-physical systems (for safety and security) • Various smart software agents • Power/light/HVAC consumption monitoring system(s)
SmU_TECHNOLOGY	• Internet-of-Things technology • Cloud computing technology • Web-lecturing technology • Web-based collaborative and communication technologies • Ambient intelligence technology • Smart agents technology

(continued)

Table 3.3 (continued)

SmU sets (components)	Examples of SmU distinctive sub-components (that go well beyond those in a traditional university)
	• Smart data visualization technology • Augmented and virtual reality technology • Computer gaming (serious gaming) technology • Remote (virtual) labs • 3D visualization technology • Wireless sensor networking technology • RFID (radio frequency identification) technology • Location and situation awareness technologies (indoor and outdoor) • Sensor technology (motion, temperature, light, humidity, etc.)
SmU_HARDWARE/ EQUIPMENT	• SMART boards and/or interactive white boards (at least 84" in size) • Ceiling-mounted projectors (in some cases, 3D projectors) • Smart panoramic video cameras (to web-tape all in-classroom activities) • Interconnected big screen monitors or TVs (to create an effect of "smart learning cave") • Interconnected laptops or desktop computers • Smart pointing devices • Voice controlled smart classroom hub (i.e., a central system to integrate and control various smart devices in a classroom) • Smart controlled and self-activated microphones • Smart speakers • Smart security video cameras • Smart classroom lock • Smart card readers • Biometrics-based access control devices (including face recognition devices) • Smart robotic controllers and actuators Smart thermostats • Smart switches
SmU_ CURRICULA	• Adaptive programs of study—major and minor programs, concentration and certificate programs—with variable structures that are adaptable to students/learners with different academic background, various components of smart pedagogy, including modes of learning content delivery, etc. • Adaptive courses, lessons, and learning modules with variable components and structure suitable for various types of teaching —face-to-face, blended, online, types of students/learners, smart pedagogy, etc.

(continued)

Table 3.3 (continued)

SmU sets (components)	Examples of SmU distinctive sub-components (that go well beyond those in a traditional university)
SmU_STAKEHOLDERS	• SmU Local (in-classroom) students • SmU remote (online) students • SmU students with disabilities • SmU life-long learners • SmU faculty (full-timers and part-timers) • SmU professional staff • SmU administrators • SmU sponsors
SmU_PEDAGOGY (or, Smart Pedagogy)	Active utilization and, if needed, adaptable combination of the following innovative types of pedagogy (teaching/learning strategies): • Learning-by-doing (including active use of virtual labs) • Collaborative learning • e-Books • Learning analytics • Adaptive teaching • Student-generated learning content • Serious games- and gamification-based learning • Flipped classroom • Project-based learning • Bring-Your-Own-Device approach • Smart robots (robotics) based learning
SmU_CLASSROOMS	• Smart classrooms that actively deploy various components of SmU_SOFTWARE, SmU_HARDWARE, SmU_TECHNOLOGY, SmU_PEDAGOGY sets (see above)
SmU_FEATURES	• Adaptation • Sensing (awareness) • Inferring (logical reasoning and analytics) • Self-learning • Anticipation • Self-optimization (re-structuring)

Smart University components	Smart University smartness levels					
	Adaptation	Sensing	Inferring	Self-learning	Anticipation	Self-optimization
SmU_SOFTWARE						
SmU_HARDWARE						
SmU_TECHNOLOGY						
SmU_CLASSROOMS						
SmU_PEDAGOGY						
SmU_CURRICULUM						
SmU_STAKEHOLDERS						
SmU_RESOURCES						

Fig. 3.1 "Smart University: Components and Features" matrix

particular SmU component and SmU smartness levels. The designers of SmU should clearly understand, for example, how to-be-deployed software systems will help SmU to

(1) adapt (for example, (a) to smoothly accommodate remote students or students with disabilities in a smart classroom, (b) to various modes of teaching, (c) to various types of learning content delivery and types of courses, etc.),
(2) sense (for example, (a) to get data about various activities or processes in a smart classroom or on smart campus, (b) to get data about student learning activities and academic performance, etc.),
(3) infer (i.e. process the obtained data, run learning analytics systems, and generate well-thought conclusions or recommendations based on big data analytics)
(4) self-learn (for example, to deploy innovative types of teaching and learning strategies or to offer classes on innovative topics that are in high demand in industry),
(5) anticipate (for example, to monitor as may areas on campus or inside buildings as possible, and predict and prevent any potentially bad events), and
(6) self-optimize (for example, to minimize consumption of electricity/heat in university classrooms and labs during night time and weekends, etc.).

Due to the limits of this chapter, it is impossible to provide details on all obtained research outcomes and findings regarding smart software systems, smart hardware, smart technology, and smart pedagogy; therefore, below we will focus on outcomes of detailed analysis of several important types of software systems for SmU.

3.3 Smart University: Software Systems' Design

Based on our vision of SmU, we believe that SmUs should deploy various types of distinctive software systems; a comprehensive list of corresponding software systems is presented in Table 3.3 above. As a part of this research project, for several most important classes of selected software systems to be deployed by SmU, the research team

(1) identified a list of desired functions (functionality) of those systems from SmU stakeholders' point of view;
(2) identified, downloaded, and analyzed approximately 10 to 20 existing software systems of designated type—usually including both open source and commercial systems—by means of (a) review of system's functions and features, (b) review of system's demo version or trial version, (c) installation and testing of actual system, and d) review of user and analysts' feedback;
(3) identified a list of main functions of system of designated type—i.e., functions to be beneficial for SmU stakeholders, and
(4) evaluated and ranked those systems.

A brief summary of our research outcomes for selected classes of software systems for SmUs is presented in multiple tables below, including

(1) pre-class learning content development systems for SmU (Tables 3.4, 3.5, 3.6 and 3.7);
(2) in-class activities recording systems for SmU (Tables 3.8, 3.9, 3.10 and 11);
(3) post-class activities' supporting systems for SmU (Tables 3.12, 3.13, 3.14 and 3.15);

Table 3.4 Pre-class learning content development systems for smart university: A list of system's desired features

#	Desired system's feature	Details
1	Screen capturing	Allows instructors to record dynamic and static visuals from computer screen
2	Audio capturing	Allows instructors to record sound, narrations for videos, VoIP calls, music, and audio that comes from the other applications on computer
3	Capturing from Webcam	Allows computer's webcam to record instructor while he/she teaches class or makes a video
4	Capturing streaming files	Allows to record streaming video and audio files onto a computer
5	Schedule recording	Allows instructor to set a time and date for an application to automatically record what's happening on computer screen (that possibly displays video from other connected resources)
6	Capturing from mobile device	Allows instructor to connect a smart phone or other mobile device to a desktop computer and record what is displayed on smart phone's screen
7	Zoom&pan	Ability to zoom in on a portion of computer screen helps an audience to focus on specific fragments of displayed learning content and better understand it; the pan effect allows instructor to smoothly move from one part of computer screen to the other one
8	Add media	Allows instructor to import video, sound, and image files from a computer into learning content files
9	Adjust audio	Supports fine-tuning of audio files
10	Add titles	Allows instructor to add title information to the beginning and/or end of video files
11	Add annotations	Add text comments and remarks to various recordings (a note: annotations are very useful for augmenting videos with additional useful information or comments that usually are not covered in the audio portion of a video)
12	Split/join video and audio files	Allows user to trim/eliminate unwanted (or low quality) video and audio fragments away from existing audio/video files and insert, if needed, other parts into final recording files

Table 3.5 Pre-class learning content development systems for smart university: A list of analyzed systems

#	Name of a system analyzed	Systems' developer (company)	System's technical platform(s)	Ref. #
Commercial systems				
1	Camtasia Studio	Techsmith	Windows/Mac	[42]
2	Adobe Presenter 11	Adobe	Windows/Mac	[43]
3	Movavi screen capture studio V7.3.0	Movavi	Windows/Mac	[44]
4	SmartPixel	SmartPixel	Windows/Mac/Android	[45]
5	Snagit 13	Techsmith	Windows/Mac	[46]
6	Screenpresso PRO	ScreenPresso	Windows	[47]
7	Bandicam	Bandisoft	Windows/Mac	[48]
8	Debut video capture software	NCH Software	Windows/Mac	[49]
9	CamVerce 1.95	Support	Windows	[50]
10	Replay Video Capture 8	Applian technologies	Windows	[51]
11	WM Recorder Bundle	WM Recorder	Windows	[52]
12	Adobe Captivate 9	Adobe	Windows/Mac	[53]
Open source (free) systems				
1	CamStudio	Cam Studio	Windows	[54]
2	Ezvid screen recorder	Ezvid	Windows	[55]
3	Screencast-O-Matic	Screen cast Matic	Windows/Mac	[56]
4	ActivePresenter	Atomi	Windows	[57]
5	Jing	Techsmith	Windows	[58]
6	Webinaria	Webinaria	Windows	[59]
7	Rylstim	SketchMan Studio	Windows	[60]
8	IceCream screen recorder	IceCream Apps	Windows/Mac	[61]
9	Flash back Express Recorder	BlueBerry Software	Windows	[62]
10	Screen Video Recorder	DVDVideo Soft	Windows	[63]

(4) Web-based audio- and video-conferencing systems for SmU (Tables 3.16, 3.17, 3.18 and 3.19);

(5) collaborative learning systems for SmU (Tables 3.20, 3.21, 3.22 and 3.23);

(6) context awareness systems for SmU (Tables 3.24, 3.25, 3.26 and 3.27).

Additionally, a summary of our research outcomes and findings for software systems, which are focused on students with disabilities at SmU, is presented in [41]. Those software systems include (a) voice-to-text (voice recognition) systems, (b) text-to-voice (voice synthesis) systems, and (c) gesture and motion recognition systems.

3.3.1 Smartness University: Pre-class Learning Content Development Systems

A brief summary of our research outcomes for pre-class learning content development systems for SmU is presented in Tables 3.4, 3.5, 3.6 and 3.7.

3.3.2 In-class Activities Recording Systems

A brief summary of our research outcomes for in-class activities recording systems for SmU is presented in Tables 3.8, 3.9, 3.10 and 3.11.

3.3.3 Post-class Activities' Supporting Systems

A brief summary of our research outcomes for post-class activities' supporting systems for SmU is presented in Tables 3.12, 3.13, 3.14 and 3.15.

Table 3.6 Pre-class learning content development systems for smart university: A list of most important features in existing systems

#	Desired system's feature	Details
1	Recording from desktop resources	Allows to record a video of what's happening on your computer screen
2	Recording of streaming audio	Allows to record video and audio streams as they come onto a computer
3	Scheduled recording	Ability to set a time and date for an application to automatically record what's happening on computer screen
4	Cropping and splice	Allows instructor to crop the unwanted or extra portions from a recording
5	Convert audio files	Allows to convert audio files into MP3, WMA, WAV formats
6	Watermarks and titles	Allows instructor to add a title information to the beginning or end of video and audio files
7	Split/join video and audio files	Allows instructor to trim unwanted portions away and insert other elements into video and audio files
8	Cursor effects	Selecting important information and highlighting it in recordings
9	Take screenshot	Capability to create screenshot images
10	Clean-up audio	Ability to fine-tune the audio portion of recording; users can alter features like volume, pitch, and reduce background noise
11	User forums	Provides students with a platform to discuss learning topics
12	Zoom in	Ability to zoom in on a portion of your screen helps your audience focus on and better understand your message.
13	Resize graphics	Ability to resize recorded video/graphics to fit window user is viewing in

Table 3.7 Pre-class learning content development systems for smart university: A list of recommended systems

#	Name of a system	Platform (s)	Price per copy (in $)	Ref. #
Commercial systems				
1	Camtasia Studio	Windows/Mac	$179 yr/$75 yr	[42]
2	Adobe Presenter 11	Windows/Mac	$149/yr	[43]
3	Movavi screen capture studio V7.3.0	Windows/Mac	$49.95 per copy	[44]
Open source (free) systems				
1	Screencast-O-Matic	Windows/Mac	Open Source	[56]
2	CamStudio	Windows	Open Source	[54]
3	Ezvid screen recorder	Windows	Open Source	[55]

Note Several co-authors developed (as a part of major NSF grants in 1999-2004) and actively used their own Web-lecturing system from 2001 to 2016—the InterLabs system; the examples of developed online learning content are available at http://www.interlabs.bradley.edu/cis573/. In this case, the Microsoft Internet Explorer web browser should be used to watch pre-recorded video lectures because the InterLabs system is based on Microsoft Media player and codecs. However, as a tangible result of this research project, the same co-authors switched to a very useful and multi-functional Screen-O-Matic system. Particularly, it significantly helps to create pre-class learning content and easily post it on YouTube to avoid any potential problems due to various technical platforms (devices) used by users (students, faculty, or learners) of developed learning content

Table 3.8 In class activities recording systems for smart university: A list of system's desired features

#	Desired system's feature	Details
1	Screen recording	Ability to capture the contents on computer screen like videos, PPT slides, animations, computer simulations, etc.
2	Live webcasting	Ability to webcast (over the Internet) classes online to remote students
3	Multi camera video	Video should be recorded and presented by multiple video cameras (views)
4	Mobile streaming	Allows faculty to broadcast live video from various mobile devices
5	Capturing of in-classroom activities	All activities (teaching, discussions, presentations, etc.) in a classroom should be captured and stored (for possible after class re-play) to provide "presence-in-classroom" effect (feeling) to remote students
6	Customization	Faculty should have an opportunity to create and edit customized instructional content
7	Sensing and automated recording	Ability to sense various activities inside a smart classroom and start recording automatically
8	Video recording management	Ability to capture video from different angles should be taken and maintained properly
9	Scheduling and automation	Basic and general-purpose activities in a smart classroom should be scheduled and/or automated (for example, an identification and registration of all in-class and remote students, automatic turn-on and set-up of all needed equipment in a smart classroom in accordance with profiles of a specific instructor or particular class, etc.)
10	Media board	Facilitates remote students to smoothly interact with local classroom

Table 3.9 In-class activities recording systems for smart university: A list of analyzed systems

#	Name of a system analyzed	Systems' developer (company)	System's technical platform(s)	Ref. #
Commercial systems				
1	Panopto	Panopto	Windows/Mac/Linux	[64]
2	Echo360 Lecture Capture	Echo360, Inc.	Windows	[65]
3	Mediasite	Sonic Foundry Inc.	Windows/Mac/Linux	[66]
4	Camtasia	TechSmith	Windows/Mac	[67]
5	Valt Software	Intelligent Video Solutions	Windows/Mac	[68]
6	Yuja Lecture Capture/Room Webcasting	YuJa Active Learning	Windows	[69]
7	Lecture Recording System	Beegeesindia	Windows	[70]
8	VIDIZMO	VIDIZMO	Windows/Mac	[71]
9	GALICASTER VIDEO PLATFORM	TELTEK	Windows	[72]
10	Adobe Presenter 11	Adobe	Windows/Mac	[73]
Open source (free) systems				
1	Opencast Matterhorn	Matterhorn	Windows/Mac	[74]
2	Kaltura	Kaltura	Windows	[75]
3	Class X	Classx	Linux	[76]
4	CamStudio	RenderSoft	Windows	[77]
5	Audionote	AudioNote	Android/Windows/Mac	[78]
6	Lecture Recordings, Lecture Notes	Everyone	Android/Windows/Mac	[79]
7	Super Notes	SuperNote	Android/Windows/Mac	[80]
8	SameView	Publication	Windows	[81]
9	INKredible	Viet Tran	Windows/IOS/Android	[82]
10	Squid	Steadfast Innovation, LLC	Windows/IOS/Android	[83]

3.3.4   Web-Based Audio- and Video-Conferencing Systems

A brief summary of our research outcomes for Web-based audio and video-conferencing systems for SmU is presented in Tables 3.16, 3.17, 3.18 and 3.19.

3.3.5   Collaborative Learning Systems for Smart University

A brief summary of our research outcomes for collaborative learning systems for SmU is presented in Tables 3.20, 3.21, 3.22 and 3.23.

Table 3.10 In-class activities recording systems for smart university: A list of most important features in existing systems

#	Desired system's feature	Details
1	Video capturing	Video of faculty is captured from different angles
2	Computer screen recording	Ability to capture the contents on computer screen like videos, PPT presentations, graphics, diagrams, photos, etc.
3	Download and publish recordings	The recorded learning content can be downloaded from or published into a repository of files with various components of learning content
4	Multi-point capture	Shows both presentation and presenter by capturing video, audio, and screen (or SMART board) in a smart classroom
5	Live streaming/Webcasting	Sessions can be made live/online to remote students
6	Voice narration	Ability to detect and record the voice of presenting or talking person
7	Capture keyboard input	Allows students to see what the professor is typing on keyboard of his/her laptop or desktop computer in smart classroom
8	Easy sharing	Recordings should be automatically be added to repository of files
9	Mobile streaming	Allow users to broadcast videos from their mobile devices onto main desktop computer in smart classroom
10	Video management	Manage camera and recording settings, view multiple camera feeds, and set alerts for tampering and motion detection
11	Watch anywhere	Plays files directly from browser on any technical platform and device like mobile, tablet, laptop, etc.
12	Media board	Allows remote students to smoothly interact with in-classroom students
13	Automated recording	Ability to sense various activities inside a smart classroom and start recording automatically
14	Remainders/alerts	Alerts are given to the students/faculty/learners before class starts
15	Presentation recording	Ability to record presentations by instructor (audio + video + PPT slides + computer simulations and other components)

3.3.6 Context Awareness Systems for Smart University

In the general case, there may be multiple types of context awareness systems to be used by SmU; they primarily deal with user's awareness of (a) learning environment, (b) learning process, (c) location on campus or inside a building, (d) safety or security inside a building or on campus, etc. As a result, it is almost impossible to integrate all types of desired user "awareness" in one system. However, below we provide (a) a united list of most desired features for learning/environment/location/safety-related context awareness systems for SmU and (b) lists of identified systems that cover various components or fragments of learning/environment/ location/safety-related awareness.

Table 3.11 In-class activities recording systems for smart university: A list of recommended systems

#	Name of a system	Platform (s)	Price per copy (in $)	Ref. #
Commercial systems				
1	Panopto	Windows/Mac/Linux	SaaS, per user $50	[64]
2	Echo360 Lecture Capture	Windows	$400 per classroom or $15 per student	[65]
3	Mediasite	Windows/Mac/Linux	$20,000 (for department CS)	[66]
Open source (free) systems				
1	Opencast Matterhorn	Windows/Mac	Open Source	[74]
2	Class X	Linux	Open Source	[75]
3	Kaltura	Windows	Open Source	[76]

Note Bradley University (Peoria, IL, U.S.A.) equipped 11 smart classrooms between 2014 and 2017. The Panopto system has been purchased, installed, and is actively used in 11 smart classrooms at Bradley to record all in-classroom activities

3.4 Towards a Smart University: Developed Components at Bradley University (Examples)

Bradley University (Peoria, IL, U.S.A.) is actively involved in research and development of SmU conceptual models, strategies, smart learning environments, smart classrooms, smart software and hardware systems, etc. in order to move from a traditional university model towards a well-thought and well-discussed smart university model. Several Bradley University pioneering initiatives in the area of design and development of smart classrooms are presented below.

Smart classrooms built (2014–2015, Westlake 316 project). From 2014 to 2016, Bradley University contracted the Crestron company (http://www.crestron.com) to set up top-quality multimedia Web-lecturing and capturing equipment for eleven (11) smart classrooms with different software/hardware configurations and set ups from a generic set of equipment for smart classroom (Fig. 3.2). For example, a smart classroom in room 316A in Westlake Hall of Bradley University is equipped with a) 84" smart board (with smart board projector), (2) HD Pro video camera and corresponding Capture HD software, (3) instructor's console with a smart control unit, (4) ceiling-mounted projector, (5) ceiling mounted microphones, (6) document camera, (7) speakers., and (8) Ponopto software system for recordings all in-classroom activities, etc.

Smart classroom of the 2nd generation (2017, Bradley 160 project). In 2017, Bradley University and, specifically, the College of Liberal Arts and Sciences (LAS), designed and developed a smart classroom of a new generation [11]—it is

Table 3.12 Post-class activities supporting systems for smart university: A list of system's desired features

#	Desired system's feature	Details
1	Video Webcasting	Allows instructors to webcast recorded class activities to students
2	Quizzes and polls	Allows instructors to quickly create quizzes and polls and assign it to a class or individual student within class
3	Mobile streaming	Allows instructors to broadcast live video from mobile devices; in this case, students can access those files using their mobile devices
4	Media uploading	Allows instructors to upload different media content
5	Interactive distance learning	Facilitates interactive (i.e., with active 2-way communication) online instructions and/or audio/video conferencing
6	Secure learning assignment submission	Allows instructors to post learning assignments on course web site; allows students to securely submit learning assignments
7	Automatic publishing	Allows instructor to easy publish various course components and learning content (recorded lectures, assignments, grades, notes, announcements, etc.) on course web site
8	Video streams' management	Allows instructor to manage camera and recording settings, view multiple camera feeds, and set alerts for tampering and motion detection
9	Scheduling and automation	Common learning activities can be scheduled and/or automated
10	Discussion threads	Allows students and instructors to discuss various after a class
11	Remainders/alerts	Facilitates students with remainders and/or alerts about the assignments
12	Post-editing	Allows instructor to edit files with recorded class activities
13	Inside-video search	Allows students to search inside the posted videos when needed

considered Phase# 1 (Sep 1, 2016—Feb 1, 2017) of the to-be-developed Center of Smart Education at LAS College and Bradley University in room 160 of Bradley Hall at Bradley University—the so-called Bradley 160 project (Fig. 3.3).

As of February 1, 2017, it is equipped with (1) a new type of SMART board—SMART Board 84—that actually works as a very big tablet with 84-inch touchable screen (the market cost is about $12,000); (2) twenty one (21) DELL 7459 AIO computers with built-in 3D video cameras and microphones (the market cost is about $1,100 per unit); each computer may serve as both desktop computer as well as flat 24" tablet with touchable screen; (3) ceiling-mounted projectors (market cost is about $1,000), (4) at least three (as of March 2017) 55" big screen TVs (market cost is about $750) for virtual presence in a classroom and communication/

Table 3.13 Post-class activities supporting systems for smart university: A list of analyzed systems

#	Name of a system analyzed	Systems' developer (company)	System's technical platform (s)	Ref. #
Commercial systems				
1	Panopto	Panopto	Windows/Mac/Linux	[64]
2	Echo360 Lecture Capture	Echo360, Inc.	Windows	[65]
3	Tegrity	McGraw Hill Education	Android/Windows/Mac/IOS	[84]
4	Camtasia	TechSmith	Windows/Mac	[67]
5	Adobe Presenter 11	Adobe	Windows/Mac	[73]
6	Corel Video Studio Pro X9.5	Corel	Windows/Mac	[85]
7	Power Director 15	Cyber link	Windows/Mac/Linux	[86]
8	Yuja Lecture Capture/Room Webcasting	YuJa Active Learning	Windows	[87]
9	Mediasite Lecture Capture	SonicFoundry	Windows/Mac/Linux	[88]
10	Mediatech Custom Classroom	Gomediatech	Windows	[89]
Open source (free) systems				
1	Sakai	SakaiProject	Windows	[90]
2	Moodle	Moodle	Windows/Mac	[91]
3	ATutor LMS	Atutor	Windows/Mac	[92]
4	CamStudio	RenderSoft	Windows	[93]
5	RCampus	Rcampus	Windows	[94]
6	Learnopia	Learnopia	Windows	[95]
7	Claroline	claroline	Windows	[96]
8	VideoPad Video Editor	NCH software	Windows/Mac	[97]
9	Jing	Techsmith	Windows	[98]
10	VSDC Free Video Editor	Videosoft	Windows	[99]

collaboration with remote/online students, (5) multiple video cameras, (6) speakers, (7) three students collaboration areas (big tables with chairs that are close to big screen TVs), and other electronics.

Support of research in Smart University area (Bradley grant REC # 1326809). In 2015, Bradley University OSP awarded one of the co-authors with a grant to support research, design, and develop conceptual models of a smart university, identify suitable software and hardware systems, smart technology, smart pedagogy, etc., in order to identify and investigate multiple aspects of Bradley University transition from a traditional university towards a smart university. A summary of up-to-date research outcomes and findings are already available in various publications by members of research team [11, 22, 23, 37, 38, 39, 40, and

Table 3.14 Post-class activities supporting systems for smart university: A list of most important features in existing systems

#	Desired system's feature	Details
1	Integrated quizzes and polls	Strengthens two-way communication and builds engagement with the system
2	Inside-video search	Provides search of the content of video lectures by automatically transcribing words used in the lecturer's visual aids
3	Digital notes and bookmarks	Facilitates student to type their notes directly into files on his/her computer or tablet
4	Record videos anywhere	Provides access to all uploaded and shared screen captures and recorded videos from anywhere on the Web
5	Secure assignment submission	Facilities for submission of assignments, their subsequent testing and marking, and the provision of feedback on assignments to students
6	Interactive distance learning	Supports regular and substantive interaction between the students and the instructor, either synchronously or asynchronously, as well as student-to-student interaction
7	Mobile streaming	Allows users to broadcast live video from their mobile devices with the touch of a button
8	Media uploading	Allows instructors to upload content produced by their institutions, researchers and even students, regardless of what tools were originally capture this content
9	Video editor	Allows instructor to edit videos after recorded and posted
10	Session management	Helps with management of learning or discussion session
11	Chat	Facilitates students to chat online
12	Multipoint capture	Allows system to capture multiple views from instructor's screen
13	Pause recording	Allows instructor to pause a recording of lectures or class activities
14	Broadcast class live	Facilitates instructors to Web-cast classes live (in real time)

Table 3.15 Post-class activities supporting systems for smart university: A list of recommended systems

#	Name of a system	Platform (s)	Price per copy (in $)	Ref. #
Commercial systems				
1	Panopto	Windows/Mac/Linux	SaaS, per user $50	[64]
2	Echo360 Lecture Capture	Windows	$400 per classroom or $15 per student	[65]
3	Tegrity	Android/Mac/Windows/IOS	Need Based	[84]
Open source (free) systems				
1	Sakai	Windows/Mac	Open Source	[90]
2	Moodle	Windows/Mac	Open Source	[91]
3	ATutor LMS	Windows/Mac	Open Source	[92]

Note Bradley University (Peoria, IL) actively uses the Sakai learning management system and the Panopto system for various post-class activities. On the other hand, some online programs on Bradley campus actively use the Pearson Embanet system for full-scale online teaching and online program management

Table 3.16 Web-based audio- and video-conferencing systems for smart university: A list of system's desired features

#	Desired system's feature	Details
1	Recording	Allows users to record video and/or audio conference and review it when needed
2	Chat/text	Allows students and faculty to chat or send instant textual messages
3	Voice calling	Allows users to make voice calls to other users available online
4	Video conferencing	Allows users to make video calls to others online users using the Internet
5	Web casting	Allows live stream video meetings on different media and/or record them for post-editing
6	Mobility	Facilitates conversations to be synchronized across various technical platforms such as make or receive voice or video calls over Wi-Fi with devices that use iOS and Android operating systems
7	Screen sharing	Facilitates faculty and students to share their desktop screen with each other and other students (usually, this feature is controlled by an instructor)
8	File sharing	Allows faculty to share different files with students
9	Group conversations	Facilitates creation of various student groups, a single group calling to multiple selected students at a time, and sharing information between them
10	Drawing tools	Facilitates user to make notes or mark certain fragments on computer screens and videos to highlight certain things on screen or video

41]. As a result, Bradley University installed a well-recognized national and international profiles in the areas of Smart Education, Smart e-Learning, Smart Classrooms and Smart University.

As an integral part of this research-focused grant, during Phase # 2 of this project (Feb 1–May 15, 2017), the Center for Smart Education will be additionally equipped with multiple identified, analyzed and tested software systems for smart education (as described above) and corresponding smart devices.

Additionally, a special emphasis in this research project is given to installation, analysis and testing of software systems to support students with disabilities in the Center of Smart Education [39, 41].

3.5 Conclusions. Future Steps

Conclusions. The performed research, and obtained research findings and outcomes enabled us to make the following conclusions:

(1) Leading academic intuitions all over the world are investigating ways to transform the traditional university into a smart university and benefit from the advantages of a smart university. Smart University concepts, principles,

Table 3.17 Web-based audio- and video-conferencing systems for smart university: A list of analyzed systems

#	Name of a system analyzed	Systems' developer (company)	System's technical platform(s)	Ref. #
Commercial systems				
1	Cisco Webex	Cisco	Windows/Mac/Linux	[99]
2	Go to Meeting	Citrix	Windows/Mac	[100]
3	ClickMeeting	ClickMeeting	Windows/Mac	[101]
4	Readytalk	ReadyTalk	Windows/Mac	[102]
5	BigMarker	BigMarker	Windows/Mac	[103]
6	Adobe connect	Adobe	Windows/Mac/Linux	[104]
7	Onstream meeting	OnStream	Windows/Mac	[105]
8	Blackboard collaborate	Blackboard	Windows/Mac	[106]
9	Ring Central	RingCentral	Windows/Mac/Linux	[107]
10	GlobalMeet	PGI	Windows/Mac	[108]
Open source (free) systems				
1	Google Hangouts	Google	Windows/Mac	[109]
2	Skype	Microsoft	Windows/Mac	[110]
3	BigBlueButton	BigBlueButton	GNU/Linux	[111]
4	Meeting Burner	MeetingBurner	Windows/Mac	[112]
5	Join.me	Join me	Windows/Mac	[113]
6	Team viewer	TeamViewer	Windows/Mac	[114]
7	Zoom	Zoom	Windows/Mac	[115]
8	Zoho Meeting	Zoho	Windows/Mac	[116]
9	Mikogo	Mikogo	Windows/Mac	[117]
10	Yugma	Yugma	Windows/Mac/Linux	[118]

technologies, systems, and pedagogy will be essential parts of multiple research, design, and development projects in upcoming years.

(2) Our vision of SmU is based on the idea that SmU—as a smart system—should implement and demonstrate significant maturity at various "smartness" levels (Table 3.1) or distinctive smart features, including (1) adaptation, (2) sensing (awareness), (3) inferring (logical reasoning), (4) self-learning, (5) anticipation, and (6) self-organization and re-structuring—the corresponding research outcomes are presented in Table 3.2.

(3) It is necessary to create a taxonomy of a smart university, i.e. to identify and classify SmU main (a) features, (b) components (smart classrooms, techno-logical resources—systems and technologies, human resources, financial resources, services, etc.), (c) relations (links) between components, (d) inter-faces, (e) inputs and outputs, and (f) limits/constraints. The premise will (a) enable SmU designers and developers to identify and predict most effective software, hardware, pedagogy, teaching/learning activities, services, etc., for the next generation of a university—smart university—and (b) help traditional

Table 3.18 Web-based audio- and video-conferencing systems for smart university: A list of most important features in existing systems

#	System's feature	Details
1	Accessibility	Easy (and intuitive) use of a system by students and faculty
2	Flexibility	Provides support (mobile applications) to various mobile communication devices –smartphones, tablets, etc. for secure and reliable business and personal communication
3	Call management	Allows user to maintain and manage calls
4	Instant messaging	Allows students and faculty to chat or send instant textual messages
5	Voice calling	Allows users to make voice calls to other online users
6	Video conferencing	Allows users to make video calls to others online users using the Internet
7	Web casting	Allows live stream video meetings on different media and/or record them for post-editing
8	Cross Platform	Facilitates conversations to sync across the devices make or receive voice or video calls over Wi-Fi with iOS and android.
9	Screen sharing	Facilitates faculty and students to share their desktop screen with each other and other students (usually, this feature is controlled by an instructor)
10	File sharing	Allows faculty to share different files with students
11	Group conversations	Facilitates creation of various student groups, a single group calling to multiple selected students at a time, and sharing information between them
12	Drawing tools	Facilitates user to make notes or mark certain fragments on computer screens and videos to highlight certain things on screen or video
13	Zoom and annotate	Zoom in on what is being shared and annotate your application or slideshow
14	Multiple monitor support	Presenters can switch between multiple monitors during a meeting; typically, the presenter will set up a specific monitor to use for desktop sharing

Table 3.19 Web-based audio- and video-conferencing systems for smart university: A list of recommended systems

#	Name of a system	Platform (s)	Price per copy (in $)	Ref.#
Commercial systems				
1	Cisco Webex	Windows/Mac/Linux	$19/month Annually	[99]
2	Go To Meeting	Windows/Mac	$19/month Annually	[100]
3	ClickMeeting	Windows/Mac	$25/month Annually	[101]
Open source (free) systems				
1	Google Hangouts	Windows/Mac	Open Source	[109]
2	Skype	Windows/Mac	Open Source	[110]
3	BigBlueButton	GNU/Linux	Open Source	[111]

Table 3.20 Collaborative learning systems for smart university: A list of system's desired features

#	Desired system's feature	Details
1	Web-based meetings at any time	Allows students/learners/faculty at different locations to work as a virtual group of project members, have online meetings/discussions, and share content or documents in real time over the Internet
2	Shared whiteboard space	Allows in-class and remote students and faculty to work together in real time over the Internet and communicate each other's thoughts and ideas and share content (using special or smart boards or tables)
3	Active online discussions and communications	Students can openly discuss and share their thoughts with their group of students, project team members or everyone in the class
4	File uploading and sharing	Faculty, students, learners, and tutors can upload different files related to student groups' activities and course learning content and share them online with a group or all classmates
5	Problem-based learning	Team-based working on student project improves student engagement and retention of learning content
6	Group calling and communication	Student group leader or moderator or tutor can call a group virtual meeting and talk to a particular group of students/learners (probably course project team members) online using various available Web-based communication tools
7	Chat/group chat	Student can talk to other student or a group of students and share ideas/thoughts/docs
8	Annotation of readings	Allows students to add notes for clarity of understanding and visually communicate thoughts/ideas/questions to other members in student group or project
9	Scheduling	Student group leader or moderator or tutor can schedule various events/meetings/sessions with various groups of students
10	Customizing content and docs to be discussed	Student group leader or moderator or tutor should be able to customize the content for a group or individual advising session as needed per each student group or individual student
11	Recording of collaborative learning sessions	All audio/video collaborative learning sessions should be recorded for possible re-play (if needed later by student group members)
12	Screen capturing	Student group leader or moderator or tutor should be able to capture all activities/processes/graphics on group's main (shared) computer screen or smart board, record, store and re-play them (if needed); it is especially important feature for "brain storming" types of collaborative group-base meetings or sessions when student bring/write ideas onto Web-based virtual "table/desk"
13	Assignments/quizzes and review/grading reports	Student group leader or moderator or tutor should be able to create assignments for various groups of students, and provide those groups with review/grading outcomes (like grading reports)

(continued)

Table 3.20 (continued)

#	Desired system's feature	Details
14	Notifications	Students should be notified about a scheduled or new upcoming event or activity
15	Reports	Ability to generate automatically various types of reports on student group activities or individual student academic performance (attendance of virtual group meetings, time spent on virtual discussions, time spent on completing test or quiz, number of attended virtual group meetings per week, etc.)

Table 3.21 Collaborative learning systems for smart university: A list of analyzed systems

#	Name of a system analyzed	Systems' developer (company)	System's technical platform(s)	Ref. #
Commercial systems				
1	Basecamp	Basecamp	Windows, Mac, Linux	[119]
2	Yammer	Yammer, Microsoft	Windows, Android, IOS	[120]
3	Blackboard	Blackboard	Windows, Mac	[121]
4	Haikyulearning	Haikyu	Windows, Mac, Android	[122]
5	OpenText FirstClass	OpenText	Windows, Mac, Linux	[123]
6	LiveText	LiveText	Windows, Mac, Linux	[124]
7	Yugma	YSL	Windows, Mac, Linux	[125]
8	Twiddla	Twiddla	Windows, Mac	[126]
9	Mindmeister	Mindmeister	Windows, Mac	[127]
10	Mikogo	mikogo.com	Windows, Mac, Linux	[128]
Open source (free) systems				
1	Edmodo	edmodo.com	Windows/Mac	[129]
2	Wikispaces	Tangient LLC	Windows, Mac	[130]
3	Wiggio	Wiggio	Windows/Mac	[131]
4	Skype, Outlook	Microsoft	Cross platform	[132]
5	Connect	McGraw-Hill Education	Windows, Mac	[133]
6	Sloodle	Sloodle	Windows, Mac	[134]
7	Google Drive, Hangouts	Google	Cross platform	[135]
8	Web Poster Wizard	ALTEC		[136]
9	OpenStudy	OpenStudy	Windows, Mac	[137]
10	Oovoo	Oovoo LLC	Windows, Mac, Android	[138]

universities to understand, identify, and evaluate paths for a transformation into a smart university. The proposed and developed conceptual modes of SmU are presented in Sect. 3.2 above.

(4) Based on our vision of SmU and outcomes of our research, SmU may have multiple components of a traditional university; however, it must have

Table 3.22 Collaborative learning systems for smart university: A list of most important features in existing systems

#	System's feature	Details
1	Web-based meetings at any time	Allows students/learners/faculty at different locations to work as a virtual group of project members, have online meetings/discussions, and share content or documents in real time over the Internet
2	Shared whiteboard space	Allows in-class and remote students and faculty to work together in real time over the Internet and communicate each other's thoughts and ideas and share content (using special or smart boards or tables)
3	Active online discussions and communications	Students can openly discuss and share their thoughts with their group of students, project team members, or everyone in the class
4	File uploading and sharing	Faculty, students, learners, and tutors can upload different files related to student groups' activities and course learning content and share them online with a group or all classmates
5	Chat/group chat	Student can talk to other student or a group of students and share ideas/thoughts/docs
6	Annotation of readings	Allows students to add notes for clarity of understanding and visually communicate thoughts/ideas/questions to other members in student group or project
7	Scheduling	Student group leader or moderator or tutor can schedule various events/meetings/sessions with various groups of students
8	Customizing content and docs to be discussed	Student group leader or moderator or tutor can to customize the content for a group or individual advising session as needed per each student group or individual student
9	Recording of collaborative learning sessions	All audio/video collaborative learning sessions should be recorded for possible re-play (if needed later by student group members)
10	Screen capturing	Student group leader or moderator or tutor should be able to capture all activities/processes/graphics on group's main (shared) computer screen or smart board, record, store and re-play them (if needed); it is especially important feature for "brain storming" types of collaborative group-base meetings or sessions when student bring/write ideas onto Web-based virtual "table/desk"
11	Notifications	Students should be notified about a scheduled or new upcoming event or activity
12	Reports	Ability to generate automatically various types of reports on student group activities or individual student academic performance (attendance of virtual group meetings, time spent on virtual discussions, time spent on completing test or quiz, number of attended virtual group meetings per week, etc.

(continued)

Table 3.22 (continued)

#	System's feature	Details
13	Search	User should be able to search and find a specific item in the system
14	Assignments/quizzes and review/grading reports	Student group leader or moderator or tutor should be able to create assignments for various groups of students, and provide those groups with review/grading outcomes (like grading reports)
15	Group calling and communication	Student group leader or moderator or tutor can call a group virtual meeting and talk to a particular group of students/learners (probably course project team members) online using various available Web-based communication tools
17	Flexibility of software	Easy to understand and use by students software tools inside the system; easy access to registered students or group of students at any time and from anywhere
18	Interactive collaborative lessons	Highly interactive group meetings and/or sessions with 2-way high quality and reliable audio/video/communication to increase students engagement into collaborative learning and brain storming (research) activities
19	Monitoring and trackability	Instructor/group leader/mentor/tutor is able to easy monitor and track student group's activities, progress in a particular project or task, and also individual student's involvement and progress in a specific task or project or assignment

Table 3.23 Collaborative learning systems for smart university: A list of recommended systems

#	Name of a system	Platform (s)	Price per copy (in $)	Ref. #
Commercial systems				
1	Basecamp	Windows, Mac, Linux	$29 per month	[119]
2	Yammer	Windows, Android, IOS	$12.50 per month/user	[120]
3	Blackboard	Windows, Mac	N/A (need based)	[121]
Open source (free) systems				
1	Edmodo	Windows, Mac	Open Source	[129]
2	Wikispaces	Windows, Mac	Open Source	[130]
3	Wiggio	Windows, Mac	Open Source	[131]

numerous additional components to support and maintain SmU distinctive features—a summary of our research outcomes is presented in Table 3.3 above.

(5) One of the most distinctive features of SmU will be multiple software systems that are usually not used by a traditional university. The obtained research data on this topic is summarized in twenty-four (24) tables presented in Sect. 3.3 above. Our research team carefully analyzed 120 + suitable software systems, carefully tested 50+ systems, and recommended 18 open-source (free) and 18

Table 3.24 Context awareness systems for smart university: A list of system's desired features

#	Desired system's feature	Details
1	Adaptation	Leaning context awareness and adaptation of learning activities, teaching style and the learning content in accordance with a) current (available) learning environment, b) academic background of current students, c) instructor's current needs and/or profile, d) students' current needs, etc.
2	Dashboard monitoring	In general, monitoring of various situations and providing metrics; particularly, monitoring of quality of learning by students, student activities, student academic performance, etc.
3	Face finding and recognition	Detects faces of different people in various types of environment—learning environment (in classroom, in a lab), buildings, on campus, etc.
4	Motion detection and recognition	To sense or detect motion of different people and objects in a classroom, lab, building, on campus, etc.
5	Gesture recognition	To identify the gestures used by instructor, TA, tutor, or students in learning environment
6	Smart surveillance	Monitoring of activities or behavioral patterns or any changes in learning environment using different types of smart devices; particularly, this feature is important for safety and security in the classrooms, labs, buildings, and on campus
7	Recording	Automatic high quality audio and video recording of various activities (situations) in classrooms, labs, buildings, and various areas on campus
8	Predictive analytics	Processed data obtained from various sensors make predictions about nest steps/actions in learning activities, or location on campus, or safety, security, etc.
9	Quick video processing and analytics	Quick and relatively easy processing of big data from the videos recorded (for example, surveillance video cameras) and processing of these data (getting analytics)
10	Quick and easy access from anywhere	Easy access by instructor (or, probably, safety officer) to real time or recorded video/audio/information from almost anywhere, but at least from a central context awareness unit (system)
11	Notifications	Send regular notifications via Email, text message, or phone call
12	Alerts	Broadcast real time immediate safety or security alerts to all people on campus and mass notifications in case of emergency
13	Smart navigation	Provides users with high accurate data about unknown locations

commercial systems for possible deployment by SmU—see Tables 3.7, 3.11, 3.15, 3.19, 3.23, 3.27 for details.

(6) Bradley University already created and implemented several smart classrooms in its curricula. A design and development of a pioneering smart classroom of the 2nd generation started in 2016. The details of developed smart classrooms are described in Sect. 3.4 above. Based on the pilot teaching of classes in smart

Table 3.25 Context awareness systems for smart university: A list of analyzed systems

#	Name of a system analyzed	Systems' developer (company)	System's technical platform(s)	Ref. #
Commercial systems				
1	NiceVision (Qognify)	Qognify	Windows, Mac	[139]
2	Blue Iris	Perspective Software	Windows	[140]
3	Sighthound	Singhthound Inc	Windows	[141]
4	EyeLine Video Surveillance	NCH Software	Windows	[142]
5	SARA	Status Solutions	Windows	[143]
6	ZoneTrigger	Omega	Windows	[144]
7	Video Insight VI monitor 6	Panasonic	Windows	[145]
8	Fibaro	Fibar Group	Windows	[146]
9	Nice Vision	Qognify	Windows	[147]
10	IP Video Surveillance Software	Qognify	Windows	[148]
Open source (free) systems				
1	Capturix (Software Set)	Capturix Technologies	Windows	[149]
2	Video surveillance software	Contaware	Windows	[150]
3	ISPY Connect	DeveloperInABox	Windows	[151]
4	Intrance Motion detector	Intrance software	Windows	[152]
5	Zone Minder	ZoneMinder	Windows, Linux	[153]
6	Active WebCam	PY Software	Windows	[154]
7	123Motion 1	MakeItEasy	Windows	[155]
8	WebCam monitor	DeskShare	Windows	[156]
9	Yawcam	Yawcam	Windows	[157]
10	Netcam Studio	Netcam Studio	Windows, Mac	[158]

classrooms, the obtained feedback from faculty and students clearly shows a keen interest from students in high-tech smart education, a significant research interest from faculty to implement various smart systems and devices into smart classroom and smart education, and, in general, proved the correctness of major design and development proposals and solution to build and actively use smart classroom in Bradley curriculum.

Next steps. The next steps (Summer 2017—December 2018) of this multi-aspect research, design, and development project deal with

(1) Implementation, analysis, testing, and quality assessment of numerous components of smart software and hardware systems, smart devices, smart technology, and smart pedagogy in everyday teaching of classes in smart classrooms.
(2) Implementation, analysis, testing, and quality assessment of numerous components of smart software and hardware systems, smart devices, and smart

Table 3.26 Context awareness systems for smart university: A list of most important features in existing systems

#	System's feature	Details
1	Adaptation	Leaning context awareness and adaptation of learning activities, teaching style, and the learning content in accordance with (a) current (available) learning environment, (b) academic background of current students, (c) instructor's current needs and/or profile, (d) students' current needs, etc.
2	Dashboard monitoring	In general, monitoring of various situations and providing metrics; particularly, monitoring of quality of learning by students, student activities, student academic performance, etc.
3	Face finding and recognition	Detects faces of different people in various types of environment—learning environment (in classroom, in a lab), buildings, on campus, etc.
4	Motion detection and recognition	To sense or detect motion of different people and objects in a classroom, lab, building, on campus, etc.
5	Gesture recognition	To identify the gestures used by instructor, TA, tutor or students in learning environment
6	Smart surveillance	Monitoring of activities or behavioral patterns or any changes in learning environment using different types of smart devices; particularly, this feature is important for safety and security in the classrooms, labs, buildings, and on campus
7	Recording	Automatic high quality audio and video recording of various activities (situations) in classrooms, labs, buildings, and in various areas on campus
8	Predictive analytics	Process data obtained from various sensors, and make predictions about nest steps/actions in learning activities, or location on campus, or safety, security, etc.
9	Quick video processing and analytics	Quick and relatively easy processing of big data from the videos recorded (for example, surveillance video cameras) and processing of these data (getting analytics)
10	Quick and easy access from anywhere	Easy access by instructor (or, probably, safety officer) to real time or recorded video/audio/information from almost anywhere, but at least from a central context awareness unit (system)
11	Notifications	Send regular notifications via Email, text message, or phone call
12	Alerts	Broadcast real time immediate safety or security alerts to all people on campus and mass notifications in case of emergency
13	Smart Search	Smart Search features in video management software allow users to set up specific search parameters to quickly scan the most relevant clips in recordings
14	Protocols	Special protocols/rules/regulations to identify problems and make improvements to prevent any significant injures, damages, loss of life, property, business, convenience, or comfort
15	Mobile supportive applications	Active use of various multiple mobile applications on smart phones to (1) inform/alert students/faculty/learners/staff about any unwanted events or activities on campus, and (2) and help them to with information to avoid/escape from those situations to secure environment
16	Active use of ONVIF standards	Active use of ONVIF—Open Network Video Interface Forum—standards in all SmU network video products to provide compatibility of and synchronization in IP video cameras and systems from different manufacturers

Table 3.27 Context awareness systems for smart university: A list of recommended systems

#	Name of a system	Platform (s)	Price per copy (in $)	Ref. #
Commercial systems				
1	NiceVision (Qognify)	Windows, Mac	Need based	[139]
2	Blue Iris	Windows, Mac	$59.95/yr	[140]
3	Sighthound	Windows, Mac, Linux	Basic license $60/Pro license $$250	[141]
Open source (free) systems				
1	Capturix (Software Set)	Windows, Mac	Open Source	[149]
2	Video surveillance software	Windows	Open Source	[150]
3	ISPY Connect	Windows	Open Source	[151]

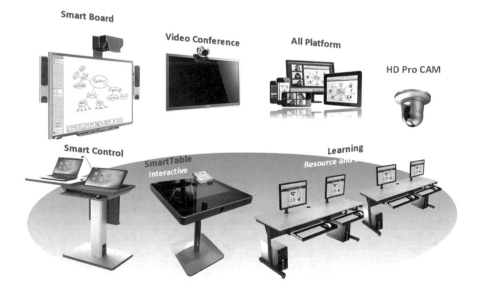

Fig. 3.2 A set of smart devices for usual smart classroom

technology at Bradley Hall (the home of majority of departments of the College of Liberal Arts and Sciences) and in some areas of the Bradley University campus.

(3) Organization and implementation of summative and formative evaluations of local and remote students and learners, faculty and professional staff, administrators, and university visitors with a focus to collect sufficient data on quality of SmU main components—features, software, technologies, hardware, services, etc.

Fig. 3.3 Smart classroom of 2nd generation at Bradley University–BR160 (with a focus on virtual presence of online/remote students in a classroom and active communication/collaboration of in-classroom instructor and students with remote/online students)

(4) Creation of a clear set of recommendations (technological, structural, financial, curricula, etc.) regarding a transition of a traditional university into a smart university.

Acknowledgements The authors would like to thank Dr. Cristopher Jones, Dean of the LAS College, and Sandra Shumaker, Executive Director, Office of Sponsored Programs at Bradley University for their strong support of our research, design and development activities in Smart University and Smart Education areas.

This project is partially supported by grant REC # 1326809 from Bradley University.

The authors also would like to thank Mr. Siva Margapuri, Ms. Mounica Yalamanchili, Mr. Harsh Mehta, Ms. Supraja Talasila, and Ms. Aishwarya Doddapaneni—the student research associates of the InterLabs Research Institute and/or graduate students of the Department of Computer Science and Information Systems at Bradley University—for their valuable contributions into this research project.

References

1. Xie, W., et al. Smart Classroom—an Intelligent Environment for Tele-Education, In: H.-Y. Shum, M. Liao, and S.-F. Chang (Eds.): PCM 2001, LNCS 2195, pp. 662–668 (2001)
2. Shi, Y., Xie, W., Xu, G., Shi, R., Chen, E., Mao, Y., Liu, F. The Smart Classroom: Merging Technologies for Seamless Tele-Education, Proc. 4th Int'l Conf. Multimodal Interfaces (ICMI 2002), IEEE CS Press, pp. 429–434 (2002)

3. Yau, S., Gupta, S., Karim, F. et al. Smart Classroom: Enhancing Collaborative Learning Using Pervasive Computing Technology, Proc. of 2003 ASEE conference, URL: http://www.public.asu.edu/~ bwang/publications/Smart Classroom-2003.pdf (2003)
4. O'Driscoll, C. et all. Deploying a Context Aware Smart Classroom, Proc. of the Arrow@DIR conference, Dublin, Ireland, http://arrow.dit.ie/engschececon/53/ (2008)
5. Huang, R., Hub, Y.,Yang, J., Xiao, G. The Functions of Smart Classroom in Smart Learning Age, Proc. 20th Int. Conf. on Computers in Education ICCE, Nanyang Technological University, Singapore, (2012)
6. Pishva, D., Nishantha, G. G. D. Smart Classrooms for Distance Education and their Adoption to Multiple Classroom Architecture, Journal of Networks, vol. 3, no. 5 (2008)
7. Gligorić, N., Uzelac, A., Krco, S. Smart Classroom: Real-Time Feedback on Lecture Quality, Proc. 2012 IEEE Int. Conf. on Pervasive Computing and Communications Workshops (PERCOM Workshops), pp. 391–394, 19–23 March 2012, Lugano, Switzerland, IEEE DOI:10.1109/PerComW.2012.6197517 (2012)
8. Slotta, J., Tissenbaum, M., Lui, M. Orchestrating of Complex Inquiry: Three Roles for Learning Analytics in a Smart Classroom Infrastructure, Proc. of the Third International Conference on Learning Analytics and Knowledge LAK'13, pp, 270–274, New York, NY, USA, ACM DOI:10.1145/2460296.2460352 (2013)
9. Koutraki, M., Maria, Efthymiou, V., Grigoris, A. S-CRETA: Smart Classroom Real-Time Assistance, IN: Ambient Intelligence—Software and Applications, Advances in Intelligent and Soft Computing, vol. 153, pp 67–74, Springer (2012)
10. Clarke, R.Y. The Next-Generation Classroom: Smart, Interactive and Connected Learning, http://www.samsung.com/in/business-images/resource/white-paper/2012/11/EBT15_1210_Samsung_ Smart_School_WP-0.pdf (2012)
11. Uskov, V.L., Bakken, J.P., Pandey, A. The Ontology of Next Generation Smart Classrooms. In: "Smart Education and Smart e-Learning", Eds: Uskov, V.L., Howlett, R.J., Jain, L.C., Springer, pp. 3–14, June 2015, 514 p., ISBN: 978-3-319-19874-3 (2015)
12. Hwang, G.J. (2014). Definition, framework and research issues of smart learning environments—a context-aware ubiquitous learning perspective, Smart Learning Environments—a Springer Open Journal, 1:4, Springer (2014)
13. Spector, J. (2014). Conceptualizing the emerging field of smart learning environments. Smart Learning Environments (1) (2014)
14. Abueyalaman, E.S.: et al Making a Smart Campus in Saudi Arabia. EDUCAUSE Quarterly 2, 1012 (2008)
15. Xiao, N. Constructing Smart Campus Based on the Cloud Computing Platform and the Internet of Things, Proc. 2nd int. conf. on Computer Science and Electronics Engineering (ICCSEE 2013), Atlantis Press, Paris, France, pp. 1576–1578 (2013)
16. Adamko, A., Kadek, T., Kosa, M. Intelligent and Adaptive Services for a Smart Campus Visions, concepts and applications, Proc. 5th IEEE int. conf. on Cognitive Infocommunications, Nov. 5–7, 2014, Vietri sul Mare, Italy, IEEE (2014)
17. Kwok, l. A vision for the development of i-campus, Smart Learning Environments—a Springer Open Journal, 2:2, Springer (2015)
18. IBM, Smart Education, available at https://www.ibm.com/smarterplanet/global/files/ au__en_uk__cities__ibm_smarter_education_now.pdf
19. Gamalel-Din, S.A.: Smart e-Learning: A greater perspective: from the fourth to the fifth generation e-learning. Egyptian Informatics Journal 11, 39–48 (2010)
20. Tikhomirov, V.P. Mir na puti k smart education. Novye vozmozhnosti dlia razvitiia [The World on the Way to Smart Education. New opportunities for Development]// Otkytoe obrazovanie [Open Education]. 2011 (3), pp. 22–28 (in Russian)
21. Neves-Silva, R., Tshirintzis, G., Uskov, V., Howlett, R., Lakhmi, J.: Smart Digital Futures. IOS Press, Amsterdam, The Netherlands (2014)
22. Uskov, V., Howlet, R., Jain, L. (eds.): Smart Education and Smart e-Learning. Springer, Berlin-Heidelberg, Germany (2015)

23. Uskov, V., Howlet, R. Jain, L. (Eds). Smart Education and e-Learning 2016, Springer, 643 p., ISBN: 978-3-319-39689-7 (2016)
24. e-Learning For Smart Classrooms, Smart Classroom Bytes journal, http://education.qld. gov. au/smartclassrooms/documents/strategy/pdf/scbyte-elearning.pdf (2008)
25. Global Smart Education Market 2016–2020. Research and Markets, http://www. researchandmarkets.com/research/x5bjhp/global_smart (2016)
26. Smart Education and Learning Market—Global Forecast to 2021, http://www. marketsandmarkets.com/Market-Reports/smart-digital-education-market-571.html
27. Coccoli, M., Guercio, A., Maresca, P., Stanganelli, L.: Smarter universities: a vision for the fast changing digital era. J. Visual Lang. Comput. **6**, 1003–1011 (2014)
28. Barnett, R.: The Future University. Routlege, New York (2012)
29. Temple, P. (ed.): Universities in the Knowledge Economy: Higher Education Organization and Global Change. Routledge, New York (2011)
30. Tikhomirov, V., Dneprovskaya, N.: Development of strategy for smart University, 2015 Open Education Global International Conference, Banff, Canada, 22–24 April (2015)
31. Hilton, M.: Exploring the Intersection of Science Education and 21st Century Skills: A Workshop Summary. National Research Council (2010)
32. Richey, R.C., Klein, J.D., Tracey, M.W.: The Instructional Design Knowledge Base: Theory, Research, and Practice. Routledge, New York (2010)
33. Ke, C.-K., Liu, K.-P., Chen, W.-C.: Building a smart E-portfolio platform for optimal e-learning objects acquisition. Math. Probl. Eng. **2013**, article ID 896027, 8 pp (2013)
34. Ruberg, L.: Transferring smart E-learning strategies into online graduate courses, SEEL2015. In: Smart Education and Smart e-Learning, Smart Innovation, Systems and Technologies, v.41, pp. 243–254. Springer (2015)
35. Uskov, V.L., Heinemann, C.: Smart Universities: literature review and creative analysis. In: Uskov, V., Bakken, J.P., Howlet, R., Jain, L. (eds) Smart Universities, Springer (accepted for publication in 2017)
36. Derzko, W.: Smart Technologies. http://archives.ocediscovery.com/discovery2007/ presentations/Session3Walter DrezkoFINAL.pdf (2007)
37. Uskov, A., Sekar, B.: Smart gamification and smart serious games. In: Sharma, D., Jain, L., Favorskaya, M., Howlett, R. (eds.) Fusion of Smart, Multimedia and Computer Gaming Technologies, Intelligent Systems Reference Library, 84, pp. 7–36, Springer. DOI 10.1007/978-3-319-14645-4_2, ISBN: 978-3-319-14644-7 (2015)
38. Uskov, V.L, Bakken, J.P., Pandey, A., Singh, U., Yalamanchili, M., Penumatsa, A.: Smart University taxonomy: features, components, systems. In: Uskov, V.L., Howlett, R.J., Jain, L. C. (eds.) Smart Education and e-Learning 2016, pp. 3–14. Springer 643 p., ISBN: 978-3-319-39689-7 (2016)
39. Bakken, J.P., Uskov, V.L., Penumatsa, A., Doddapaneni, A.: Smart universities, smart classrooms, and students with disabilities. In: Uskov, V.L., Howlett, R.J., Jain, L.C. (eds) Smart Education and e-Learning 2016. Springer, pp. 15–27, June 2016, 643 p., ISBN: 978-3-319-39689-7 (2016)
40. Serdyukova, N.A., Serdyukov, V.I., Uskov, V.L., Ilyin, V.V., Slepov, V.A.: A formal algebraic approach to modeling smart university as an efficient and innovative system. In: Uskov, V.L., Howlett, R.J., Jain, L.C. (eds.) Smart Education and e-Learning 2016, pp. 83–96. Springer, June 2016, 643 p., ISBN: 978-3-319-39689-7 (2016)
41. Bakken, J.P., et al.: Smart University: software systems for students with disabilities, In: Uskov, V., Bakken, J.P., Howlet, R., Jain, L. (eds.) Smart Universities. Springer (accepted for publication in 2017)
42. Camtasia Studio software system, https://www.techsmith.com/camtasia.html
43. Adobe Presenter 11software system, http://www.adobe.com/mena_en/products/presenter. html
44. MovaviSudio V7software system, http://www.movavi.com/screen-capture/
45. Smartpixel video screen capture and editor software, http://www.smartpixel.com/
46. Snagit screen capture software system, https://www.techsmith.com/snagit.html

47. Screenpresso software system, http://www.screenpresso.com/
48. BandiCam, https://www.bandicam.com/
49. Debut Video Capture Software, http://www.nchsoftware.com/capture/index.html?ref=cj
50. CamVerce 1.95, http://innoheim.com/camverce_more.php
51. Replay Video Capture 8 software, http://applian.com/replay-video-capture/?utm_source=Shareasale&utm_medium=affiliate
52. WM RECORDER 16, http://wmrecorder.com/home/
53. Adobe Captivate 9, http://www.adobe.com/products/captivate.html
54. CamStudio 2.7.2, http://camstudio.org/
55. Ezvid wiki & screen recorder, http://www.ezvid.com/
56. Screencast-O-Matic software system, http://screencast-o-matic.com/home
57. Active Presenter Screen Recording software, http://atomisystems.com/activepresenter/
58. Jing software system, https://www.techsmith.com/jing.html
59. Webinaria create software system, http://www.webinaria.com/
60. Screen Recorder from Rylstim, http://www.sketchman-studio.com/rylstim-screen-recorder/
61. Icecream Screen Recorder system, http://icecreamapps.com/Screen-Recorder/
62. FlashBack Express software system, http://www.flashbackrecorder.com/express
63. Screen Video recorder, http://www.wordaddin.com/screenvcr/
64. Panopto https://panopto.com
65. Echo360 Lecture Capture software system, https://echo360.com/connect/contact/
66. Mediasite software system, http://www.sonicfoundry.com/solutions/education/lecture-capture/
67. Camtasia https://www.techsmith.com/camtasia.html
68. Valt Software http://ipivs.com/products/valt-software/
69. Yuja Lecture Capture/Room Webcasting http://www.yuja.com/lecture-capture/
70. Lecture Recording System http://www.beegeesindia.com/lecture-recording-system/
71. VIDIZMO http://www.vidizmo.com/solutions/education/lecture-capture/
72. Gallicaster, https://wiki.teltek.es/display/Galicaster/Galicaster+project+Home
73. Adobe Presenter 11 http://www.adobe.com/products/presenter.html
74. Opencast Matterhorn software system, http://www.opencast.org/software
75. Kaltura software system, http://corp.kaltura.com/products/features/Lecture-Capture
76. ClassX http://classx.sourceforge.net/index.html
77. CamStudio http://camstudio.org/
78. Audionote http://appcrawlr.com/ios/audionote-notepad-and-voice-rec
79. LectureRecording http://appcrawlr.com/android/lecturerecordings
80. Super Notes http://clearskyapps.com/apps/SuperNote/OnlineHelp.html
81. SameView http://docplayer.net/8244387-Smart-remote-classroom-creating-a-revolutionary-real-time-interactive-distance-learning-system.html
82. INKredible, http://inkredibleapp.com/
83. Squid, http://squidnotes.com/
84. Tegrity, http://www.mhhe.com/tegrity/product.html
85. Corel Video Studio Pro X9.5, http://www.videostudiopro.com/en/products/videostudio
86. Power Director 15, http://www.cyberlink.com/downloads/trials/powerdirector-ultra/download_en_US.html
87. Yuja Lecture Capture/Room Webcasting, http://www.yuja.com/lecture-capture/
88. Mediasite lecture capture, http://www.sonicfoundry.com/solutions/education/lecture-capture/
89. Mediatech custom classroom, http://gomediatech.com/wp/custom-classrooms/
90. Sakai, https://www.sakaiproject.org/
91. Moodle, https://moodle.org/
92. Atutor, http://www.atutor.ca/
93. RCampus, http://www.rcampus.com/login.cfm?&fltoken=1459290664125&
94. Learnopia, http://www.learnopia.com/search/
95. Claroline, http://www.claroline.net/

96. Videopad, http://www.nchsoftware.com/videopad/
97. Jing, https://www.techsmith.com/jing.html
98. VSDC Free Video Editor, http://www.videosoftdev.com
99. Cisco Webex, http://www.webex.com/
100. Go To Meeting, http://www.gotomeeting.com/
101. ClickMeeting, https://clickmeeting.com/solutions/education
102. ReadyTalk, https://www.readytalk.com/products-services
103. BigMarker, http://www.bigmarker.com/
104. Adobe Connect, http://www.adobe.com/products/adobeconnect.html
105. OnStream meeting, http://www.onstreammedia.com/onstream_meetings.php
106. Blackboard Collaborate, http://www.blackboard.com/
107. Ring Central, http://www.ringcentral.com/
108. GlobalMeet, https://www.globalmeet.com/
109. Google Hangouts, https://hangouts.google.com/
110. Skype, https://www.skype.com/en/
111. Bigbluebutton, http://bigbluebutton.org/
112. Meeting Burner, https://www.meetingburner.com/
113. Join me, https://www.join.me/
114. Team Viewer, https://www.teamviewer.com/en/
115. Zoom, https://zoom.us/
116. Zoho meeting, https://www.zoho.com/meeting/
117. Mikogo, https://www.mikogo.com/
118. Yugma, https://www.yugma.com/download/allproducts-jvm.php
119. Basecamp, https://basecamp.com/
120. Yammer, https://www.yammer.com
121. Blackboard,   http://www.blackboard.com/learning-management-system/collaborative-learning-platform.aspx
122. HaikyuLearning, http://www.haikulearning.com/
123. OpenText   FirstClass,   http://www.opentext.com/what-we-do/products/specialty-technologies/firstclass
124. LiveText, https://www.livetext.com/
125. Yugma, https://www.yugma.com/index.php
126. Twiddla, https://www.twiddla.com/
127. Mindmeister, https://www.mindmeister.com/signup/educational
128. Mikogo, https://www.mikogo.com
129. Edmodo, https://www.edmodo.com
130. WikiSpaces Classroom, https://www.wikispaces.com/content/classroom
131. Wiggio, https://wiggio.com/
132. Skype, https://www.skype.com/en/download-skype/skype-for-computer/
133. Connect,   http://www.mheducation.com/highered/platforms/connect/instructor-backtoschool2016/testimonials.html
134. Sloodle, https://www.sloodle.org/
135. Google Drive, https://drive.google.com
136. Web Poster Wizard, http://poster.4teachers.org/
137. OpenStudy, http://openstudy.com/
138. Oovoo, http://www.oovoo.com/
139. NiceVision (Qognify), http://www.qognify.com/nicevision/
140. Blue Iris, http://blueirissoftware.com/
141. Sighthound, https://www.sighthound.com/products/sighthound-video
142. EyeLine, http://www.nchsoftware.com/surveillance/
143. SARA (Situation Awareness and Response Assistant), http://www.statussolutions.com/technologies/automated-alerting
144. ZoneTrigger, http://www.zonetrigger.com/motion-detection/

145. VI Monitor 6 Video Management Software, http://www.video-insight.com/products-video-management-software.php
146. Fibero, http://www.fibaro.com/en/the-fibaro-system/motion-sensor
147. Nice Vision, http://www.qognify.com/nicevision/
148. IP Video Surveillance, http://www.qognify.com/ip-video-surveillance-software/
149. Capturix (software set), http://www.capturix.com/
150. Video surveillance software, http://www.contaware.com/
151. ISPY Connect, http://www.ispyconnect.com/
152. Intrance Motion detector 2.0, http://www.tucows.com/preview/329773/Intrance-Motion-Detector
153. ZoneMinder, https://www.zoneminder.com/
154. Active WebCam, http://www.pysoft.com/ActiveWebCamMainpage.htm
155. Motion 1, http://www.tucows.com/preview/405108/123Motion
156. WebCam monitor, http://www.deskshare.com/wcm.aspx
157. Yawcam, http://www.yawcam.com/
158. Netcam Studio, http://www.netcamstudio.com/

Chapter 4
Smart University: Software Systems for Students with Disabilities

Jeffrey P. Bakken, Vladimir L. Uskov, Suma Varsha Kuppili,
Alexander V. Uskov, Namrata Golla and Narmada Rayala

Abstract Smart universities, smart classrooms and smart education are the wave of the future in a highly technological society. One of the distinctive features of a smart university is its ability of adaptation to and smooth accommodation of various types of students/learners such as regular students and life-long learners, in-classroom/local and remote/online students/learners, regular students and special students, i.e. students with various types of disabilities including physical, visual, hearing, speech, cognitive and other types of impairments. This chapter presents the outcomes of an ongoing research project aimed at systematic identification, analysis, and testing of available open source and commercial text-to-voice, voice-to-text and gesture recognition software systems—those that could significantly benefit students with disabilities. Based on obtained outcomes of completed research and analysis of designated systems we identified and recommended top text-to-voice, voice-to-text and gesture recognition software systems for implementation in smart universities.

Keywords Software systems · Students with disabilities · Text-to-speech · Voice-to-text · Gesture recognition · Smart university

4.1 Introduction

Smart universities (SmU) and smart classrooms (SmC) can create multiple innovative opportunities for students to learn material and communicate to classmates in a variety of highly technological ways. In addition, they can give students who

J.P. Bakken (✉)
The Graduate School, Bradley University, Peoria, IL, USA
e-mail: jbakken@fsmail.bradley.edu; jbakken@bradley.edu

V.L. Uskov · S.V. Kuppili · A.V. Uskov · N. Golla · N. Rayala
Department of Computer Science and Information Systems,
and InterLabs Research Institute, Bradley University, Peoria, IL, USA
e-mail: uskov@fsmail.bradley.edu; uskov@bradley.edu

© Springer International Publishing AG 2018
V.L. Uskov et al. (eds.), *Smart Universities*, Smart Innovation,
Systems and Technologies 70, DOI 10.1007/978-3-319-59454-5_4

would normally not have access to these learning materials opportunities to interact with digital learning content as well as the instructors and other in-classroom and/or remote/online students. Although not designed or even conceptualized to benefit students with disabilities, this concept would definitely have an impact on the learning process and access to learning content for students with different types of disabilities.

4.1.1   Literature Review

4.1.1.1   Smart Classrooms: Literature Review

Pishva and Nishantha in [1] define a SmU as an intelligent classroom for teachers involved in distant education that enables teachers to use a real classroom type teaching approach to teach distant students. "Smart classrooms integrate voice-recognition, computer-vision, and other technologies, collectively referred to as intelligent agents, to provide a tele-education experience similar to a traditional classroom experience" [1].

Glogoric, Uzelac and Krco [2] addressed the potential of using Internet-of-Things (IoT) technology to build a SmU. "Combining the IoT technology with social and behavioral analysis, an ordinary classroom can be transformed into a smart classroom that actively listens and analyzes voices, conversations, movements, behavior, etc., in order to reach a conclusion about the lecturers' presentation and listeners' satisfaction" [2].

Slotta, Tissenbaum and Lui [3] described an infrastructure for SmUs called the Scalable Architecture for Interactive Learning (SAIL) that "employs learning analytic techniques to allow students' physical interactions and spatial positioning within the room to play a strong role in scripting and orchestration".

Koutraki, Efthymiou, and Grigoris [4] developed a real-time, context-aware system, applied in a SmU domain, which aims to assist its users after recognizing any occurring activity. The developed system "…assists instructors and students in a smart classroom, in order to avoid spending time in such minor issues and stay focused on the teaching process" [4].

Given all the research publications that focus on SmUs, no literature was located on "SmC's software systems and students with disabilities" topic. This is the reason that this topic is in the center of our research activities.

4.1.1.2   Smart Universities: Literature Review

Coccoli in [5] argue that "…primary focus of SmU is in the education area, but they also drive the change in other aspects such as management, safety, and environmental protection. The availability of newer and newer technology reflects on how the relevant processes should be performed in the current fast changing digital era.

This leads to the adoption of a variety of smart solutions in university environments to enhance the quality of life and to improve the performances of both teachers and students. Nevertheless, we argue that being smart is not enough for a modern university. In fact, all universities should become smarter in order to optimize learning. By "smarter university" we mean a place where knowledge is shared between employees, teachers, students, and all stakeholders in a seamless way" [5].

Aqeel-ur-Rehman et al. in [6] present the outcomes of their research on one feature of future SmU—sensing with RFID (Radio frequency identification) technology; it should benefit students and faculty with identification, tracking, smart lecture room, smart lab, room security, smart attendance taking, etc.

Lane and Finsel in [7] emphasize the importance of the big data movement and how it could help to build smarter universities. "Now is the time to examine how the Big Data movement could help build smarter universities—in situations that can use the huge amounts of data they generate to improve the student learning experience, enhance the research enterprise, support effective community outreach, and advance the campus's infrastructure. While much of the cutting-edge research being done with Big Data is happening at colleges and universities, higher education has yet to turn the digital mirror on itself to innovate the academic enterprise" [7]. Big data analytics systems will strongly support inferring characteristic of a SmU.

Al Shimmary et al. in [8] analyzed advantages of using RFID and WSN technology in development of SmU. "The developed prototype shows how evolving technologies of RFID and WSN can add in improving student's attendance method and power conservation". RFID, WSN as well as Internet-of-Things technology are expected to significant parts of a SmU and strongly support sending characteristics of SmU.

Doulai in [9] presents a developed system for a smart campus. This system "… that offers an integrated series of educational tools that facilitate students' communication and collaboration along with a number of facilities for students' study aids and classroom management. The application of two widely used technologies, namely dynamic web-based instruction and real-time streaming, in providing support for "smart and flexible campus" education is demonstrated. It is shown that the usage of technology enabled methods in university campuses results in a model that works equally well for distance students and learners in virtual campuses".

Yu et al. in [10] argue that "… with the development of wireless communication and pervasive computing technology, smart campuses are built to benefit the faculty and students, manage the available resources and enhance user experience with proactive services. A smart campus ranges from a smart classroom, which benefits the teaching process within a classroom, to an intelligent campus that provides lots of proactive services in a campus-wide environment". The authors described 3 particular systems—Wher2Study, I-Sensing, and BlueShare—that provide sensing, adaptation, and inferring smart features of a SmU.

One area that so far has had a limited attention is "students with disabilities and SmU". Although features, components, and systems of SmU taxonomy have been discussed in [11], only one publication could be located that discussed SmU, SmC, and students with disabilities [12]. Given that 10% of all school/college/university students have some kind of disabilities, this is definitely an area that needs a more thorough investigation.

4.2 Students with Disabilities and Software Systems

Categories of students with disabilities. Students of schools/colleges/universities may experience a variety of different categories of disabilities; they include but are not limited to:

(1) Deaf/hearing impairments
(2) Learning disabilities
(3) Physical disabilities
(4) Psychological/neurological disorders
(5) Speech or language impairments
(6) Visual impairments
(7) Cognitive impairments

Software systems for students with disabilities. Software systems allow students with disabilities equal access in the classroom and learning environments. Often these systems also help them learn more efficiently and effectively and in many cases allow them to interact better with their professor and classmates. Where traditional classrooms do not specifically address software systems and how students with disabilities could be impacted, the implementation of specific advanced software systems in SmU and learning environments would definitely approach learning barriers from the perspective of universal accessibility: providing greater learning opportunities for all students in the SmU classroom—including students with disabilities.

A list of possible software systems that may benefit students with various types of disabilities are listed in Table 4.1.

Bradley University and students with disabilities. Bradley University (Peoria, IL USA) is a top-ranked private university that offers 5,400 undergraduate and graduate students opportunities and resources of a larger university and the personal attention and exceptional learning experience of a smaller university. Bradley offers more than 185 undergraduate and 43 graduate academic programs in business, communications, education, engineering, fine arts, health sciences, liberal arts and sciences, and technology.

The Center for Learning and Access (CLA) at Bradley University is the University's primary academic support service responsible for helping students acquire skills essential to achieve academic and personal success (https://www.bradley.edu/offices/student/cla/). Under the CLA umbrella, the Office of Access Services currently serves approximately 310 students (or, about 6% of the total

Table 4.1 Types of students with disabilities and software systems that may be beneficial

Category of disabilities	Software systems that may benefit these types of students
Deaf/hearing impairments	Voice-to-text software systems, assistive listening systems, touch screen technology, eye gaze software, auditory tools (headset worn by user to limit distractions), writing software (Inspiration), closed captioning, video camera, real-time captioning
Learning disabilities	Text-to-voice, voice-to-text, gesture recognition and facial recognition software systems, touch screen technology, systems to improve auditory processing abilities, systems to develop basic math skills and mathematical reasoning, systems to improve organizational and memorization skills, systems to improve reading skills, talking word processors, word prediction software, spelling software, writing software (Inspiration)
Physical disabilities	Text-to-voice, voice-to-text, gesture recognition software systems, sip-and-puff systems, touch screen technology, switch access (clicker), eye gaze software, screen magnification software, talking word processors
Psychological/ neurological disorders	Text-to-voice, voice-to-text, gesture recognition and facial recognition software systems, touch screen technology, spelling software, writing software (Inspiration)
Speech or language impairments	Text-to-voice, voice-to-text, gesture recognition and facial recognition software systems, touch screen technology, speech synthesizers, talking word processors, word prediction software, spelling software, writing software (Inspiration)
Visual impairments	Text-to-voice software systems, voice-to-text software systems, screen magnification and readers software systems, talking word processors, screen review systems, audio books, spelling software
Cognitive impairments	Text-to-voice, voice-to-text, gesture recognition and facial recognition software systems, touch screen technology, switch access (clicker), eye gaze software, screen magnification software, talking word processors, word prediction software, spelling software

student number) that have provided appropriate documentation and registered for services.

In accordance with information from Bradley's Center for Learning and Access (CLA) the current distribution of students with disabilities at Bradley by various designated categories is follows: (1) with a health impairment—19 students, (2) with a hearing impairment—6 students, (3) with learning disabilities—84 students, (4) with a physical disability—5 students, (5) with psych/neuro impairments —186 students, including 11 students with ASD (Autism Spectrum Disorder) and 61 students with ADHD (Attention Deficit Hyperactivity Disorder), (6) with a speech impairment—2 students, and (7) with a visual impairment—8 students.

The software systems currently in use at CLA by various categories of students with disabilities at Bradley University are summarized in Table 4.2.

Table 4.2 Software systems used by students with disabilities at Bradley University

#	Software system	Technical platform	Cost	Category of students served
1	Kurzweil Reading Edge	Windows/Mac	–	3, 5, 6, 7
2	CAR (Central Access Reader)	Windows/Mac	Free	3, 5, 6, 7
3	Inspiration	Windows/Mac	V9.2—$40; bundle pack —$60; household—$95	3, 5, 7
4	Jaws	Windows	Pro—$1,100; home— $900; 90-day trail—$180	7
5	Firefox Add on "Text to voice"	Windows/Mac/Linux	Free	3, 6
6	Dragon Naturally speaking	Windows/Mac	$75	2, 3, 4, 7
7	Zoom text	Windows	$600	7

The CLA specialists identified a list requirements for software systems to be used by students with various categories of disabilities—those features and functions should provide users with significant benefits; a list of CLA of most important requirements is presented in Table 4.3.

Table 4.3 CLA's list of most important requirements to software systems for students with disabilities

Requirement	Details (software systems should be able to ...)
Maximal number of students to get benefits	Serve and be useful and beneficial for as many students with various categories of disabilities as possible.
Graphic user interface and pre-defined commands	Help users to navigate the system easily without any discomfort; the system should have a list of pre-defined commands, for example, to access folders and files, send email, etc.
Voice-to-text functionality	Accurately convert user's speech into text, structure text into notes, create emails, support punctuation, support spell checking using built-in vocabulary, word prediction functionality, optical character recognition (OCR) functionality (ability to scan the printed information or camera captured image's text and convert it into digital text which can be read by the software system), multi-lingual user interface language packs (MUI) functionality (ability to download and use different languages in graphic user interface), provide quick and accurate voice training, provide editing of built-in dictionary, support punctuation, email the outcome text file, etc.

(continued)

Table 4.3 (continued)

Requirement	Details (software systems should be able to ...)
Text-to-voice functionality	Accurately convert user's text into voice (or, synthesize audio), use various types of "voices" (kid's, female, male voices, etc.), read web-based data, math data, data in tables, etc.
Input/output formats	Handle (i.e. work with) various types of input data, especially, PDF, TXT, DOC, HTML, etc. and provide output data in various formats such as DOC, TXT, MP3, MP4, etc.
Web-based content	Read data from the Web such as the content of web pages, web applications, web simulations, etc.
Titles and text on images	Read the text present in an image.
Math data	Recognize math notation (math equations), or, in other words, have a special math reader.
Data in tabular form	Read data presented in tabular form.

The outcomes of our research as well as CLA requirements clearly shows that (1) text-to-voice (or, text recognition), (2) voice-to-text (or, speech recognition; also including captioning of all lectures and video materials), and (3) gesture (and, face) recognition systems are among most actively systems that may be used by students with disabilities in SmU. This is the main reason that during initial part of our project we focused research activities primarily on these types of software systems.

4.3   Project Goal and Objectives

The performed analysis of above-mentioned and multiple additional publications and reports relevant to (1) SmU, (2) university-wide smart software and hardware systems and technologies, (3) SmC, (4) smart learning environments, (5) smart educational systems, and (6) students with disabilities undoubtedly shows that SmU-related topics will be in the focus of multiple research, design and development projects in the upcoming 5–10 years. It is expected that in the near future SmU concepts and hardware/software/technological solutions will start to play a significant role and be actively deployed and used by leading academic institutions in the world.

Project Goal. The overall goal of this ongoing multi-aspect research project is a) to research and analyze various open source and commercial software systems in the areas of text-to-voice, voice-to-text, and gesture recognition, and b) identify top systems that could be recommended for implementation and active use in SmU and/or SmC to aid students with disabilities (and possibly students without disabilities).

The premise is that these software systems will make the curriculum more accessible for students with and without disabilities and will help traditional universities to understand the impact this software could have on the learning of students with disabilities and how this software could aid universities to a possible transformation from a traditional university into a smart one.

Project Objectives. The objectives of this project are

(1) close collaboration with subject matter experts and identification of most desired features and functions for software systems to be used by students with disabilities in SmU and SmC;
(2) extensive research and identification of available software systems in text-to-voice, voice-to-text, and gesture recognition areas;
(3) identification and thorough analysis of available software systems in each designated area, including at least 10 commercial and 10 open-source systems,
(4) identification of a list of most important (i.e. most useful for students with disabilities) features (functions) of existing software systems in each designated area;
(5) perform analysis of most powerful (in terms of functionality) existing software systems in each area;
(6) ranking of analyzed systems, i.e. identification of top 3 commercial and top 3 open-source systems among analyzed systems in each area, and
(7) develop lists of open-source and commercial software systems in each designated area that are recommended for in-depth testing by actual students with various categories of disabilities and subject matter experts in smart classrooms and smart universities (and, probably, traditional universities).

The obtained research and analysis outcomes are presented below.

4.4   Research Outcomes: Analysis of Text-to-Voice Software Systems

There are many available text-to-voice software systems that could be implemented in a smart classroom within a SmU. This software will allow the user to convert text to voice so they can hear what information the text is trying to convey if they have issues with reading and comprehending text. Instead of students focusing on reading the text they can focus on comprehending it. For example, the act of reading for some students is a cognitive process. These students see words and have to figure out what letters are in the words, what the letters sound like, and what the actual word is so all there energy is spent on the task of reading, not comprehending the material. Using this software will make the material more accessible to the student with these difficulties. For other students, the actual act of reading is automatic and they can focus on comprehending what they are reading.

After investigating the desired features of text-to-voice software systems (Table 4.4) that, in our mind, should be available for students with disabilities in SmU, the next steps in our research and analysis project were:

(1) Identification and thorough analysis of about 10 commercial and 10 open-source text-to-voice available software systems,
(2) identification of a list of most important (i.e. most useful for students with disabilities) features (functions) of existing text-to-voice software systems,

(3) examples of obtained analysis outcomes of powerful (in terms of functionality) text-to-voice existing software systems, and our ranking of those systems,

(4) our recommendations, i.e. top 3 commercial and top 3 open-source text-to-voice software systems to be implemented and actively used in SmU.

The obtained research and analysis outcomes are summarized and presented in Tables 4.5, 4.6, 4.7, 4.8, 4.9, 4.10, 4.11, 4.12 and 4.13.

4.5 Research Outcomes: Analysis of Voice-to-Text Software Systems

There are many available voice-to-text software systems that could be implemented in a smart classroom within a SmU. This software will allow the user to convert their voice to text if they have issues with written expression. Instead of students focusing on the actual writing process they can focus their attention on producing a high quality product. For example, the act of writing for some students is a cognitive process. These students think of a word, have to think of the letters that make up this word, and then have to think of how the letter looks so they can retrieve it from memory and write it down. This process is very time consuming and by the time they have written a few words they have lost their thoughts on what they initially had planned to write. Using voice-to-test software systems will allow the student with a disability more access and the ability to produce higher quality written products. For other students, the actual act of writing is automatic (i.e., letter

Table 4.4 A list of desired features of text-to-voice software systems for SmU

#	Desired system feature	Feature details
1	Quick response	The system should convert text-to-voice instantly
2	Proof reading	Student or faculty should be able to listen to their notes or assignments, in order to improve the quality of information
3	Access on mobile-devices	It should allow users to convert text-to-voice anywhere
4	Drag-and-drop	This option should allow users to drag their external files to the software, so that it reads aloud for them
5	Multi-linguistic	The software should support several popular languages
6	Highlight word	The word that is read aloud should be highlighted
7	Pronunciation editor	Manually modify the pronunciation of a certain word
8	Batch convertor	Convert multiple documents to MP3, WAV, WMA, etc. files
9	Type-and-talk	A mute student should be able to communicate easily by simply typing what he/she wants to say
10	High quality	Speech should be of high quality with clear pronunciation and minimal errors

Table 4.5 Analyzed 10 commercial and 10 open source text-to-voice software systems

#	Systems analyzed	Company-developer	Technical platform	Ref.
Commercial systems				
1	Natural Reader	NaturalSoft Limited	Windows/Mac	[13]
2	Text Speech Pro	NetHint	Windows/Mac	[14]
3	Read The Words	True Logic	Windows/Linux	[15]
4	TextAloud3	Nextup	Windows/Linux/Mac	[16]
5	Verbose	NCH Software	Windows/Linux/Mac	[17]
6	Voki	Voki	Windows/Mac	[18]
7	Oddcast TTS	Oddcast	Windows/Linux/Mac/Android	[19]
8	Ultra Hal	Zabaware	Windows/Linux	[20]
9	Neo Speech	NeoSpeech	Windows/Mac	[21]
10	Texthelp Read&Write	Texthelp Ltd.	Windows/Mac/IOS/Android with add on to Google Chrome	[22]
Open source (free) systems				
1	Balabolka	Balabolka	Windows/Mac	[23]
2	Text-to-Speech Reader	Speech Logger	Windows/Linux/Mac	[24]
3	Text-to-Speech (TTS)	Poon Family	Windows	[25]
4	Microsoft Word TTS	Microsoft	Windows/Mac	[26]
5	ClipSpeak	Code Plex	Windows/Mac	[27]
6	WordTalk	WordTalk	Windows	[28]
7	Imtranslator	Smart Link	Windows/Linux	[29]
8	iSpeech	Apple	Mac	[30]
9	Google Translate	Google	Windows/Linux/Mac	[31]
10	Power Talk	Atis4all	Windows/Linux	[32]

Table 4.6 A list of most important for SmC/SmU features in existing text-to-voice software systems

#	Existing important features	Details of existing important features
1	High quality	The high speech quality allows users to clearly and better understand information that is produced
2	Pronunciation editor	Allows to manually modify pronunciation of a certain word
3	Batch converter	Allows to convert multiple documents to MP3 at same time
4	Reading speed	Can change the reader and reading speed at any point of time
5	Read documents	Students can listen to any kind of documents like electronic textbooks, PDF files, Microsoft Word documents and web pages
6	Highlight text	Allow students with learning difficulties to hear the material and simultaneously see the words highlighted on screen
7	Spell checking	Can read text as typed, by word or by sentence, and delivers critical real-time feedback to help students with dyslexia improve coordination of sounds and letter combinations

(continued)

Table 4.6 (continued)

#	Existing important features	Details of existing important features
8	Optical Character Recognition (ORC)	Works with scanner to convert printed characters into digital text
9	Natural voice	Includes a great selection of natural-sounding voices
10	Floating bar	Integrate add-in toolbars in MS Word, Outlook and PowerPoint which offers a simple way of reading texts directly on page
11	Multi-linguistic	Supports several popular languages like English, French, German, Spanish etc.
12	Saving formats	Allows saving audio in a file of different formats like MP3, WAV etc.
13	Response time	Conversion to speech is done instantly without much waiting time
14	Mobile-devices	Allow students to listen even on mobile-devices from remote locations
15	Read as typed	Allows a student without speech to easily communicate with faculty or in a group by just typing what he/she wishes to say

Table 4.7 *Natural Reader* [13] commercial text-to-voice system: the analysis outcomes

#	System's characteristics	System's details
1	Main most important system's features and functions	• It can read to the user various types of textual information such as Microsoft Word files, webpages, PDF files, and e-mails • It can be synchronized with iPhone, iPad and Android apps • Textual files can be converted into audio files • Easy-to-use software system with natural-sounding voices; a wide variety of speakers to choose from; it has a selection of over 11 languages and over 50 voices, including children's voices • Speech specs (speed, frequency, etc.) can be easily adjusted • It converts any text or document into natural-sounding voices and even to MP3 or WAV files • The floating bar is a handy tool where user can listen to text in other applications • It allows multi-tasking processing, for example, listening of eBooks while walking, running etc. • For users with low vision, it can display text in large fonts and highlight text while reading aloud • It has high contrast color interface

(continued)

Table 4.7 (continued)

#	System's characteristics	System's details
2	Strengths and opportunities	• It allows students to listen to textbooks, class notes, assignments, emails, etc. • It can convert multiple documents into MP3 files at same time with Batch Converter tool • It considerably assists students with dyslexia, reading challenges or visual impairments • It saves eye strain due to ergonomic GUI • It makes proof reading effective • It has very useful features—spell checking and word predicting
3	Possible weaknesses and threats	• It cannot handle some text features and symbols, for example, bullets or dashes
4	Technical platform	• Windows, Mac OS
5	Prices	• $70 (personal) • $130 (professional) • $200 (ultimate) • Various packages are available for schools—details are available at http://www.naturalreaders.com/exploring.html
6	Colleges/universities that currently use this system	• Hamburg University of Technology • University of Pennsylvania
7	System's ranking	**Our ranking of this system: 1** Main reasons: • It works in 11 different languages and has over 50 voices available • It has spell checking and word predicting features • It works with e-books and/or e-textbooks from various vendors

Table 4.8 *Read the Words* [15] commercial text-to-voice system: the analysis outcomes

#	System's characteristics	System's details
1	Main most important system's features and functions	• It supports three different languages and 15 different voices • Virtual Reader avatar helps students to read along and improve their reading skills • It illustrates how words are spoken by native speakers • It can upload any Microsoft Office document, Adobe PDF, TXT or HTML document • It easily converts textual files into MP3 files • It helps non-native speaker to speak English correctly
2	Strengths and opportunities	• 2 levels of quality for audio—highest and high • Relatively easy integration with most popular applications • Highly accurate text interpretation • It is useful for students learning English, French and Spanish

(continued)

Table 4.8 (continued)

#	System's characteristics	System's details
3	Possible weaknesses and threats	• Recording speed cannot be adjusted
4	Technical platform	• Windows OS, 100 MB hard disk (at least), 256 MB RAM (at least)
5	Prices	• $100 per year or $20 per month
6	Schools/universities that currently use this system	• Hodges University • Monroe Public School • Springfield Elementary School • St. Charles Parish Public Schools
7	System's ranking	**Our ranking of this system: 2** Main reasons: • Grammatical errors can be fixed by listening to the typed notes • Podcast feature enables students to broadcast their paper to everyone instantly

Table 4.9 *TextHelp Read&Write* [22] commercial text-to-voice system: the analysis outcomes

#	System's characteristics	System's details
1	Main most important system's features and functions	• System reads texts from TXT and PDF files • It supports typing echo feature • It includes fact finder where the user can be able to search information from the provided search engines • It contains a picture dictionary which provides the relevant pictures for the words searched • It contains a built-in dictionary • It contains a translator for a selected text to be translated into selected other language • The scan feature is able to scan 5-8 image formats such as the.JPG,.PNG,.GIF,.BMP to text. • Daisy reader can be used to read any digital e-books • Uses PDF aloud feature to read PDF files • Word prediction feature is available which will be able to provide suggestions for the next required word • It also includes the speech to text software

(continued)

Table 4.9 (continued)

#	System's characteristics	System's details
2	Strengths and opportunities	• Based on the type of disability, the user can use needed feature: Reading, Writing or Research • Students can make use of typing echo where the software reads the words as soon as the student completes writing it • Spell checking is available—it provides word suggestions for the misspelled words • Screenshot reader helps users to take screen shot of any text which is then read out to them • Verb Checker helps users to check for specific verbs, later receive the examples about how the verb is used • Voice note helps students to record speech, replay it, save the audio into a file, and, if needed, insert it into a document • Picture dictionary can help users who have problem remembering the context; in this dictionary every word is presented as a picture
3	Possible weaknesses and threats	• Cannot read data from the DOC file
4	Technical platform	• Supports Windows and Mac operating systems
5	Price (if any)	• $145/copy
6	Colleges/universities/companies that currently use this system	• Colorado College • Los Angeles City College
7	System's ranking	**Our ranking of this system: 3** Main reasons: • Can read the text from PDF files • Helps students with reading and writing disabilities • Also includes speech to text software

Table 4.10 *Balabolka* [23] open-source text-to-voice system: the analysis outcomes

#	System's characteristics	System's details
1	Main most important system's features and functions	• It allows to save the converted file in WAV, MP3, MP4 or WMA • Every line of text can be converted to an audio file if necessary • It has a timer and it can be controlled by hot keys • It can read the clipboard context and view text from documents saved as DOC, EPUB, HTML, PDF, RTF, etc. • It can extract text from PDF files directly • It offers fine tuning of text-to-speech voices • It can alter how fast the voice reads, what pitch it reads in and can even use custom dictionaries to improve pronunciation

(continued)

Table 4.10 (continued)

#	System's characteristics	System's details
2	Strengths and opportunities	• It supports around 32 languages • The HELP support files are available in 7 different languages • It allows to customize the window appearance with downloadable skins • It allows to use regular expressions like "Find" and "Replace" • It can run in full screen mode or in a window
3	Possible weaknesses and threats	• Importing documents does not retain its format • The quality of outcomes (deliverables) may vary
4	Technical platform	• Windows XP/Vista/7/8/10 • Mac OS
5	Price (if any)	• Open Source
6	Colleges/universities that currently use this system	• University of Edinburgh, Scotland • The Open University, UK
7	System's ranking	**Our ranking of this system: 1** Main reasons: • The audio is of high quality and users can alter voices, pitch, cadence and speed • It can use language modules from Microsoft Office for spell checking

Table 4.11 *Text-to-Speech Reader* [24] open-source text-to-voice system: the analysis outcomes

#	System's characteristics	System's details
1	Main most important system's features and functions	• It supports multiple languages • This is ideal system to listen class notes, articles etc. • It can adjust speed of speech • It helps students with learning disabilities and visual impairments to read text aloud in natural sounding voices • It highlights text that is read aloud • It has a user-friendly interface • It allows "copy and paste" procedure
2	Strengths and opportunities	• It reads text aloud with natural sounding voices • It is very useful for mute users; they can simply type in and click on play button • The user interface is easy to understand • It helps to avoid contact with computer screen for reading long textual information • It allows to listen to hours of textual podcasts with close to zero data consumption • It is a multi-linguistic system
3	Possible weaknesses and threats	• It needs continuous internet connection • It cannot drag and drop files

(continued)

Table 4.11 (continued)

#	System's characteristics	System's details
4	Technical platform	• Windows, Mac OS • Any standard Web browser like Chrome, Safari can be used • Decent Internet connection is required
5	Price (if any)	• Open Source
6	Colleges/universities that currently use this system	• Kurzweil Educational Institutions • Albany Hills State School, Australia • Hamilton State School, Australia
7	System's ranking	**Our ranking of this system: 2** Main reasons: • Students with visual impairments can listen to very long textual data files—this is a great benefit for those students • The system saves a lot of data from mobile traffic • It has a very simple and intuitive GUI

Table 4.12 *Text-to-Speech TTS* [25] open-source text-to-voice system: the analysis outcomes

#	System's characteristics	System's details
1	Main most important system's features and functions	• User can type, copy and paste, or open text files • "Click to Speech" begins enunciating the contents of user's text, highlighting the current part as it is being read • Operation in 32 different languages is available for users • It allows to listen in the background while working on other applications • It allows speech speed control • It can work with any webpage
2	Strengths and opportunities	• It makes communication easier for people with speech disorders, vision impairments, and dyslexia • It is multi lingual system • It highlights text when reading • It allows multi-tasking processes
3	Possible weaknesses and threats	• It cannot read PDF files • It lacks speed-reading control feature • It has no spell checker available
4	Technical platform	• Windows 8.1, 10, Windows 10 Mobile, Windows Phone 8.1 • Architecture × 86, × 64, ARM
5	Price (if any)	• Open Source
6	Colleges/universities that currently use this system	• University of Kent, UK • University of Michigan
7	System's ranking	**Our ranking of this system: 3** Main reasons: • Users can listen audio in the background while working on other applications • Highlighting the text when read aloud is an useful advantage • Wide range of languages is available for users

Table 4.13 Our recommendations: existing top 3 commercial and top 3 open-source text-to-voice software systems to be implemented and actively used in SmC/SmU

#	System	Company-developer	Details	Ref.
Commercial systems				
1	Natural Reader	NaturalSoft Limited	$70/copy (personal), $130 (professional), $200 (ultimate)	[13]
2	Read The Words	True Logic	$100 per year or $20 per month	[15]
3	Texthelp Read&Write	Texthelp Ltd.	$145/copy	[22]
Open source (free) systems				
1	Balabolka	Balabolka	Windows/Mac	[23]
2	Text-to-Speech Reader	Speech Logger	Windows/Linux/Mac	[24]
3	Text-to-Speech (TTS)	Poon Family	Windows	[25]

formation, word spellings, punctuation, etc.) and they can focus on the content of the message or assignment they are involved in writing.

Based on our current and past research project and obtained research outcomes, a generalized list of desired features of voice-to-text software systems for SmU is presented in Table 4.14.

After investigating the desired features of voice-to-text software systems (Table 4.14) that, in our mind, should be available for students with disabilities in SmU, the next steps in our research and analysis project were:

Table 4.14 A list of desired features of voice-to-text software systems for SmU

#	Desired system feature	Feature details
1	Dictate continuously	It should help faculty to dictate notes continuously in a normal, conversational pace without slowing down pace or over-enunciating words
2	Robust documentation	It should allow users to create documents with punctuation marks
3	Accent support	It should allow faculty from different locations to communicate easily
4	Hands-free	It should help students with disabilities such as repetitive strain injury (RSI), dyslexia, vision impairment, etc.
5	Recognition speed	The text should appear on screen as it is dictated, without any delay
6	Accuracy	The text should be accurate without any major errors
7	Mobility	Documents should be easily integrated with cloud technology
8	Web search	Students should be able to search the Web by just dictating
9	Multi-lingual support	System should be able to listen to text in native language voices and recognize voices with quality
10	Easy-to-use	It should help users to dictate, and, when finished, simply copy-paste dictated text where needed

Table 4.15 Analyzed 10 commercial and 10 open source voice-to-text software systems

#	Systems analyzed	Company-developer	Technical platform	Ref.
Commercial systems				
1	Dragon Naturally Speaking Premium (Home Edition)	Nuance	Windows/Mac/Android/iOS	[33]
2	Dragon Professional Individual	Nuance	Windows/Mac/Android/iOS	[34]
3	Braina Pro	Brainasoft	Windows/Android	[35]
4	Tazti Speech Recognition Software	Voice Tech Group Inc.	Windows	[36]
5	SpeechGear Compadre Interact	SpeechGear	Windows/Mac	[37]
6	Dictation Pro	Desk Share	Windows/Mac/Linux	[38]
7	e-Speaking	e-speaking	Windows	[39]
8	Voice Finger 2.6.2	Robson Cozendey	Windows	[40]
9	Text Shark	textshark	Windows	[41]
10	Speechmatics	Speechmatics	Windows	[42]
Open source (free) systems				
1	Windows Speech Recognition	Microsoft	Windows	[43]
2	Apple Dictation	Apple	Mac	[44]
3	TalkTyper	TalkTyper	Windows/Mac	[45]
4	Jasper	Jasper Project	Windows/Mac/Linux	[46]
5	Dictation 2.0	Digital Inspiration	Windows/Mac	[47]
6	Speechnotes	SpeechNotes	Windows/Mac/Linux	[48]
7	Digital Syphon Sonic Extractor	Digital Syphom	Windows/Mac	[49]
8	Balabolka	Ilya Morozov	Windows	[50]
9	Speech Logger	Speehclogger	Windows/Mac	[51]
10	Google Docs—Speech Recognition	Google	Any platform that supports Google Chrome browser	[52]

(1) identification and thorough analysis of about 10 commercial and 10 open-source voice-to-text available software systems,

(2) identification of a list of most important (i.e. most useful for students with disabilities) features (functions) of existing voice-to-text software systems,

(3) examples of obtained analysis outcomes of powerful (in terms of functionality) voice-to-text existing software systems, and our ranking of those systems,

(4) our recommendations, i.e. top 3 commercial and top 3 open-source voice-to-text software systems to be implemented and actively used in SmU.

The obtained research and analysis outcomes are summarized and presented in Tables 4.15, 4.16, 4.17, 4.18, 4.19,4.20, 4.21,4.22 and 4.23.

Table 4.16 A list of most important for SmC/SmU features in existing voice-to-text software systems

#	Existing important features	Details of existing important features
1	Dictate continuously	It helps faculty to dictate notes continuously in a normal pace, without slowing down pace
2	Collaboration	Students can collaborate on group projects by sending e-mails or instant messages entirely by voice
3	Enhanced diction	It can save information like name, relationships and send e-mails to friends and professors easily
4	Accent support	It allows users from different locations to document reports
5	Flawless report	It generates reports and documents with almost zero errors
6	Edit and format	Users can edit and format documents and reports all by voice
7	Web search	Students who cannot type can easily search the Web for any information and save the required data
8	Instant translator	Faculty can communicate with deaf students easily
9	Compatibility	The system is compatible with device's screen of the user using even outside the smart classroom
10	Voice control	Users can control their computers by performing various functions such as (a) launching application, (b) selecting menu items, (c) switch between windows, and other functions using voice commands
11	Mobility	Users can easily use cloud technology and access documents from anywhere
12	Robust documentation	It allows the user to write documents with punctuation marks like "period" for ".", "question mark" for "?", etc.
13	Multi-lingual support	It supports multiple language inputs
14	Word suggestions	It provides suggestions for words pronounced similarly
15	Import audio	Students can import different audio lecture files and convert them into a textual file

Table 4.17 *Braina Pro* [35] commercial voice-to-text system: the analysis outcomes

#	System's characteristics	System's details
1	Main most important system's features and functions	• It uses natural language interface and speech recognition to interact with its users and allows them to use English language sentences to perform various tasks on their computer • It can find information on the Internet, take dictation, find and open files, set alarms and reminders, perform math calculations, etc. • It adapts to the user's behavior over time to better anticipate needs • It is a multi-functional software that provides a single window environment to control user computer and perform wide range of tasks using voice commands • It can remember notes, automate various computer tasks, read e-books, etc.

(continued)

Table 4.17 (continued)

#	System's characteristics	System's details
2	Strengths and opportunities	• "Easy to learn and easy to use" graphic user interface • Easy to set up; no need of voice training • User can ask system to learn information from file using "Learn from File" feature
3	Possible weaknesses and threats	• Currently it can only understand simple English language at present • It gets confused if complex sentences are used
4	Technical platform	• Operating system: Window 10/8.1/8/7/Vista/XP • Dependencies: msvcr100.dll • Internet connection required to use some features
5	Price (if any)	• $40 per year
6	Companies that currently use this system	• TechRadar • Wikipedia • Brainasoft, the Intel Software Partner
7	System's ranking	**Our ranking of this system: 1** Main reasons: • It allows user to customize voice commands and replies • Users can schedule events like class times and when homework is due just by giving commands

Table 4.18 *Dragon Naturally Speaking Premium (Home Edition)* [33] commercial voice-to-text system: the analysis outcomes

#	System's characteristics	System's details
1	Main most important system's features and functions	• It allows users to dictate words three times faster than typing with up to 99% accuracy of word recognition • It allows user to speak, and up to 160 words per minute are transcribed and appear on computer screen; it can read back transcribed text to ensure that the material is truly what the user intended to say • It allows users with blindness/low vision, dyslexia, dyspraxia, mobility or dexterity impediments are able to show proof of what they know because they can speak out their ideas to compose an essay or writing an answer • It allows students who have dyslexia to complete assignments orally and still produce written output for grading purposes • It allows users to send email and instant messages to collaborate on group or classroom projects to students, faculty, parents, administrators—entirely by voice • It provides users with an easier way to control a computer that is less physically and cognitively taxing

(continued)

Table 4.18 (continued)

#	System's characteristics	System's details
2	Strengths and opportunities	• Users who would normally write incomplete or simple sentences have the ability to use more complex sentence structures to appropriately describe their ideas • Teachers can dictate feedback more quickly because students work is completed electronically • It has 99% accuracy of word recognition; the outcome text will be spell-checked • It has an integrated dictionary that takes care of all spell checking; as a result users can focus on their thoughts and do not worry a lot about the process of typing and spell checking • It allows user to speak, and up to 160 words per minute are transcribed and appear on computer screen; it can read back transcribed text to ensure that the material is truly what the user intended to say
3	Possible weaknesses and threats	• Outdated computers may not be able to cope with this system • The initial task to set up the program for a new user is quite complex—users need to read a lengthy passage • It takes time to "teach" the system to recognize/learn user's voice style and to help it recognize how certain words are pronounced by that particular user; in some cases, it may be a frustrating experience for some users
4	Technical platform	• Operating systems—Windows 7, 8, 8.1 (32bit and 64bit), Windows Server 2008 64bit, 2012 64bit • RAM—2 GB for 32-bit Windows 7, 8 and 8.1, 4 GB for 64-bit Windows 7, 8 and 8.1 and Windows Server 2008 R2, Windows Server 2012 • CPU—min 2.2 GHz Intel dual core or equivalent AMD processor • Free hard disk space—4 GB • A sound card supporting 16-bit recording • A Nuance-approved microphone • An Internet connection for automatic product activation
5	Price (if any)	• About $100 per copy • About $130—Dragon for Mac Student/Teacher Bundle
6	Colleges/universities that currently use this system	• Yale Center for Dyslexia and Creativity (YCDC) • Arizona State University
7	System's ranking	**Our ranking of this system: 2** Main reasons: • It is exclusively developed for educational purposes and is widely used by educational institutions

Table 4.19 *Dragon Professional Individual* [34] commercial voice-to-text system: the analysis outcomes

#	System's characteristics	System's details
1	Main most important system's features and functions	• It adapts to user's unique voice, environment and becomes more accurate as user dictates his/her information • It is designed with next generation speech engine leveraging Deep Leaning Technology • User can tailor the vocabulary with the terms used every day • It allows users to add formatting rules by voice, such as bold or underline • It is available in English, French, Dutch, Japanese, Spanish, Italian and German languages
2	Strengths and opportunities	• It "learns" the words and phrases that users use the most to minimize corrections • It optimizes accuracy for speakers with accents or in slightly noisy environment such as a classroom • Smart Format Rules automatically adapt to how student/lecturer want abbreviations, dates, phone numbers, etc. to appear • It allows faculty to speak as long as possible without pausing to wait for the software to transcribe the already pronounced information • It allows to import and export to/from popular cloud-based document-sharing tools like Dropbox and note-taking apps like Evernote
3	Possible weaknesses and threats	• It works with most common Web browsers but not Edge • If there is a considerable noise in the environment, it is likely to get far more errors • There should be minimum background noise • It is relatively expensive software
4	Technical platform	• RAM: Minimum 4 GB • CPU: Intel dual core or equivalent AMD processor • Free hard disk space: 8 GB • Operating systems: Windows 7, 8.1, 10(32-bit and 64-bit); Windows Server 2008 R2 and 2012 R2 • Built-in microphone or a Nuance-approved microphone • An Internet connection for product download and automatic product activation
5	Price (if any)	• Starts at $300 per copy
6	Colleges/universities that currently use this system	• Lamas College, Iowa • Youth Villages (Health and Human Services)
7	System's ranking	**Our ranking of this system: 3** Main reasons: • It can transcribe from .MP3, .AIF, .WAV, .MP4 audio files to text easily • It significantly saves time and eliminates reliance on costly transcription services; for example, out of 250 words dictated only one was incorrectly transcribed

Table 4.20 *Google Doc–Speech Recognition* [52] open-source voice-to-text system: the analysis outcomes

#	System's characteristics	System's details
1	Main most important system's features and functions	• Voice typing of the document with the help of microphone • The Google documents are saved to user's Google drive thereby making it easier to view documents later • Corrections can be made while voice typing process takes place • It supports many languages alongside different dialects of the same language • Punctuation can be added to the text by using special pre-defined phrases such as the period ('.'), comma, question mark, exclamation point…etc. • Commands can be used to edit, select or format the text • The users can stop or resume voice typing by using special pre-defined commands • It can create tables • User can dictate continuously without the use of the Internet connection, but need a proper connection for saving of files • Accuracy is pretty favorable without many mistakes
2	Strengths and opportunities	• It is a very helpful system for users to prepare documentation anywhere and anytime • It can be installed by anyone as long as they have Google Chrome browser • No tension of losing the information in document due to any technical errors; this is because the system automatically regularly saves document • It does not require lengthy voice training from the user • Spelling mistake can be corrected • Speaking can be stopped or resumed whenever needed • It is a free add-on for any operating system that supports Google Chrome browser
3	Possible weaknesses and threats	• If there is no proper Internet connection, saving of a document automatically is impossible • If there is loud background voice or noise, then there might be high chance of errors • If the microphone is far from the user, then accuracy (or, quality) of outcome text may decrease
4	Technical platform	• Any operating system that supports Google Chrome browser
5	Price (if any)	• Free
6	Colleges/universities that currently use this system	• Google • Many local schools in the United States use this system
7	System's ranking	**Our ranking of this system: 1** Main reasons: • It supports many languages and dialects • Commands can be given for punctuation signs • Editing and formatting the document can also be done • It proved to be a very feasible system

Table 4.21 *Windows Speech Recognition* [43] open-source voice-to-text system: the analysis outcomes

#	System's characteristics	System's details
1	Main most important system's features and functions	• Users can type and use simple commands with their voices • The system can be trained to better understand your language • It uses speech profile to store information about user's voice • It can navigate your computer through commands • It has about 97.8% accuracy • Users can dictate text within documents • Opening, closing, switching and scrolling programs' functions are very easy • Users can dictate their e-mails and the system will type them • Filling forms can also be done using voice commands • The accuracy of speech recognition increases through active use • Individual speech profiles can be created on a per-user basis • Users can say "How do I … (specify a task)", and the system will act accordingly
2	Strengths and opportunities	• Permanent on-going adaptation to both your speaking style and accent continually improves speech/voice recognition accuracy • The new interface provides a simple and efficient experience for dictating and editing text, correcting mistakes and controlling user's computer by voice • It has support for multiple languages—English (United States), English(United Kingdom), German(Germany), French(France), Spanish(Spain), Japanese, Traditional Chinese and Simplified Chinese • It can help people who has trouble using their fingers and/or hands • It can help users who have cognitive disabilities • No extensive training is required for users • It increases productivity; for example, faculty can dictate their feedback/evaluation more quickly because student/learner work is completed electronically • It is free and in-built for Windows operating system • It has long term benefits for students/learners
3	Possible weaknesses and threats	• If there is a background noise or some other sound in the room, the number is errors will increase • More distant (at a distance) microphones from users will tend to increase the number of errors • It has an initial period of adjusting to each user's voice • It may not work properly in a cubicle environment
4	Technical platform	• Windows 7 or 8 or 10 OS
5	Price (if any)	• Free (since it is in-built Windows OS)
6	Colleges/universities/companies that currently use this system	• Many leading companies like Microsoft, IBM, Symantec, Oracle, SAP

(continued)

Table 4.21 (continued)

#	System's characteristics	System's details
		• Many sister companies of Microsoft such as NextBase, Fox Aoftware, Netwise, PlaceWare, Groove Networks, Metanautix
7	System's ranking	**Our ranking of this system: 2** Main reasons: • It allows users to dictate emails and documents easily • It increases an independence for students with physical disabilities and possibly other disabilities

Table 4.22 *TalkTyper* [45] open-source voice-to-text system: the analysis outcomes

#	System's characteristics	System's details
1	Main most important system's features and functions	• It is a free software in Web browser • It can get started right away after user gives microphone permission to talk to the application • Allows to translate all the text into another language • Various options are available to correct the outcome text • It can copy to clipboard, add punctuation, print, and clear all texts • It can send an e-mail, and tweet your text • English Language students/learners can speak English, play it back, and correct it until it "sounds right" and expresses their ideas correctly
2	Strengths and opportunities	• Users can bypass poor typing skills, dysgraphia, dyslexia and physical disabilities • It doesn't require any additional downloads onto user computer • Writing an e-mail to classmates or lecturer is easier as student needs to just click the e-mail button after dictating the text
3	Possible weaknesses and threats	• It is not completely hands-free system; some editing must be done with mouse and keyboard • It needs a constant connectivity to the Internet • If paused for a short time (even a fraction of second), the system may cut user off)
4	Technical platform	• Internet
5	Price (if any)	• Free
6	Colleges/universities that currently use this system	• Clinton County Regional Educational Service Agency (www.ccresa.org) • The Australian Disability Clearing House on education (www.adcet.edu.au) • TeachersFirst (http://www.teachersfirst.com)
7	System's ranking	**Our ranking of this system: 3** Main reasons: • Users with physical disabilities, dyslexia, dysgraphia and poor typing skills may get a great advantage to improve content and overcome problems related to writing and/or typing

Table 4.23 Our recommendations: existing top 3 commercial and top 3 open-source voice-to-text software systems to be implemented and actively used in SmC/SmU

#	System	Company-developer	Details	Ref.
Commercial systems				
1	Braina Pro	Windows/Android	$40 per year	[35]
2	Dragon Naturally Speaking Premium (Home Edition)	Windows/Mac/Android	$100 per copy	[33]
3	Dragon Professional Individual	Windows/Mac/Android	$300	[34]
Open source (free) systems				
1	Google Docs—Speech Recognition	Any platform that supports Google Chrome browser	Open Source	[52]
2	Windows Speech Recognition	Windows	Open Source	[43]
3	TalkTyper	Windows/Mac/Linux	Open Source	[45]

4.6 Research Outcomes: Analysis of Gesture Recognition Systems

Gesture recognition software systems, in general, will allow the user to communicate with the machine naturally, using human-machine interface (HMI) and without any mechanical devices. For example, using the gesture recognition technology, it is possible to point a finger at the computer screen so that the computer cursor on a screen will move accordingly. Potentially, this technology could make conventional input devices such as mouse, keyboards and even touch-screens redundant. For the individual with any type of motor difficulties this could make a huge contribution to them having access to content in the Smart Classroom.

Currently, there are several gesture recognition software systems available that potentially in the future could be implemented in a smart classroom within a SmU. Unfortunately, most of them are not mature enough to be recommended at this moment for an implementation and active use in SmU.

Based on the outcomes of extensive literature review and creative analysis of existing gesture recognition systems, we arrived with a generalized list of desired features of gesture recognition systems suitable for SmU; those features are presented in Table 4.25.

After investigating the desired features of gesture recognition software systems (Table 4.24) that, in our mind, should be available for students with disabilities in SmU, the next steps in our research and analysis project were:

(1) identification and thorough analysis of about 10 commercial and 10 open-source available gesture recognition software systems,

Table 4.24 A list of desired features of gesture recognition software systems for SmU

#	Desired system feature	Feature details
1	Provide alternatives	Alternatives to mouse and keyboard input should be provided for users with hearing or visual impairments or physical disabilities
2	High-speed recognition	The system should have no difficulty to recognize high speed hand movements/signals and input from multiple users at the same time
3	User-friendly system	The system should be easy to learn, easy to use and easy to understand by students with disabilities
4	Transcribe to text	It should be able to understand quickly and with a good quality the sign language's elements/components and transcribe them into a text
5	Interact with digital content	Tracking cameras should be able to detect hand movement to allow user or multiple users to interact with digital content on a screen
6	Engage every user	It should allow multiple users to participate at once, making sure that nobody is left behind
7	Multimodality	It should be adaptable to users with wide range of communication (gesture) styles
8	Immediate feedback	It should be able to provide immediate feedback (reaction of the system)
9	Portability of a system	It should be easy to move a system to different locations and assemble it
10	Technical platform independent system	The system should be work on multiple operating systems

(2) identification of a list of most important (i.e. most useful for students with disabilities) features (functions) of existing gesture recognition software systems,

(3) examples of obtained analysis outcomes of powerful (in terms of functionality) gesture recognition existing software systems, and our ranking of those systems,

(4) our recommendations, i.e. top 3 commercial and top 3 open-source gesture recognition software systems to be implemented and actively used in SmU.

The obtained research and analysis outcomes are summarized and presented in Tables 4.25, 4.26, 4.27, 4.28, 4.29, 4.30, 4.31, 4.32, 4.33.

Table 4.25 Analyzed 10 commercial and 10 open source gesture recognition software systems

#	Systems analyzed	Company-developer	Technical platform	Ref.
Commercial systems				
1	Kinect	Microsoft	Windows/Mac	[53]
2	Intel RealSense	Intel	Windows/Linux/Mac	[54]
3	IISU	SoftKinetic	Windows/Linux/Android	[55]
4	G-Speak	Oblong	Windows/Linux/Mac	[56]
5	Kinems	Kinems	Windows/Mac/Linux	[57]
6	Elliptic Labs Software	Elliptic Labs	Windows	[58]
7	Leap Motion	Leap Motion	Windows/Mac	[59]
8	UbiHand	Petr Musilek	Windows	[60]
9	Kinect Education	KinectEDucation	Windows	[61]
10	ArcSoft 3D hand gesture recognition	ArcSoft	Windows/Linux/Mac	[62]
Open source (free) systems				
1	HandVu	Moves Institute	Windows/Linux/Mac	[63]
2	FUBI (Full Body Interaction Fr.)	Augsburg	Windows/Linux	[64]
3	Wiigee	Ninttendo	Windows/Linux	[65]
4	ControlAir	Apple	Mac	[66]
5	GestTrack3D	GestureTek	Windows	[67]
6	iGesture	Globis	Windows/Linux/Mac	[68]
7	OpenCV	IntoRobotics	Windows	[69]
8	KinectCAD	Catia	Windows	[70]
9	Accelerometer Gesture Recognizer	David Uberti	Windows	[71]
10	Gesture Clustering toolkit (GECKo)	Jacob O. Wobbrock	Windows	[72]

Table 4.26 A list of most important for SmC/SmU features in existing gesture recognition systems

#	Existing important features	Details of existing important features
1	High-speed recognition	The system is able to recognize high-speed hand signals
2	Multiple-users	The system is able to handle multiple users at the same time
3	Transcribe to text	The system is able to take sign language and transcribe it into text
4	Engage every student	The system is able to handle multiple participating students at once
5	Friendly GUI	Dome systems have friendly GUI
6	Camera tracking	Tracking cameras are able to detect user hand movement and allow the user to interact with digital content
7	Sign language recognition	The system is able to detect/recognize elements/components of sign language; it helps mute students to communicate with a system and digital content
8	Multimodality	System supports user's different communication styles
9	Interactivity	Students with disabilities are immersed in a variety of interactives with a physical (in-classroom) and online (remote) students
10	Platform independent	Some systems work on different operating systems

Table 4.27 *Microsoft Kinect* [53] commercial gesture recognition system: the analysis outcomes

#	System's characteristics	System's details
1	Main most important system's features and functions	• Faculty can enhance traditional lesson plans, special education, physical education, school communication and collaboration and after-school programs with immersive full-body experiences that help students get engaged in learning, stay on task and inspire creativity and camaraderie with peers • It tracks as many as six complete skeletons and 25 joints per person • It allows the sensor to see in the dark; the IR capabilities produce a lighting-independent view • It has four microphones to capture sound, record audio, as well as find the location of the sound source and the direction of the audio wave
2	Strengths and opportunities	• It will track every user and so is very adaptable for wheelchair users
3	Possible weaknesses and threats	• For students with Specific Learning Disability (SLD), the program is quite visually messy • Although it can track wheelchair users it is not 100% ready for this • It requires skilled faculty to use; faculty development is required • Visually impaired students cannot use the system
4	Technical platform	• Xbox Kinect Sensor • Kinect SDK • 64-bit ($\times$ 64) processor • 4 GB Memory (or more) • Dual-core 3.1 GHz or faster processor • USB 3.0 controller dedicated to the Kinect for Windows v2 sensor • Microsoft Kinect v2 sensor, which includes a power hub and USB cabling • Projector, Smart Board, or large screen or a big TV screen
5	Price (if any)	• The Kinect SDK is free to download; however, the Kinect Sensor Bar will cost around $60–$100
6	Colleges/universities that currently use this system	• Los Angeles Unified School District (California) • Chicago Public Schools (Illinois) • Houston Independent School District (Texas) • Scottsdale Unified School District (Arizona) • Flagstaff Unified School District (Arizona) • Fairfax County Public Schools (Virginia) • Loudoun County Public Schools (Virginia) • University of Washington
7	System's ranking	**Our ranking of this system: 1** Main reasons: • Schools which used this system reported that it showed a trend of improved executive function, which is the portion of the brain responsible for planning, problem-solving and working memory

Table 4.28 *Intel Real Sense* [54] commercial gesture recognition system: the analysis outcomes

#	System's characteristics	System's details
1	Main most important system's features and functions	• It provides facial recognition, hand gestures, background removal, depth enabled photo, scene perception, 3D scanning and other functions • It supports Augmented Reality • The camera is advanced—it measures depth and enable the computer to read facial expressions and gestures and swap out backgrounds
2	Strengths and opportunities	• Device can be accessed by logging in with student's or faculty's face and it has a good security • It enables person detection and tracking, skeleton tracking, gestures, object recognition etc.
3	Possible weaknesses and threats	• The camera to be used is intended solely for use with Intel RealSense SDK for Windows • The camera drivers need to be downloaded separately
4	Technical platform	• Microsoft Windows 8.1 or 10 OS 64-bit • 4th generation (or later) Intel Core processor • 8 GB free hard disk space • An Intel RealSense camera SR300 (Front-Facing) or R200 (Rear-Facing) • The RealSense Camera F200 has three cameras in one —a 1080p HD camera, an infrared camera and infrared laser projectorC# (Microsoft.NET 4.0 Framework is required) • Java (JDK 1.7.0_11 or later) • Microsoft Visual Studio 2010-2015 or newer
5	Price (if any)	• Intel RealSense SDK is free to download • Intel RealSense Developer Kit R200 will cost $99 • Intel RealSense Developer Kit SR300, which is the next version, will cost $149
6	Companies that currently use this system	• Acer • Dell • Lenovo • HP • Fujitsu
7	System's ranking	**Our ranking of this system: 2** Main reasons: • Intel RealSense SDK can create the next generation of natural, immersive and intuitive software applications • It has many capabilities like facial recognition, hand gestures, background removal, depth enables photo, scene perception, 3D scanning, etc. • The key features involve depth video recording and replay with frame by frame navigation

Table 4.29 *IISU* [55] commercial gesture recognition system: the analysis outcomes

#	System's characteristics	System's details
1	Main most important system's features and functions	• It offers full-body skeleton tracking as well as precise hand and finger tracking • IISU's toolbox provides access to live data and performance analytics during development
2	Strengths and opportunities	• It offers a robust solution from individual finger tracking up to full-body skeleton tracking • The middleware enables developers to easily and rapidly produce gesture based applications for both Close Interaction (3 feet) and Far Interaction (10 feet) experiences • Setup time will be reduced to zero and will not require the multiple calibration steps necessary with other gesture recognition systems
3	Possible weaknesses and threats	• The different plug-ins permits the faculty and students to control the monitors using hand gestures only while sitting comfortably • The free version can only actively track one person • The updates have been discontinued since the release of specialized DepthSense Libraries
4	Technical platform	• Windows 7, Windows 8, Linux and Android 4.1. • Plugins are available for (a) Adobe Flash—access all IISU SDK functions in Action Script 3, (b) Unity3D (V3.0 +)—retrieve and send data in real time using mono-native support of the Unity platform • It supports powerful legacy camera such as the original Kinect, the Asus Xtion, the Panasonic D-Imager and the SoftKinetic DepthSense 331 and 325
5	Price (if any)	• It is free for non-commercial use for three months • Commercial license for IISU can be bought for a one-time fee of $1,500
6	Colleges/universities that currently use this system	• Seneca School, Canada • GURU training Systems • Disney's "The Sorcerer's Apprentice"
7	System's ranking	**Our ranking of this system: 3** Main reasons: • IISU's features are turned to use minimal CPU and memory resources with rapid refresh rates at 25/30/50/60 fps • It is optimized for higher performance

Table 4.30 *Hand Vu* [63] open-source gesture recognition system: the analysis outcomes

#	System's characteristics	System's details
1	Main most important system's features and functions	• It is intended to track and record hand gestures • It discovers and reports three key pieces of information about the tracked hand, including the x and y coordinates and its "posture". The system has 26 hand postures recognitions • The software collection implements a vision-based hand gesture interface. It detects the hand in a standard posture, then tracks it and recognizes key postures—all in real-time • The output is accessible through library calls
2	Strengths and opportunities	• The system can work under various degrees of background lightening conditions • It has 26 hand standard postures recognitions • It has about 90% average recognition rate • It is available free of charge • It is well documented with full reference manual • It has user friendly graphic interface
3	Possible weaknesses and threats	• Tracking is not very effective with fast hand motion • The tracking will get off more frequently and recognition rates might suffer if there is a very hard contrast in the background • It requires sensor cameras that usually very expensive; it does not work with miniature cameras
4	Technical platform	• Windows, Mac OS
5	Price (if any)	• Free download
6	Colleges/universities that currently use this system	• Pennsylvania State University • University of Electronic Science and Technology, China • Montana State University
7	System's ranking	**Our ranking of this system: 1** Main reasons: • It has about 90% average recognition rate (this is a very good technical outcome) • It can be integrated into more complex software systems that can add significant benefits to users with disabilities

Table 4.31 *FUBI (Full Body Interaction Framework)* [64] open-source gesture recognition system: the analysis outcomes

#	System's characteristics	System's details
1	Main most important system's features and functions	• It recognizes full body gestures and postures in real time from the data of a depth sensor • It records gesture performances and can generate valid XML file • It supports gestural interaction by using gesture symbols • It provides buttons and swiping menu to implement freehand GUI interaction • The download comes with Visual Studio 2010 and 2013 solutions • It is written in C ++ and additionally includes a C# - wrapper
2	Strengths and opportunities	• It distinguishes between four gesture categories: (a) static postures (configuration of several joints, no movement); (b) gestures with linear movement (linear movement of several joints with specific direction and speed); (c) combination of postures and linear movement (combination of above two with specific time constraints); (d) complex gestures (detailed observation of one or more joints over a certain amount of time) • It supports gestural interaction by using gesture symbols
3	Possible weaknesses and threats	• It requires the installation of additional files and a compliant middleware to support full body tracking • Testing and fine tuning of gestures could be an issue; it is a time consuming process
4	Technical platform	• Modern Windows operating systems • It requires the installation of OpenNI (OpenNI binaries) and a complaint middleware supporting full body tracking
5	Price (if any)	• Freely available under the terms of the Eclipse Public License—v 1.0
6	Colleges/universities that currently use this system	• Augsburg University • Ontario College of Art & Design • MukiBaum Treatment Centers
7	System's ranking	**Our ranking of this system: 2** Main reasons: • High percent (90% +) were recognized in third-party research of this product • It has a freehand GUI interaction component

Table 4.32 *Wiigee* [65] open-source gesture recognition system: the analysis outcomes

#	System's characteristics	System's details
1	Main most important system's features and functions	• It's main goal is to allow the training and recognition of arbitrary gestures using the Nintendo remote controller by utilizing state of the art probability theory methods • It delivers reliable results in a fast and efficient way • It is able to handle multiple wiimotes
2	Strengths and opportunities	• User can connect to Wii Remote controller like any other Bluetooth device • Wii Remote controller does not need special handling anymore
3	Possible weaknesses and threats	• It requires to purchase additional hardware—wiimote's hardware; it includes infrared camera, LED lights, vibration motor, etc. • It runs only on one technical platform—Linux
4	Technical platform	• Linux OS • Wiimotes
5	Price (if any)	• Free; http://wiigee.org/download/download.htm
6	Colleges/universities that currently use this system	• University of Oldenburg
7	System's ranking	**Our ranking of this system: 3** Main reasons: • Utilizes state of the art probability theory that delivers reliable results quickly in an efficient way

Table 4.33 Existing top 3 commercial and top 3 open-source gesture recognition software systems among analyzed systems

#	System	Company-developer	Details	Ref.
Commercial systems				
1	Kinect	Microsoft	$60–100	[53]
2	Intel RealSense	Intel	$99 or $149	[54]
3	IISU	SoftKinetic	$1500 one-time fee	[55]
Open source (free) systems				
1	HandVu	Moves Institute	Open Source (free)	[63]
2	FUBI	Augsburg	Open Source (free)	[64]
3	Wiigee	Ninttendo	Open Source (free)	[65]

4.7   Research Outcomes: Strengths and Weaknesses of Tested Text-to-Voice and Voice-to-Text Systems

The next step of research and analysis was to

(1) download the actual trial or demo versions of selected ranked software systems,
(2) test and evaluate those systems against CLA requirements (Table 4.3), and
(3) summarize lists of strengths and weaknesses of analyzed systems.

The outcomes of systems' testing and lists of identified strengths and weaknesses of each system (using evaluation criteria from Table 4.3) are presented in Tables 4.34, 4.35, 4.36, 4.37.

Based on outcomes of the performed SWOT (Strengths-Weaknesses-Opportunities-Threats) analysis and obtained testing outcomes of designated

Table 4.34 Strengths and weaknesses of ranked commercial text-to-voice software systems

Rank	System ref	Strengths	Weaknesses
1	Natural Reader [13]	• Accepts files in DOC, DOCX, TXT and HTML formats • Reads text in an image • Reads text in a tabular form • Typing echo is available • Spell checker is available • Built-in dictionary is available • Supports OCR functionality • Provides natural voices • Supports 7 languages • Highlights the text while reading • Word prediction functionality is available	• Does not support reading of math equations • Cannot read the e-mails
2	Read the Words [15]	• Accepts files in DOC, DOCX, TXT and HTML formats • Reads text in a tabular form • Supports 5 languages • Saves audio in audio files • Incorporates the created audio files into emails	• Does not support reading of math equations • Merges data in the tabular form • No spellchecker • No OCR functionality • No e-mail reader • No built-in dictionary • No highlighting text functionality while reading • No word prediction functionality

Table 4.35 Strengths and weaknesses of ranked open source text-to-voice software systems

Rank	System ref	Strengths	Weaknesses
1	Balabolka [23]	• Accepts files in DOC, DOCX, TXT and HTML formats • Reads titles/text in an image • Reads text in a tabular form • Supports 30 languages • Saves audio in audio files • Highlights the text while reading	• Does not support reading of mathematical equations • Spell checker is not able to detect typos • No built-in dictionary • Merges data in the tabular form • No optical character recognition (OCR) functionality • Cannot read the e-mails No word prediction functionality
2	Text to Speech Reader [24]	• Accepts files in TXT and PDF formats • Reads text in a tabular form • Highlights the text while reading • Supports 10 languages	• Does not support DOC and DOCX formats • Cannot read text in an image • Merges data in the tabular form • Does not support reading of mathematical equations • Spell checker is not able to detect typos • No built-in dictionary • No optical character recognition (OCR) functionality • Cannot read the e-mails • No word prediction functionality • Do not save audio into an audio file

systems, we recommend the following systems to be considered for an implementation, testing by actual students with disabilities of various categories, and active use in SmU, and, probably, traditional universities:

(1) **text-to-voice systems:** *Natural Reader* [13] (about $ 70/copy) and *Read The Words* [15] (about $ 40/year) commercial systems, and *Balabolka* [23] and *Text to Speech Reader* [24] open source systems;

(2) **voice-to-text systems:** voice-to-text systems: *Braina* [35] (about $ 30/year) and *Dragon Naturally Speaking Home Edition* [33] (about $ 100/copy) commercial systems, and *Google Docs–Speech Recognition* [52] and *Windows Speech Recognition* [43] open source systems.

Table 4.36 Strengths and weaknesses of ranked commercial voice-to-text software systems

Rank	System ref	Strengths	Weaknesses
1	Braina [35]	• Output format is in DOC, DOCX and TXT formats • Transcribes the entire audio files into text • Supports 40 languages • Does not require lengthy voice training • Editing of a dictionary is available • Supports punctuation • Can e-mail the outcome text file • Can access files and/or folders • Note-taking feature is available • Enables users with predefined and customized commands • Provides reading of math equations • Can also convert text to speech • Can set alarms and reminders	• Cannot e-mail the output text files • Spell checking is not available
2	Dragon Naturally Speaking Home Edition [33]	• Output format is in DOC and DOCX formats • Supports about10 languages • Does not require lengthy voice training • Spell checking is available • Can e-mail the output text file • Enables users with predefined and customized commands • Can also convert text to speech • Mobile dictation is available	• Does not support reading of mathematical equations • Cannot access files and/or folders • Does not support punctuation

Table 4.37 Strengths and weaknesses of ranked open source voice-to-text software systems

Rank	System ref	Strengths	Weaknesses
1	Google Docs–Speech Recognition [52]	• Output format is in DOC and DOCX formats • Supports 70 + languages • Does not require much voice training • Editing of a dictionary is available • Spell checking is available • Supports punctuation • Can e-mail the text document • Pre-defined commands are available	• Does not support reading of mathematical equations • No customization of commands • Cannot access files and/or folders
2	Windows Speech Recognition [43]	• Output format is in DOC and DOCX formats • Supports 6 languages • Supports punctuation • Pre-defined commands are available • Can access files and/or folders • Editing of a dictionary is available • Spell checking is available	• Does not support reading of mathematical equations • Requires MUI language pack for additional languages • Requires lengthy voice training • No customization of commands. • Cannot e mail the text directly

4.8   Conclusions. Future Steps

To be successful in a college/university environment, students with disabilities need more support than students without disabilities. We believe the implementation of specific software systems in SmU and SmC is a key for this to happen. Software systems that address speech-to-text, text-to-speech, and gesture recognition will help students with disabilities to be more successful in the educational setting. In addition, students without any disabilities may benefit as well.

We are suggesting that SmC be equipped with various software systems so that all students (a) will have better access to the content being delivered, (b) be able to adequately interact with the professor and classmates, and (c) feel they are an integral part of the innovative learning environment—SmC in SmU.

Although not all university professors have knowledge or experience with students with disabilities, all of them should try to include them in the learning environment.

Conclusions. The performed research helped us to identify new ways of thinking about "students with disabilities in smart classroom and smart university environment" concept. The obtained research findings and outcomes enabled us to make the following conclusions:

1. Smart universities and smart classrooms can significantly benefit students with disabilities even though they are not the focus.
2. Many technologies that are geared towards students without disabilities will actually impact the learning of students with disabilities.
3. Some students with disabilities may need specialized technology to have equal access in the classroom.
4. Some technologies and software systems focusing on the success of students with disabilities may help students without disabilities to be successful.
5. There are a variety of commercial and open-source software systems in the areas of text-to-voice, voice-to-text and gesture recognition to aid students with disabilities.
6. Given the variety of commercial and open-source software systems an in-depth hands-on assessment of these systems by actual students with disabilities and subject matter experts should be conducted.
7. Each of the commercial and open-source software systems in the areas of text-to-voice, voice-to-text and gesture recognition have different features and capabilities.
8. More research and testing in real-world scenarios needs to be completed addressing commercial and open-source software systems in the areas of text-to-voice, voice-to-text and gesture recognition to decide which of them would have the most benefits for students with disabilities.
9. More research needs to be completed where actual students with disabilities experience and evaluate commercial and open-source software systems in various learning environments and scenarios.
10. More research needs to be completed that directly focuses on students with disabilities in smart classroom and smart university environment.

Next steps. The next steps of this multi-aspect research, design and development project deal with

1. Implementation, analysis, testing and quality assessment of numerous components of text-to-speech, speech-to-text, and gesture recognition software systems in Bradley Hall (the home of majority of departments of the College of Liberal Arts and Sciences) and in some areas of the Bradley University campus.
2. Implementation, analysis, testing and quality assessment of text-to-speech, speech-to-text, and gesture recognition software systems (a) in everyday teaching of classes in smart classrooms and (b) with actual students with disabilities.
3. Organization and implementation of summative and formative evaluations of local and remote students and learners with and without disabilities, faculty and professional staff, subject matter experts, administrators, and university visitors with a focus to collect sufficient data on quality of implemented text-to-speech, speech-to-text, and gesture recognition software systems.

4. Creation of a clear set of recommendations (technological, structural, financial, curricula, etc.) regarding a transition of a traditional university into a smart university pertaining to software and students with and without disabilities.

Acknowledgements The authors would like to thank Dr. Cristopher Jones, Dean of the LAS College, and Sandra Shumaker, Executive Director, Office of Sponsored Programs at Bradley University for their strong support of our research, design and development activities in smart university and smart education areas.

The authors would like to thank Lynne Branham, Interim Director and Students with Disabilities Counselor, and Dr. Susan Rapp, Associate Director and Students with Disabilities, Center for Learning and Access, Bradley University, for active participation in project related activities and collaboration with project team members.

The authors also would like to thank Ms. Aishwarya Doddapaneni, Ms. Supraja Talasila, Mr. Siva Margapuri, and Mr. Harsh Mehta—the research associates of the InterLabs Research Institute and/or graduate students of the Department of Computer Science and Information Systems at Bradley University—for their valuable contributions into this research project.

This project is partially supported by grant REC # 1326809 from Bradley University.

References

1. Pishva, D., Nishanthia, G.G.D.: Smart classrooms for distance education and their adoption to multiple classroom architecture. J. Netw. **3**(5) (2008)
2. Gligorić, N., Uzelac, A., Krco, S.: Smart classroom: real-time feedback on lecture quality. In: Proceedings 2012 IEEE International Conference on Pervasive Computing and Communications Workshops (PERCOM Workshops), pp. 391–394, 19–23 March 2012, Lugano, Switzerland. IEEE doi:10.1109/PerComW.2012.6197517 (2012)
3. Slotta, J., Tissenbaum, M., Lui, M.: Orchestrating of complex inquiry: three roles for learning analytics in a smart classroom infrastructure. In: Proceedings of the Third International Conference on Learning Analytics and Knowledge LAK'13, pp. 270–274, New York, NY, USA. ACM. doi:10.1145/2460296.2460352 (2013)
4. Koutraki, M., Maria, Efthymiou, V., Grigoris, A.: S-CRETA: smart classroom real-time assistance. In: Ambient Intelligence—Software and Applications. Advances in Intelligent and Soft Computing, vol. 153, pp 67–74. Springer (2012)
5. Coccoli, M., et al.: Smarter Universities: a vision for the Fast Changing Digital Era. J. Visual Lang. Comput. **25**, 103–1011 (2014)
6. Aqeel-ur-Rehman, Abbasi, A.Z., Shaikh, Z.A.: Building a Smart University using RFID technology. In: 2008 International Conference on Computer Science and Software Engineering (2008)
7. Lane, J., Finsel, A.: Fostering Smarter Colleges and Universities Data, Big Data, and Analytics. State University of New York Press (2014). http://www.sunypress.edu/pdf/63130.pdf
8. Al Shimmary, M.K., Al Nayar, M.M., Kubba, A.R.: Designing Smart University using RFID and WSN. https://www.researchgate.net/publication/221195787_Building_a_Smart_University_Using_RFID_Technology (2008)
9. Doulai, P.: Smart and flexible campus: technology enabled university education. In: Proceedings of The World Internet and Electronic Cities Conference (WIECC), Kish Island, Iran, 1–3 May 2001, pp. 94–101 (2001)

10. Yu, Z. et al.: Towards a smart campus with mobile social networking. In: Proceedings on the 2011 International Conference on Cyber, Physical and Social Computing, Oct 19–21 2011, Dalian, China, IEEE, pp. 162–169 (2011)

11. Uskov, V.L, Bakken, J.P., Pandey, A., Singh, U., Yalamanchili, M., Penumatsa, A.: Smart University taxonomy: features, components, systems. In: Uskov, V.L., Howlett, R.J., Jain, L. C. (eds.) Smart Education and e-Learning 2016. Springer, pp. 3–14, June 2016, 643 p. ISBN: 978-3-319-39689-7 (2016)

12. Bakken, J.P., Uskov, V.L, Penumatsa, A., Doddapaneni, A.: Smart Universities, Smart Classrooms, and students with disabilities. In: Uskov, V.L., Howlett, R.J., Jain, L.C. (eds.) Smart Education and e-Learning 2016. Springer, pp. 15–27, June 2016, 643 p., ISBN: 978-3-319-39689-7 (2016)

13. Natural Reader software system, http://www.naturalreaders.com/priceorder.html#voices

14. Text Speech Pro software system, http://www.textspeechpro.com/download.html

15. Read The Words software system, http://www.readthewords.com/

16. Text Aloud3 software system, http://sites.fastspring.com/nextup/product/textaloud

17. Verbose software system, http://www.nch.com.au/verbose/index.html

18. Voki software system, http://www.voki.com/site/products

19. Oddcast TTS software system, http://www.oddcast.com/home/demos/tts/tts_example.php?sitepal

20. Ultra Hal software system, https://www.zabaware.com/reader/

21. Neo Speech software system, http://www.neospeech.com/products

22. Texthelp Read&Write system, https://www.texthelp.com/en-us/products/read-and-write-family/

23. Balabolka software system, http://www.cross-plus-a.com/balabolka.htm

24. Text to Speech software system, https://www.microsoft.com/en-us/store/p/text-to-speech-tts/9wzdncrdfm3b

25. TTS Reader software system, http://ttsreader.com/#about

26. Microsoft Word TTS software system, https://support.office.com/en-us/article/Using-the-Speak-text-to-speech-feature-459e7704-a76d-4fe2-ab48-189d6b83333c

27. ClipSpeak software system, https://clipspeak.codeplex.com/

28. WordTalk software system, http://www.wordtalk.org.uk/Home/

29. Imtranslator software system, http://text-to-speech.imtranslator.net/

30. iSpeech software system, https://www.ispeech.org/text.to.speech

31. Google Translate software system, https://translate.google.com/

32. Power Talk software system, http://fullmeasure.co.uk/powertalk/#requirements

33. Dragon Naturally Speaking Premium (Student/Teacher Edition), http://www.nuance.com/for-business/by-industry/education/dragon-education-solutions/index.htm

34. Dragon Professional Individual, http://www.nuance.com/for-business/by-product/dragon/dragon-for-the-pc/dragon-professional-individual/index.htm

35. Braina Pro software system, https://www.brainasoft.com/braina/

36. Tazti Speech Recognition software system, http://www.tazti.com/downloads.php

37. SpeechGear Compadre Interact Software system, http://www.speechgear.info/products/interact-as

38. Dictation Pro software system, http://www.deskshare.com/dictation.aspx

39. e-Speaking Software system, www.e-speaking.com

40. Voice Finger software system, http://voicefinger.cozendey.com/

41. Text Shark software system, http://www.textshark.com/products

42. Speechmatics software system, https://www.speechmatics.com/

43. Window Speech Recognition, http://windows.microsoft.com/en-us/windows/set-speech-recognition#1TC=windows-7

44. Apple Dictation, https://support.apple.com/en-us/HT202584

45. Talktyper system, https://talktyper.com

46. Jasper software system, https://jasperproject.github.io/

47. Dictation 2.0 software system, https://dictation.io/

48. Speechnotes software system, https://speechnotes.co/
49. Digital Syphon Sonic Extracter, http://www.digitalsyphon.com/technologies_sonicextract.asp
50. Balabolka software system, http://balabolka.en.softonic.com/
51. Speech Logger software system, https://speechlogger.appspot.com/en/
52. Google Docs Speech Recognition software system, https://chrome.google.com/webstore/detail/speech-recognition/idmniglhlcjfkhncgbiiecmianekpheh?hl=en
53. Kinect software system, https://www.microsoft.com/en-us/education/products/xbox-kinect/default.aspx
54. Intel RealSense software system, https://software.intel.com/en-us/intel-realsense-sdk
55. IISU software system, www.iisu.com
56. G-Speak software system, http://www.oblong.com/g-speak/
57. Kinems software system, http://www.kinems.com/
58. Elliptic Labs software system, http://www.ellipticlabs.com/technology/
59. Leap Motion software system, http://www.arcsoft.com/technology/gesture.html
60. UbiHand software system, http://dl.acm.org/citation.cfm?id=1179805
61. Kinect Education, http://www.kinecteducation.com/
62. ArcSoft 3D hand gesture recognition software system, http://www.arcsoft.com/technology/gesture.html
63. HandVu software system, http://www.movesinstitute.org/~kolsch/HandVu/HandVu.html#people
64. FUBI software system, http://www.idownloadblog.com/2015/02/05/controlair-control-your-mac-gestures/
65. Wiigee software system, http://www.wiigee.org/
66. Control Air software system, https://www.informatik.uni-augsburg.de/en/chairs/hcm/projects/tools/fubi/
67. GestTrack3D software system, http://www.gesturetek.com/3ddcpth/introduction.php
68. iGesture software system, http://www.igesture.org/
69. OpenCV software system, https://www.intorobotics.com/9-opencv-tutorials-hand-gesture-detection-recognition/
70. KinectCAD software system, https://sourceforge.net/projects/kinectcad/
71. AGR software system, https://sourceforge.net/projects/agr/
72. GECKo software system, http://depts.washington.edu/aimgroup/proj/dollar/gecko.html

Chapter 5
Building a Smarter College: Best Educational Practices and Faculty Development

Nobuyuki Ogawa and Akira Shimizu

Abstract In a smarter university/college, various matters should be interconnected for effective functioning. These include hardware, software, systems, institutional policies, including admission policy, curriculum policy, and diploma policy, faculty development, and information sharing. Our past practices of innovative smart education are highly evaluated and, as a result, financially supported by the Ministry of Education, Culture, Sports, Science and Technology, Japan, and National Institute of Technology, Gifu College (NIT, Gifu College). The efforts of smart education are being promoted at an accelerated rate. All of the faculty members are interconnecting the various matters. Active learning practiced at NIT, Gifu College is characterized by the use of educational systems with ICT equipment. From this point of view, in this chapter, active learning has almost the same meaning as smart education. Based on this idea, we describe various practices for building a smarter college.

Keywords Smart education · Smarter university/college · ICT · Active learning · Faculty development · Curriculum

5.1 Introduction

5.1.1 Globalization of Smart Education

Uskov et al. [1] proposed that "… a future smart university is expected to have distinctive features that go well beyond features of a traditional university, sufficiently taking in *Adaptation, Sensing, Inferring, Self-learning, Anticipation, Self-*

N. Ogawa (✉)
Department of Architecture, National Institute of Technology, Gifu College,
Gifu College, Motosu, Japan
e-mail: ogawa@gifu-nct.ac.jp

A. Shimizu
General Education, National Institute of Technology, Gifu College, Motosu, Japan
e-mail: ashimizu@gifu-nct.ac.jp

© Springer International Publishing AG 2018
V.L. Uskov et al. (eds.), *Smart Universities*, Smart Innovation,
Systems and Technologies 70, DOI 10.1007/978-3-319-59454-5_5

Organization". These concepts are the core concepts of "smartness" that represent the foundation of smart education. In this chapter, it is clarified which concept among the smartness levels each content of smart education has.

Regarding a smart university, Hwang [2] describes the importance of the real-world learning environment, stating: "It can be seen that applying intelligent tutoring or adaptive learning techniques to real-world learning scenarios has become an important and challenging issue of technology-enhanced learning. That is, incorporating intelligent tutoring or adaptive learning techniques to context-aware ubiquitous learning has become one of the important issues of technology-enhanced learning" [2].

Coccoli et al. [3] state: "By 'smarter university' we mean a place where knowledge is shared between employees, teachers, students, and all stakeholders in a seamless way. … To cope with this reality, technology is no longer sufficient. We suggest that a paradigm change is necessary to transform a smart university into a smarter university, hence more efficient, more effective, and with a higher participation of both students and teachers, collaborating to achieve the common objective of better learning". Both the former and latter authors point out the importance of seeing our real society in addition to the technology, such as "the technology and the real environment" and "the technology and sharing knowledge among the university's faculty members" [2, 3].

Active learning (AL) encourages learners to actively and independently solve a problem. On such occasions, it is important to interact well with the other members within a group and a teacher as a facilitator. In smart education, various types of interactions are supported by ICT equipment, software, and other kinds of systems. Recording the interactions in the learning process will cause the system to send feedback or information helpful to learners and teachers in the form of a portfolio. For example, e-Learning using educational videos enables learners to do self-learning at home through the idea of flipped learning. Also, the use of ICT equipment and some kinds of systems for problem solving learning, such as flipped learning in class at a university/college, will effectively be able to bring out learners' active nature. Thus, AL becomes more finely tuned, more intelligent, and more effective when combined with smart education. At NIT, Gifu College, we are promoting this type of AL associated with smart education.

As globalization proceeds in the field of education, the increasing importance of the smart university concept is considered as a global standard. In 2013, Japan's National Institute for Educational Policy Research, which is under the Ministry of Education, Culture, Sports, Science and Technology (MEXT), proposed the "21st century competencies": in the future, Japanese students are expected to acquire, not only knowledge, but also the capability to find problems for themselves and solve them through thinking. The former Ministry of Education, now MEXT, substantially revised the standards for establishing universities in 1991 as the Japanese government promoted the decentralization of power and announced a policy to ease the restrictions.

5.1.2 Foundation for Smart Education in Japan

In the revisions, the related laws, including the School Education Act and the standards for establishing universities, were drastically amended. This allowed individual schools to flexibly develop unique education and research based on its own educational philosophy and objectives. This takes place while responding appropriately to the advancement of learning and the demands of society. The related legal revisions led to the elimination of the details of the standards including curriculum [4]. Under the revisions, the requirements of the standards were eased, but a policy in instituted that stated that universities themselves should assure quality of education and research. Consequently, universities were required to conduct self-inspections and assess the quality of their education and research.

As a result of these revisions, the number of private universities in Japan increased from 372 in 1990 to 605 in 2012. Also, the percentage of students who go on to a 4-year university increased from 25.5% in 1991 to 50.2% in 2009, surpassing the 50% line for the first time. In recent years, however, the university advancement rate has remained unchanged at approximately 50%. The total university enrollment also tends to decrease, following the decline in the number of high school students due to a declining birthrate in Japan. From a standpoint of both enhancing international competitiveness amid globalization and quality assurance of university education, MEXT announced "a University Reform Action Plan" in June of 2012. This clearly indicated its intention to exclude universities that cannot adjust to socictal changes. The plan includes a policy to intensively provide subsidies to universities that improve the quality of education and promote the development of human resources focused on global outlook and/or contribution to the local community. They must also severely deal with universities that have problems with managing status and/or educational environment, etc. Also, because of the decline in the number of high school students due to the falling birthrate, MEXT has presented the following rough plan: universities are highly expected to promote the abolition of departments and graduate schools related to teacher training, humanities, and social science. They must also promote the abolition of their reorganization to fields with high societal demand.

In Japan, it is estimated that the number of 18-year-olds, which stood at 2.07 million in 1991, will continue to decline to 1.01 million in 2030. The declining number of children will cause the integration and abolition of universities in the future. Just as with the worldwide trend to positively introduce e-Learning and active learning (AL) into education amid increasing globalization, it is a natural tendency to introduce ICT-driven education and e-Learning in Japan under the present situation. MEXT is also strongly promoting the use of AL and e-Learning in elementary, secondary, and higher education. In order to develop an environment where school education appropriate for the 21st century can be realized, MEXT budgeted the amount required for attaining the targeted level in the Second Basic Plan for the Promotion of Education. This was endorsed by the Cabinet decision on June 14, 2013. Based on the "four-year environmental improvement program for

IT-based education (2014 to 2017)", 167.8 billion yen will be budgeted every year up to the fiscal year 2017, allotting 671.2 billion yen each year for four years.

5.1.3 Development to Smart Education in NIT, Gifu College

Colleges in the National Institute of Technology (NIT) accept junior high school graduates who have just completed compulsory education. We focus on applicants entering the best suited department by visualizing the educational content of each department and publicizing the differences among all the departments. This has made it possible to start specialized education at the first year and allow students to acquire university-level expertise, receiving a high evaluation from several external evaluation organizations, including OECD. As a success of the educational activities in NIT colleges, the graduates of departments have received a high evaluation from society whether through gaining employment after graduation or transferring into the third year at four-year universities. Those who have entered graduate schools after completing the advanced courses of NIT colleges have also received a high evaluation from the graduate schools they entered. Furthermore, the examinations conducted by the National Institution for Academic Degrees, Japan, and JABEE, such as the institutional certified evaluation and accreditation for the advanced course of NIT colleges, have recently pointed out the need to develop a curriculum while being more conscious of the specialized fields of respective departments. Accordingly, the research performances in his/her specialized field of the faculty of the departments of NIT colleges are screened more strictly than before. The sophistication and complexity of the technologies, society needs, globalization, global environmental issues, and the rapid progress of technological innovation have produced an idea that we should foster engineers with wide range of skills, though. Considering the social situation, NIT is promoting its educational and organizational reforms throughout the whole organization.

Our past challenges such as advanced education using ICT equipment, e-Learning education under the credit transfer agreement practiced in a consortium, and developmental continuation of education constructed financially supported by the "Support Program for Contemporary Educational Needs (GP)" led to awards from several academic societies for education. Furthermore, NIT, Gifu College acquired the AP. While financially supported by MEXT for five years, we started to promote college wide educational reform at an accelerated rate for a smart college. And now, entering the third year of support, it has been decided that the period of our AP project would be extended from five years to six years. This will enable us to more fully develop items that were scheduled for the latter three years. This can further increase the quality assurance level of the graduating students. Our challenge of smart education is characterized by the practice of AL using ICT-driven equipment as well as some kinds of systems. Also, AL as described in this chapter is AL in a broad sense of the term, including project-based learning, learning-by-doing, adaptive learning, and flipped classrooms. We are practicing smart education by using ICT-driven equipment and some kinds of systems. In short, the objectives of our AP project practiced during this period, are to have students acquire independent-minded, self-directive, and self-improvement minds as well as their ability through learning at college, holding the practice of AL, and the visualization of the learning outcomes as key words, while ensuring continuity after the completion of the project. The advancement and internationalization of the education of NIT colleges, which had been the education goals of NIT, Gifu College before the acquisition of the AP, are related to the AP themes of other fields.

5.2   Smart Education: Literature Review

5.2.1   Use of Data Mining

The wide use of the Internet has caused an explosive increase in the amount of information available to society. It has also caused a tremendous increase in the amount of information learners can access and, accordingly, provided them with a greater range of interest and choices. On the other hand, it is becoming difficult for teachers to meet all the needs of all learners. This is because teachers are limited in knowledge beyond their lectures. By automatically supplying learning materials, distributing reference materials, collecting reports, and conducting tests using ICT, though, we will be able to build an educational system which responds to the learner-centered personalized needs. Also, analyzing big data will make it possible to create a system related to smart education. This is where teachers will support learners to match each student's ability.

Technology to extract knowledge is developing rapidly; it is conducted by exhaustively applying some kinds of data analysis techniques. Statistics, pattern recognition, and artificial intelligence by use of data mining can all be used to deal with such large amounts of data [5–7]. Data mining analysis of learners' study history/portfolio of, for example, LMS, will make it possible to give feedback to students and teachers. This corresponds to "sensing" when it comes to smartness levels. Relational databases and its operation language (SQL) appeared in the 1980s, making it possible to perform dynamic data analysis on demand. Data volume increased explosively in the 1990s and provided a spark for people to use data warehouses for data accumulation. This trend created an idea of using data mining as a way to deal with massive amounts of data in a database. It was then slowly applied as a way of statistical analysis, the search technology in the field of artificial intelligence, and so on. In the 2010s, a variety of practical services using big data analysis, where data mining is conducted with an enormous quantity of data, have appeared and are provided.

5.2.2 Data Processing on Students' Physiological Responses

By gathering students' data, in addition to giving feedback to learners and teachers based on the analysis of data from LMS and so on, research is being conducted where, after measuring learners' automatic biological information, the result of data analysis is fed back to learners and teachers [8–10]. Here, we find research to evaluate educational materials with an entirely new approach. This is where students' physiological responses are employed to evaluate the quality of digital educational materials scientifically and objectively, not by questionnaires or forms of the same nature. This kind of coping using biometrics corresponds to "sensing" of the smartness level. The authors measured the physiological responses of both students learning with interactive educational materials and students learning with non-interactive educational materials based on blood-flow engineering. They then explained the difference between them. The data analysis was based on bioinstrumentation data. By incorporating the data into a system, we will be able to directly provide feedback to the students as well as the faculty. Thus, we will provide a smart education environment. For example, when the operational bioinstrumentation system detects that the students are inactive, a teacher can immediately provide feedback by giving them appropriate or different educational materials. Besides, as a different method for utilization of the measured results of automatic biological reactions, the system will be able to analyze where and how learners have setbacks in light of their pattern of learning. It can then provide detailed advice to them for dealing with their setbacks. This matter corresponds to "sensing" of the smartness level. Asanka et al. [11] took a look at how this can utilize eye blinking and a learner's attitude.

5.2.3 Challenges in Education in a Virtual World

Problem-based learning (PBL) in a virtual world has some challenges. The following references are worth noting in regards to the measurement of the eye blinking in a virtual world and a learner's attitude. Barry et al. [12] shows that the introduction of a virtual world like metaverse into education will make it possible to realize smart education where distance in space is followed by techniques. We can create an environment where students feel as if they were learning in the same classroom through metaverse as an educational environment, even though they really are distant from each other [13, 14].

5.2.4 Application to Smart University

At NIT, Gifu College, we are advancing the above-mentioned research. Their research results have already been introduced into smart education classes. Their practical realization will make it possible to introduce smart education, specifically, to mostly shift individualized instructions by teachers to each student to ICT. This will bring a substantial change to future models for education and the role of teachers. In knowledge-acquisition learning, through an integrated management system where an ICT environment and a software environment are associated each other, each student studies by him/herself or with the support of a teacher. In class, students take advantage of information, have problem consciousness, seek a solution, and present his/her idea. Teachers play a role as a guide for learning. Specifically, they act as a facilitator who motivates students for learning and activate a discussion. They also need flexibility to deal with learners' diverse interests as well as their different desires to learn. These are regarded as the embodiment of future education, which MEXT and the ruling party suggested, and are the guidelines that universities/colleges all around the world aiming for "smart universities/colleges", including NIT, Gifu College, should follow.

5.3 Goals and Objectives of Smart Education

5.3.1 NIT, Gifu College's Transformation into a Smarter College

On April 1st, 2004, the National Institute of Technology (NIT) in Japan was inaugurated to manage 55 national colleges (KOSEN) including NIT, Gifu College [15]. In Japan, almost all prefectures have one KOSEN. It is an institution of higher education with a unique feature that the campuses exist throughout Japan. In 2014, ten years after its foundation, the Institute has developed the third-stage medium-term programs and goals where the promotion of AL, which would highly contribute to the realization of smart college using ICT-driven equipment, is described. On the other hand, NIT, Gifu College, a higher education institution celebrating its 53rd birthday in 2016, has been working on unique challenges related to smart college since before the NIT in Japan was inaugurated. For more than fifteen years, NIT, Gifu College has been developing ICT-driven educational content to cultivate students' voluntary learning. Also, we have developed a system to give incentives for voluntary learning and its evaluating method.

MEXT's "Support Program for Contemporary Educational Needs (GP)" once sponsored our activity. These implementations led to use receiving several awards from academic societies for education. Our ICT-driven education that has been practiced for the past fifteen years was highly evaluated. As a result, our program was selected as a project of the AP by MEXT in 2014. Our AP project began college-wide smart education and was funded by MEXT for six years.

5.3.2 Curriculum

The acquisition of the AP has triggered the introduction of students' voluntary learning into the classes of all teachers; this previously had been performed only in a few classes. In our AP project, we will introduce AL into all the formal subjects at NIT, Gifu College. We also visualize the learning outcomes of both non-curriculum and AL of curriculum, based on the practical engineering credit system. Though AL includes various kinds of activities, such as flipped learning, they correspond to "adaptation" of the smartness level. In addition, AL places emphasis on thinking for oneself and solving problems. These matters are related to "self-learning" of the smartness level. We will promote active educational improvement and the visualization of the learning outcomes by doing the following:

(1) to consolidate in a server and distribute both educational content for teachers. This ensures the educational quality based on Model Core Curriculum (MCC) established by the National Institute of Technology and learning assistance content for students.
(2) to improve the classroom ICT environment.

    (a) projectors with an electronic blackboard system,
    (b) wireless LAN environment,
    (c) terminal environment, and
    (d) systems and software environment.

5.3.3 Creation of an Environment for Smart Education

We think it's vital for first year students to become accustomed to AL in order to understand its meaning and to experience and master the skills of creating new things by themselves and/or with friends in classes over a short time period; of course, these are true for students in upper grades as well. We are now developing the supporting system of students' voluntary learning; all of the teachers learn how to use the electronic blackboard system, the LMS server, tablet PCs, and software for creating teaching materials.

5.3.4 Development of Educational Materials

When starting this AP project, we consulted with some senior graduates of NIT, Gifu College. Each has worked long and/or is currently working in Japanese industries as leading engineers. There is emphasis on the knowledge and skills that they think are important for when students enter the industry after graduation. Specifically, in the 2014 questionnaire, they picked up forty-five important items among those included in MCC and recommended visualization of educational outcomes of these items. More specifically, they suggested subjects, keywords, related matters, backgrounds, reasons, motivation, the introductory level, the intermediate level, the advanced level, and familiar products and cases in use, using diagrams as well as texts when needed. From this academic year, we will create the learning content and learning support content of each item based on the suggestions of the graduates. We will also create the content while being conscious of the learning level of respective content and the relations among the subjects.

5.3.5 Promoting and Visualizing Smart Education

It is important to construct both hardware and software interdependently in order to realize smart education: the introduction of equipment and the advances in technology of classroom environment in terms of hardware as well as the systematic analyses of the constitution of educational content, the establishment of the related education system, and the creation of the educational content that goes along with the system in terms of software.

The aim of our AP project is to promote students' active way of learning by practicing AL [16–27]. Therefore, we are creating an environment for practicing ICT education useful for AL and developing teaching materials based on MCC established by the NIT to ensure educational quality [15]. The AP project of NIT, Gifu College is composed of two themes, Theme I and Theme II. We are now promoting both Theme I, which is the promotion of AL, and Theme II, which is the visualization of the learning outcomes. As for the initiatives for Theme II, we are visualizing the educational and learning outcomes of NIT, Gifu College for off-campus people. It also examines them every time NIT, Gifu College undergoes accreditation by evaluation organizations such as the National Institution for Academic Degrees and Quality Enhancement of Higher Education and the Japan Accreditation Board for Engineering Education (JABEE) [28, 29].

In our AP project, we are proceeding with:

(1) the improvement of educational methods by introducing AL into all classes of the formal subjects of NIT, Gifu College, and
(2) the construction of a system where the educational and learning outcomes are visualized, evaluated, and examined quantitatively under the practical engineering credit system, or a point system where educational outcomes are visualized as independent learning outcomes.

Theme I, the promotion of AL, acts as a catalyst for accelerating the improvement of educational methods of respective subjects, encouraging teachers to improve their teaching methods as well as content, and cooperating with other teachers concerned. The efforts made by respective teachers are shared and visualized through our faculty development (FD) and debrief sessions. This is helpful in spreading the efforts within NIT, Gifu College. Theme II, the visualization of the learning outcomes, focuses, not only on the learning outcomes of respective subjects and the improvement of the educational outcomes, but also on the students' independent learning process toward the objectives from entrance to graduation. Under Theme II, we also visualize the process of developing the ability necessary to acquire competencies, such as cooperativeness and humanity as well as literacies of planning, independence, etc., so that students can attain their objectives.

5.4 The Creation of the Environment: Methods and Outcomes

As described earlier, NIT, Gifu College is leading the way in the promotion of AL among all the National Institutes of Technology. More precisely, we are improving the environment for AL including ICT-driven equipment, e-Learning and teaching materials and promoting educational practice with them as the AP, funded by MEXT for five years.

In the academic years of 2014, 2015, and 2016, we introduced projectors with an electronic blackboard system into all 25 classrooms for all five years of all the five departments through bids at the expense of the AP budget. Furthermore, the wireless LAN device was set up for use in the 25 classrooms so that the introduced LMS systems, such as Moodle and Blackboard, could be used in classes. STORM Maker, software for making teaching materials, was introduced to develop teaching materials to store in the LMS. The special characteristic of STORM Maker, which has an automatic voice synthesis function, simplifies the process of making content that is based on materials. Therefore, we can easily create teaching materials with voice for e-Learning with the work of entering character, without recording narration voice. Both male and female voices can be synthesized, depending on the use and characteristics of teaching materials. Moreover, we introduced 163 Tablet computers (Toshiba), 50 notebook PCs (Fujitsu) and 20 Microsoft Surfaces. All of them were introduced for lending and being set up to connect to all the access points of the wireless LAN for e-Learning in classes.

5.4.1 The Foundation for Introducing AL

As the actual performances that have led NIT, Gifu College to a hub for promoting AL, there are two pillars:

(1) The practice of ICT-driven education performed mainly in the multimedia educational building.
(2) The practice of the credit transfer system education using e-Learning.

5.4.1.1 The Practice of ICT-Driven Education Performed Mainly in the Multimedia Educational Building

The first pillar is the practice of ICT-driven education performed mainly in the multimedia educational building. In NIT, Gifu College, from 2001 until the present, ICT-driven education has been practiced in the multimedia educational building where the fourth-year classrooms of all five departments reside. The installation of ICT-driven equipment for teachers, as well as desktop personal computers for all students with desks having a storage feature for locking away each personal computer, made it possible to perform ICT-driven education in all classes (Fig. 5.1).

ICT-driven equipment was introduced into teachers' workspaces as well as students' workspaces in every classroom. Since the ICT-driven education system has been introduced, ICT-driven education can be applied in various subjects, including social studies, languages, science, and engineering. Every student can use his or her own personal computer and enjoy multimedia education at his or her own will. In NIT, Gifu College, ICT-driven equipment of the multimedia educational building has been internally budgeted for replacement every five years. In 2016, we introduced new ICT-driven equipment useful for AL. This will be mentioned later.

5.4.1.2 The Practice of the Credit Transfer System Education Using E-Learning

Additionally, supported by the "Support Program for Contemporary Educational Needs (GP)" since 2004, we developed the content and system for e-Learning. Also, the program to provide two different kinds of consortium with lectures by using the e-Learning system has successfully been continued until today. When studying by e-Learning, students are supported by e-mail and teachers face-to-face. They are supposed to study the content within the server for themselves, though. Their studies with simulation software, interactive software, videos, texts, etc.

Fig. 5.1 ICT-driven equipment installed in the fourth-year classrooms

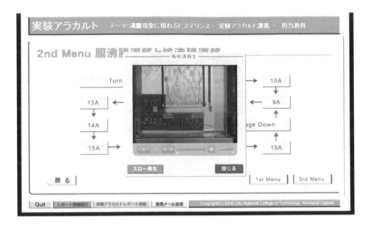

Fig. 5.2 An example of a presentation using the e-Learning system

correspond to "self-learning" of the smartness level. The number of NIT colleges participating in the consortium is increasing year after year. To this day, nearly half of all of the NIT colleges have already participated in the consortium (Fig. 5.2).

NIT, Gifu College has been strongly promoting ICT-driven education and e-Learning. As described above, in a broad sense, we have been practicing AL for a long time. Also, like other colleges, NIT, Gifu College has been practicing AL by PBL for the instruction of practical experiments for the past years. Under the groundwork of the above-mentioned AL, NIT, Gifu College began to strongly promote AL. This includes flipped learning for classroom lectures in 2012. The states of our practice, both before and after the acquisition of the AP, are described below.

5.4.2 AL Before the Acquisition of AP

5.4.2.1 The Learning Environment of AL at NIT, Gifu College Library (2nd Floor)

From this point forward, we will describe the learning environment of AL that was developed in a room on the 2nd floor of the library of NIT, Gifu College. In order to do flexible education of AL, it is important that the equipment in the classrooms is flexible.

As shown in Fig. 5.3, considering these points, NIT, Gifu College has developed a learning environment by installing some custom-made trapezoidal tables. These can be combined in various forms according to the type of group work required, in addition to movable downsized whiteboards for group discussion in the 2nd floor room of the library. During breaks between classes and after school, students can

Fig. 5.3 The learning environment of AL in NIT, Gifu College's library

use the classroom freely. They can have an active discussion using freely arranged desks for completing a report or studying for an exam.

5.4.2.2 Best Practices of AL

NIT, Gifu College is presenting a plan to introduce AL into all classes of NIT, Gifu College and establish it within the term of the third-stage medium-term programs and objectives. In 2012, NIT, Gifu College started AL with flipped learning in the classes of some subjects underlying engineering. More specifically, these classes were mathematics, physics, chemistry, applied physics, and applied mathematics. These subjects took a central role in promoting AL in NIT, Gifu College.

In AL, where the utilization of knowledge is important, such class activities as group discussion, debate, and group work, are effective. Also, cooperative learning done in groups and presentations are indispensable.

In order to do flexible education of AL, it is beneficial that the equipment of the classrooms to be used is versatile. In 2013, considering these points, NIT, Gifu College developed ICT environment by installing electronic blackboards, tablet computers, a file server for teaching materials, and the like in the laboratory classrooms of applied physics, and chemistry.

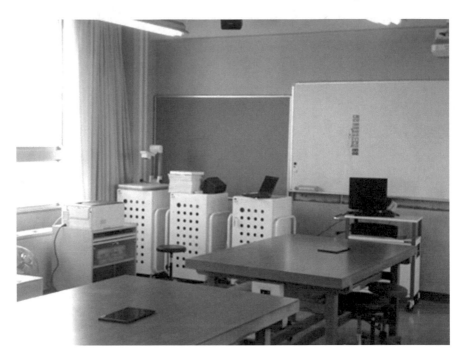

Fig. 5.4 ICT installment in the laboratory classroom of applied physics

The classroom has also been used for classroom lectures of applied physics and applied mathematics and a new type of flipped learning where classroom lectures, practical experiments, and ICT-driven education are combined has been practiced (Fig. 5.4).

5.4.3   AL After the Acquisition of AP

5.4.3.1   ICT-Driven Equipment Introduced by the AP Funds

So far, at the expense of the AP budget, we have introduced 163 Tablet computers (Toshiba), 50 notebook personal computers (Fujitsu) and 20 Microsoft Surfaces. Besides this, the following four items were introduced through bids for the purpose of developing various systems for smart education with ICT-driven equipment. The name of the company written at the end reveals which company won a bid and is in charge of constructing a respective system:

(1)   an electronic blackboard system that uses ICT (Kameta Inc.),
(2)   software for making teaching materials, STORM Maker (Otsuka Shokai Co., Ltd.),

(3) lease and maintenance operations of wireless LAN switch (Nippon Telegraph And Telephone West Corporation), and

(4) LMS server (Moodle) and DB server+FileMaker (Nippon Telegraph and Telephone West Corporation).

We believe that it's vital for freshman students to become accustomed to AL in order to understand its meaning as well as to experience and master the skills of creating new things by themselves and/or with friends in classes over a short time period; of course, these are true for students of the upper grades as well. Therefore, in 2014, when our AP project started, we replaced the blackboards in the back of the first-year classrooms of all five departments with whiteboards and introduced Epson-manufactured projectors. Each has the function of electronic blackboard that is described in the above item (1).

In the 2015 academic year, we introduced the same ICT environment into the second and third classrooms of the five departments. Moreover, in the 2016 academic year, we are planning to introduce the same ICT environment into the fourth and fifth classrooms of the five departments. Thus, the same projectors with the electronic blackboard functions will have been installed into all classrooms of all departments.

The introduced electronic blackboards make it possible to draw and write on its whiteboard with a dedicated electronic pen without connecting a personal computer. Digital data of drawing and writing can be recorded and stored in a file server connected to the network. Moreover, the linkage function of a tablet makes it possible for teachers to arbitrarily select students' tablets up to 50 units from the teachers' tablet and display them on up to four screens. Using the projector control toolbar displayed on the projection screen of the electronic blackboard, teachers can easily select and control students' tablet screen by operating on the screen (Fig. 5.5).

It was basically left to respective departments in which personal computers the assigned licenses are to be installed. The assignment of software for making teaching materials, STORM Maker, is as follows: Two licenses for the departments of liberal arts and natural science, respectively. Three licenses for the specialized five departments, respectively.

The special characteristic of STORM Maker is that we can easily develop content based on materials. It contains the functions to create original Power Point material with animations, write an article on it using the notebook function, and automatically translate original educational material into educational material with a movie using the speech-synthesis function. (Figure 5.6) The software makes it possible to automatically synthesize voice from text, which corresponds to "adaptation" of the smartness level.

AL, which makes students learn on their own initiative, will help them gain a deeper understanding. It is considered important that students present their knowledge and opinions in a positive manner through presentations, discussions,

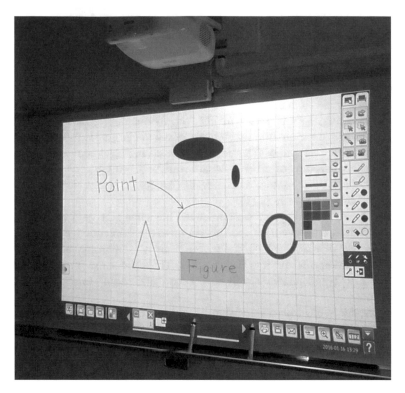

Fig. 5.5 Drawing on an introduced electronic blackboard with an electronic pen

and debates. The intention of the software is not only for teachers to make content, but also for students to make content for future use. In addition to the fact that students' creative activities can also be considered AL, the teaching materials they produce are practiced in classes of AL. When students make educational materials, teachers are hoped to facilitate their content. Doing so will assure the quality and quantity of the teaching materials and increase the quality of AL where students do content creative activities.

The wireless LAN device was set up for use in all 25 classrooms every year (from the first to the fifth year) in all five of the departments. The system was developed by providing two access points for the wireless LAN in each classroom and by controlling them using MAC addresses to prevent injustice access. The same setting was conducted for more than 163 tablet computers (Toshiba). They can be connected to all of the access points for the wireless LAN of 25 classrooms. Therefore, though tablet computers are stored in the storage cabinets near the first-year classrooms, they can be used in other classrooms as well. More specifically, more than 160 tablet computers (Toshiba) were introduced so that we could use them in four classrooms at the same time. All of them were set up for connecting to all the access points of the wireless LAN of the above-mentioned 25

Fig. 5.6 A screen of materials selection for making educational materials using STORM maker

classrooms. Though tablet computers are stored near the first-year classrooms, they can be used in the classrooms of the second, third, fourth and fifth years by simply carrying them. In NIT, Gifu College, we are promoting AL while utilizing two kinds of LMS; LMS (Blackboard) of started by NIT, Japan and LMS (Moodle) introduced by the above-mentioned AP. LMS can be used in classes by using equipment we introduced. It can also be used at home. When learners use it for his/her own learning, LMS corresponds to "self-learning" of the smartness level. The same goes for the 50 notebook computers (Fujitsu) and 20 Microsoft Surfaces we introduced. They can be used in every classroom with the wireless LAN system.

The LMS server was introduced so that personal computers and tablet computers within and outside the campus, as well as smartphones, could obtain access. This enabled students to access it at home in addition to in the classroom. The use of these kinds of mobile devices corresponds to "adaptation" of the smartness level. In the 2014 academic year, students used the LMS server for conducting student evaluation while teachers used it for submitting a report on AL. "DB server +FileMaker" will be used for DB processing to visualize the learning outcomes and visualization itself.

"DB server+FileMaker" will be used for DB processing to visualize the learning outcomes and visualization itself. It will also be used in an effort to visualize students' learning activities performed outside the formal curriculum. Moreover, we intend to build a "student analytics system" by using LMS and a DB server we introduced, which corresponds to "inferring" of the smartness level.

5.4.4 Replacement in the Multimedia Educational Building

In NIT, Gifu College, ICT-driven equipment of the fourth-year classrooms of all five departments, which we mentioned in "4.1 The foundation for introducing AL", has been replaced every five years since it was first introduced. In the fourth replacement done in the 2015 academic year, we changed each of the fourth-year classrooms of all five departments into a flexible classroom environment. Now, using tablet/notebook computing bought using the AP budget allow the practice of AL. These connect to the classroom wireless LAN. The expenses needed for the replacement were not from the AP budget, but from NIT, Gifu College operating cost. Additional costs were covered by MEXT as well as the NIT. We expected, however, that the change of the fourth-year classrooms into a flexible classroom environment for practicing AL with tablet/notebook computers would make it almost impossible to practice programming/CAD. Before that, the students were learning these by using high-specification desktop computers with high memory capacity, high-speed arithmetic processing capacity, and advanced drawing performance. These were installed in the fourth-year classrooms. In order to deal with this situation, we increased the number of the computer rooms of the Information Processing Center from 3 to 5 in the replacement (Fig. 5.7).

Through the replacement of the Information Processing Center, the following were introduced and distributed into 5 computer rooms: 242 Client PCs (DELL: OptiPlex 3020 SFF), 4 Servers (DELL: PowerEdge R430), 1 NAS (Dell Storage NX3230), 10 Printers (Canon Laser Beam Printer Satera LBP8710), and 5 Document Cameras (EPSON: ELPDC12).

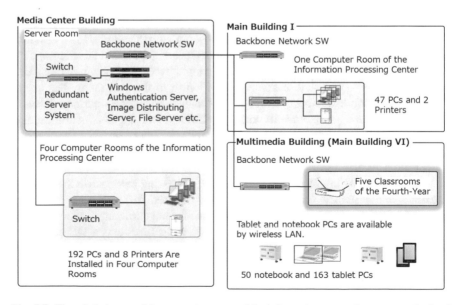

Fig. 5.7 The whole image of the computer rooms of the information processing centre and related facilities after the replacement

For the sake of managerial convenience, the system we introduced was set up as "thin client" and incorporated in the server system. Moreover, clients were set up to be under the same server system and work in the same way with the notebook PCs bought with the AP budget. This made it possible to conduct a smart management of software introduced in clients. For example, though software systems such as Adobe and CAD are managed under the license server, the connection of the license server with the "thin client" system will make it possible to use software with appropriate licenses in appropriate client PCs. We built a system where the server automatically shuts down in the event of a power outage. This corresponds to "self-organization" of the smartness level. The server system was designed to be load-balanced in addition to a doubling for redundancy. This also corresponds to "self-organization" of the smartness level.

5.5 Validation of Research

5.5.1 Faculty Development

As described earlier, NIT, Gifu College has established the system of associating hardware with software. It also created an educational environment for smart education. All teachers are practicing it; however, operation of the system depends on human factors. In other words, even though using the same instrument/system in a variety of types of smart education can be practiced, it depends on how teachers use them. We consider it important to share information among faculty regarding these matters. A teaching example of a teacher can provide the faculty with helpful ideas. Sharing information in good examples among the faculty will promote the practice of smart education at a college-wide level, leading to a true smart university/college.

It is necessary to consider the following two things when practicing AL according to the MCC curriculum:

(1) How to use the e-Learning system and ICT-driven equipment, and
(2) Educational methods of AL and e-Learning.

Teachers have different degrees of knowledge and skills regarding these two items. This means it is important to improve the teachers' knowledge and skills through FD in order to promote AL and e-Learning within NIT, Gifu College. In later paragraphs, we will describe the upward spiral of ICT-driven education through FD in NIT, Gifu College.

In the 2014 academic year, we established the office for promoting AL as a college wide organization and since then have been practicing AL. The office members consist of the representative teachers of all the departments including Mechanical Engineering, Electrical and Computer Engineering, Electronic Control Engineering, Civil Engineering, Architecture, liberal arts, and natural science. This makes it possible to exchange information smoothly between the office and each department. This system will be maintained in the future. Also, the members of the office learn the newly introduced e-Learning system, ICT-driven equipment, and the approaches of AL in advance. Each member then conveys new information to his/her department. Moreover, the office is playing a leading role in promoting AL by implementing the following two kinds of FD activities at every faculty meeting: (A) teaching methods of AL and (B) How to use the e-Learning system and ICT-driven equipment. In NIT, Gifu College, we regard "teaching methods of smart education and AL" and "the use of ICT-driven equipment" as important cores. We also have the idea that a variety of AL can be practiced by combining teaching methods, the e-Learning system, and methods of using ICT-driven equipment. Actually, some teachers, inspired by the FD sessions held at faculty meetings, have created and practiced their own methods of AL.

Table 5.1 shows the FD sessions related to e-Learning and ICT-driven education held in NIT, Gifu College in the 2015 academic year. Since the beginning of the 2016 academic year, we have been holding the FD sessions under the two new subjects: one is an integrated subject of "teaching methods of smart education and AL" and "the use of ICT-driven equipment", and the other is a newly born subject, "the FD sessions where examples of AL performed within college include field works". It has been decided that, in this academic year and from the next academic year, the FD sessions will be held continuously. This is so that the teachers will be able to acquire more advanced skills. By doing so, trainers can conduct detailed instructions and respond to different teachers with different skills.

It is important to enhance teachers' skills for teaching and student counseling/guidance along with systematic curriculum to make college wide organizational deployment of education work more effective. It is necessary for each teacher to improve, not only his or her teaching ability, but also his or her skills of using ICT-driven equipment in order to practice education based on e-Learning and AL. This requires the improvement of the coordinated training system, sharing and publicity of model education examples, and the system of properly evaluating teachers' education examples. Also, it is necessary to introduce the system of teacher evaluations and student counseling/guidance conducted within a campus (peer review). In NIT, Gifu College, we consider it important to do continuous improvement performed by sharing and evaluating the education examples that, we believe, make high-quality e-Learning and AL penetrate within the college. These methods work well by both a shared understanding of the objectives of the curriculum and the curriculum itself by teachers and the effort for improving teachers' skills for teaching and student counseling/guidance. Considering the importance of organizational deployment of education curriculum, it is necessary to hold organizational workshops (FD) concerning the objectives of the curriculum, education content, and methods. It is important to keep in mind that the characteristics and creativity of education should not be impaired.

Table 5.1 The content of the FD sessions of NIT, Gifu College held in the 2015 academic year

Dates	Type of activity performed for promoting AL	People that the sessions targeted (headcount)	Results, expected effect	Problems	Measures against the problems described in the left column
Faculty meetings: Apr. 1, Jun. 3, Aug. 5, Sep. 18, Nov. 18, Feb. 10, Mar. 14	NIT, Gifu College, collegewide level (FD on the teaching of AL and ICT-driven equipment)	All teachers (about 80 people)	The FD lecture sessions are effective because they are held when all teachers get together	There are different needs because the teachers have different degrees of skills	A wide variety of subjects are treated
The FD meetings: May 7, Oct. 14	NIT, Gifu College (collegewide level)	All teachers (about 80 people)	Useful lecture by visiting lecturers	General topics, not concrete content	Concrete content is treated at FD regarding AL conducted at faculty meetings
(1) May 26–28 workshop of blackboard (basic) (2) Jun. 1–3 workshop of blackboard (intermediate) (3) Jun. 8–10 workshop of blackboard (advanced)	NIT, Gifu College (collegewide level)	All teachers (about 80 people)	To acquire how to use blackboard (basic, intermediate, advanced) and practice AL in class)	Some teachers cannot attend workshop because of other school affairs. The teachers have different degrees of skills	The same content was presented for three days, considering the teachers' schedule. The participants were free to select the level among three (basic, intermediate, advanced)
Workshop of moodle was held three times in Jun.	NIT, Gifu College (collegewide level)	All teachers (about 80 people)	To acquire how to use moodle and utilize it in class	Some teachers cannot attend workshop because of other school affairs	The same content was presented for three days, considering the teachers' schedule
Jul. 23 (Akashi) Oct. 14 (Gifu) Dec. 3 (Kyoto) Mar. 1 (Maizuru). The 2015 AL promotion study team of the third block	The third block committee members of AL promotion study team (the colleges that belong to the third block)	The number of colleges that belong to the third block multiplied by two committee members of each college	NIT at Akashi and Gifu Colleges, leading colleges of AL, are supposed to lead the other colleges within the third block to the positive practice of AL	Each college has a different perspective and degree of penetration of AL, which makes it difficult to have a common understanding of AL	To respond to diverseness among colleges, first, it is necessary to assess the position of each college by conducting a survey of the teachers who belong to the third block

(continued)

Table 5.1 (continued)

Dates	Type of activity performed for promoting AL	People that the sessions targeted (headcount)	Results, expected effect	Problems	Measures against the problems described in the left column
Sep. 24: workshop of projectors which have the functions of electronic blackboards and tablet PC (the fosterage of trainers: for the members of AL promotion WG)	For the members of AL promotion WG of NIT, Gifu College	For the members of AL promotion WG (seven people)	How to use projectors which have the functions of electronic blackboards and tablet PC. How to conduct AL classes using equipment	The teachers have different degrees of skills	Workshop for the members of AL promotion WG was held, for fostering trainers who would instruct the faculty members. The instructions for teachers by trainers are supposed to be conducted within each department
Sep. 25, 28, 29: workshop of projectors which have the functions of electronic blackboards and tablet PC (for all teachers)	NIT, Gifu College (collegewide level)	All teachers (about 80 people)	How to use projectors which have the functions of electronic blackboards and tablet PC. How to conduct AL classes using equipment	The instruction of how to use equipment is insufficient for actual use. Some teachers cannot attend workshop because of other school affairs	In the workshop held at a classroom of each department after school, teachers actually operated projectors, which have the functions of electronic blackboards and tablet PC
Sep. 1–4: workshop of cybozu (collaborative software)	NIT, Gifu College (collegewide level)	All teachers and college staff	To acquire knowledge of procedure, methods for managing various information within college	Some teachers and college staff cannot attend workshop because of other school affairs	The same content was presented for four days, considering the teachers' schedule

5.5.2 Feedback on Smart Education

5.5.2.1 Teachers' Feedback

The teachers practicing AL have shown the following good points and other comments.

- A group solved a problem by their freewheeling thinking.
- It seems that group work made students consider how to make use of their knowledge and that one student's idea stimulated the other participants.
- Students coped flexibly with the new method.
- At the beginning of a class, I had students expect a solution. With the passage of time, the final solution began to appear like fog clearing away. In some cases, the solution was the same as the initial expectation and, in other cases, it wasn't. In all cases, most students seemed to have interest in the process.
- Some students could submit a task, which shows what they had learned through homework and classwork in a wonderful way as soon as a class ended.
- Students were requested to complete common answer sheets by teaching one another among the group members from the step of preparation; however, they didn't perform research beforehand or teach one another. Those who could solve a problem for themselves continued to do so at their own pace. Those who couldn't solve a problem, though, didn't even ask questions. They didn't initiate a discussion for completing a common answer sheet for the presentation.
- I hoped to get common answer sheets, which are almost the same level within a group soon after the class ended; however, there seemed to be little difference between the sheets of the students who work hard and the sheets of the students who don't work hard.
- The situation where the students teach one another among the group members makes it possible for a teacher to contact them individually based on their level of understanding.
- Few students fall asleep during the time. And we can urge sleepy-looking students to sleep beforehand.
- Group work itself is effective and the instruction that a representative of a group writes on the blackboard led several groups to a discussion on how they arrange their thoughts in order to present their ideas in a simplified way. Also, the class was very stimulating for a group idea which was given low evaluations at the beginning yet turned out to be very useful in the process of treating handling advanced content.

5.5.2.2 Students' Feedback

The teachers that practiced AL have shown the students' attitude. Most of them are positive.

- When I told the students the development processes of all groups, they acted as a stimulus and the students worked harder in competition with the other groups.
- Group work activated the class and created a condition where the students made a lot of remarks.
- A student who was not interested in the activity at first (i.e., not knowing the aim and what was going on) began to have interest in the activity as the class advanced and his level of understanding deepened. Finally, he seemed to be satisfied with the result.
- The students who followed a teacher's lecture didn't actively ask questions about the content of the presentations by the other students. Also, they seemed to be unsatisfied when their teacher didn't give explanations.
- The students who don't work hard habitually in class didn't make a preparation and take part in a discussion. Also, they relied on the other students at the time of presentation and submitting a report. In short, they didn't change their attitude.
- Nearly every time, more than a few students in the class fell asleep during a lecture. They seemed uninterested in the content of the lecture. Today, however, they tried to solve a problem through discussion with fellow group members. They seemed to have more motivation than usual. Seeing them, I thought that most students could understand the problem. Today's trial seemed to be successful.
- I think some students increased their interest in mathematics. They were doing calculations rather amusingly.
- At the beginning, they were solving a problem silently in an unfamiliar situation, but they began to exchange their ideas soon after.
- An hour or so after the class started, some students began to talk in whispers through lack of concentration.
- Some students may not have taken part in the discussion, but most students seemed to do so actively.
- Most groups made an active discussion. The description of their ideas on the blackboard was useful for processing them.

5.5.2.3 Teachers' Comments

Some teachers presented their comments, including the future feedbacks and problems, after they practiced AL.

- The tasks that require students to think for themselves and solve problems in a group setting are helpful for making their knowledge useful. It will stimulate and activate problem-solving skills and communication skills.
- It is important to improve AL by providing feedback following its trial.
- Many teachers are hoping to learn the content of the advanced examples of AL (flipped learning). We are planning to hold a workshop where an extramural lecturer will make a lecture with the aim that we will spread AL in NIT, Gifu College.

- If we let students take the initiative, from beginning to end in AL class, they will be at a loss for what to do first. I think that teachers should teach basic knowledge. It might be an idea that students learn basic knowledge preliminarily as homework, but it is unreasonable to carry out the idea considering the fact that students have different degrees of ability.
- When students made presentations, the teacher tried correcting mistakes and making comments. But some students didn't pay attention to the teacher, remaining unfocused on studying.
- If students perform preparation, group work, and presentations seriously, I think their studies proceed smoothly.
- It was difficult to make the students concentrate on a new method as I instructed. I think it is important to make them have an idea that they must study at their own initiative and employ Teaching Assistants.
- They made few discussions within a group. Questions and answers were not active at the presentation. These are probably because of their poor communication skills.
- It was effective in one class, but not effective in the other class. I am thinking of increasing time spent doing group work and getting cooperative work on track in the latter class. There may be some students who want to solve a problem individually, but I am thinking of leading those diligent students to teach other students voluntarily.
- It is difficult to know students' reactions in an ordinary class, but it is possible to check the level of understanding of each student, as a teacher gets a direct response from each student in an environment where students can speak freely.
- Since the time a teacher speaks unilaterally decreases sharply, students and observing parents might be mistaken into believing there is insufficient effort on behalf of the teacher. Therefore, a teacher should always walk around the classroom during a class, monitoring students as they work.
- I think how to solve a problem is difficult. I think it necessary to continue the method and produce results, but I have not grasped the direction yet.
- Group work is very effective. Only a question will develop into a deep thinking activity. But it takes some time. So, a teacher must consider the balance of the method with the progress schedule.

5.6 Discussion

5.6.1 A Summary of Smartness Levels

In this chapter, we describe our practice of smart education in NIT, Gifu College. Smart education was classified based on its smartness level. In each section, we wrote our efforts related to respective smartness levels [1]. The key concepts are as follows:

(1) Adaptation: software for synthesizing automatic voice, mobile device access designing, flipped learning, etc.;
(2) Sensing: feedback on learning history, coping with physiological responses, coping with learners' setbacks;
(3) Inferring: a system that visualizes students' learning activities performed outside the formal curriculum;
(4) Self-learning: learning materials such as simulation software, interactive software, videos and texts; LMS; active learning which urges learners to think for themselves and solve problems;
(5) Anticipation: an automatic registration of student information data operated by combining some systems, a mail system as a crisis measure; and
(6) Self-Organization: a system where the server automatically shut down in the event of a power outage; load-balancing in addition to a doubling for redundancy.

5.6.2 Curriculum in Smart Education

A true smart university/college is where smart education is practiced at a university/college-wide level. In order to realize this situation, it is important to share information on smart education among the faculty and hold the FD sessions on related matters in addition to a smart system environment mentioned above. More specifically, this is a system environment of smart education created by associating hardware and software. Also, it is necessary to create university/college educational concepts, including admission policy and diploma policy, and the education system of respective departments based on the educational concept, in consideration of smart education. Moreover, it is necessary to prepare the curriculum based on a common viewpoint in relation to the practice of smart education.

Though the education of the NIT, which NIT, Gifu College belongs to, is highly evaluated by various domestic and overseas quarters. The NIT is intended to pursue the sophistication of its education, centering on the improvement of ICT environment. Taking world movements into consideration, the respective NIT colleges have improved its educational content. This led to certification by some examination bodies such as JABEE; however, before the release of Model Core Curriculum (MCC) by the head office of NIT on March 23, 2012, there was no uniform standard regarding curriculum for all NIT colleges. Through the trend of the revisions to standards for establishing universities directed by the Japanese government, the NIT started working to establish MCC ahead of universities after conducting a hard survey of the current situation of curriculum, afterwards, repeatedly conducted investigations, and published MCC in 2012.

5.6.3 MCC in Smart Education

MCC is intended to clarify the model of a practical, creative engineer. NIT colleges are fostering and the policies of educational content and methods assure the quality of education in itself. They further promote the educational reform and improvement with individuality and characteristics of respective NIT colleges. In the schematization of NIT education, in addition to showing "Core (a minimum standard)", the minimum skill level and content to be studied for all the students of NIT, and "Model", a guideline for further advancement of NIT education, is presented to respond to more advanced social requests. The curriculum is promoting both "Model" and "Core", so the name "Model Core Curriculum (MCC)" is used. For "Model", in addition to only presenting the method of thinking, it is important to share pioneering challenges performed in some colleges among all NIT colleges and introduce them into the other colleges based on the realities of respective colleges. For this purpose, NIT colleges are complementing MCC by sharing education examples of engineering design created on a periodic basis. MCC doesn't indicate concrete curriculum itself organized and practiced by school as a narrowly defined curriculum. It represents students' attainment targets, or outcomes, as a guideline for curriculum organization. The way of thinking conforms to domestic and international trends and, when it comes to aiming to achieve the targets, respective NIT colleges are indicated to devise a way to combine various educational activities such as concrete subjects, practical studies of every kind, and extracurricular activities based on their own conditions and policies. MCC never uniformly determines concrete subject names or content.

5.6.4 ICT Managed Classification of MCC in the Promotion of Smart Education

MCC is organized from the viewpoint of the advancement of NIT to respond to social needs. The direction of NIT is as follows:

(1) The fostering of engineers who can be active internationally in response to the globalization of society and industry,
(2) The fostering of innovative human resources who can contribute to the sustainable social progress, and
(3) The expansion into the composite, integrated fields that respond to the needs of the local communities and industries.

MCC clearly specifies the targets for students to attain from the viewpoint of 10 items: mathematics, natural science, art and social science, basis of engineering, Specialized Engineering Categorized by Field, Engineering Experiments and Practical Skills Categorized by Field, Substantiation of Specialized Skills, Versatile Skills, Attitude/Orientation, Comprehensive Learning Experience, and Creative

Thinking Power. The teachers are proceeding with the following work and integrating them with each other in order to meet our targets, following the curriculum created based on MCC:

(1) Improvement of lecture and teaching method (Ex. group work, workshop-type learning),
(2) Cooperation among teachers,
(3) Improvement of educational evaluation and checkup method (Ex. interview and oral examination, portfolio of students and teachers),
(4) Development of teaching materials, and
(5) Activities of Faculty Development (FD) and Staff Development (SD).

In order for NIT, Gifu College to be a higher education institution that contributes to local industries, it is essential to have a viewpoint of industry-college-government cooperation, as well as regional cooperation. The Targets of MCC are to be attained by combining the students' attained level with various kinds of methods and subjects. This is cooperated among teachers. The students' attained level is evaluated by what and how far they have learned. In MCC, the students' attained level has been established in accordance with the dimensions of cognitive processes of the Bloom's Taxonomy (cognitive domain) Table (the revised edition):

(1) Knowledge-Memory Level (to be able to remember, recognize, recall related knowledge)
(2) Comprehension Level,
(3) Application Level (apply knowledge, theory and information for applied cases and problem-solving, execute, practice),
(4) Analyzing Level,
(5) Evaluation Level (to be able to judge based on criteria and standards, adjust, find, observe, verify, criticize, judge), and
(6) Creativity Level (newly construct elements to organize the whole, to be able to newly reorganize elements, produce, plan and design, create).

The study targets were established for each item of MCC, which will be linked with the ICT systems such as e-Learning, the syllabus system, and the portfolio system. For example, in the item of "Natural Science" (Large Classification), there exists the item of "Physics" (Middle Classification) as one of the many items. In the item of "Physics" (Middle Classification), there exists the item of "To be able to synthesize and resolve forces" (Small Classification). The item of "Physics" doesn't mean the subject of physics itself, but represents collectivity of the attainment targets regarding physics-related fields. When trying to make students rise to the "(3) Application Level" in the upper years, with respect to the item of "To be able to synthesize and resolve force", the curriculum of the department is supposed to organize with the following in mind: In the curriculum of a construction-related department, for example, students are expected to rise to the "Knowledge-Memory Level" in physics class of the early years, so that they can solve problems of "synthesis and resolution of forces". On the other hand, they are expected to rise to the "Application Level" in the structural dynamics class of the upper years so that they can perform practical calculations of load when lifting heavy goods. Which subject of curriculum each item of MCC corresponds to and how far it is taught is checked and managed by the syllabus, e-Learning, and portfolio systems.

By combining the syllabus system with the e-Learning system and adding the MCC check system function to the combined system, these systems work in conjunction with each other, developing a system with an automatic registration function. This kind of system corresponds to "anticipation" of the smartness level. As another example of this, NIT, Gifu College has a mail system as a crisis measure. In the case of a typhoon, for example, the system immediately and automatically provides the students with a notice as to whether school will be closed or not.

5.7 Conclusions

MEXT grants support promoting of university reform with the aim of activating Japanese higher education and contributing to the development of highly skilled personnel. It is intended to develop educational and research projects working on leading educational reform in Japan by introducing competitive support. Our efforts made for more than 15 years include the following: ICT-driven education, e-Learning, distance education using e-Learning under the credit transfer agreement practiced in a consortium with more than 20 colleges and universities within the prefecture, distance education using e-Learning under the credit transfer agreement practiced in a consortium with about 30 higher education institutions such as NIT colleges, universities, graduate universities, and the Open University of Japan.

In that process, NIT, Gifu College applied for and successfully acquired the "Support Program for Contemporary Educational Needs (GP)" of MEXT (three-year financial support) in 2004. Also, in 2014, our application was picked up as a project of the AP by MEXT with three-years of financial support. Moreover, this year, it was decided that the period of our AP project would be extended from five years to six years. Therefore, we are going to continue our research and education on advanced practices until the academic year 2019.

Making use of subsidies from MEXT necessary for the projects to promote educational reform of NIT colleges, NIT, Gifu College is promoting college wide AL, as well as visualization of education from the viewpoint of both students and teachers. Here, in this article, we described our efforts to visualize extra curriculum education for promoting AL. In NIT, Gifu College, we consider it important to visualize, not only feedback by means of visualization of education, such as e-portfolio in formal curriculum, but also efforts in extra curriculum education.

The efforts encourage students to act based on their own ideas, help students acquire generic skills, active nature, initiative, independence, and cooperativeness. In our AP project, we visualize the outcomes of extra curriculum education by analyzing, classifying, and grading them. Moreover, we created a relational database system in order to visualize each student's data with ease.

Making effective use of the budget granted by the central government during the extended period, we will strongly introduce beneficial approaches related to students' activeness. Also, we are strongly promoting the FD sessions for the use of AL for all NIT colleges, as per an attempt to expand NIT, Gifu College's efforts. Moreover, we are developing the improvement of a learning-support ICT environment, including the Learning Management System (LMS) with Blackboard into the efforts of many NIT colleges for expanding it among all NIT colleges. Based on the above, we added the following three items to the original plan of our AP project:

(1) research on students' abilities conducted by using progress reports on generic skills and the provision of individual portfolio,
(2) the great expansion of the Learning Commons environment in NIT, Gifu College. which the universities aiming for smart university are promoting,
(3) the support of the FD programs regarding the use of AL, LMS, etc., and
(4) more accelerative promotion of the reduction of the burden on NIT, Gifu College's staff and effective use of our educational resources, mainly by introducing a terminal management software and human resources, to provide support; we added this item because of the ICT environment, which is not fully utilized now due to lack of manpower.

The above-mentioned additional four items were highly evaluated by MEXT. It was determined to extend the period of our AP project from five years (the initially scheduled period) to six years.

Acknowledgements We have been supported by the Ministry of Education, Culture, Sports, Science and Technology for our GP project of the "Support Program for Contemporary Educational Needs (GP)" (2004 to 2006) and for our AP project of the "Acceleration Program for University Education Rebuilding" (starting in 2014). We are very grateful to the organization for making our projects possible by the financial support.

References

1. Uskov, V.L., Bakken, J.P., Pandey, A., Singh, U., Yalamanchili, M., Penumatsa, A.: Smart University taxonomy: features, components, systems, smart education and e-Learning of the series. Smart Innov. Syst. Technol. **59**, 3–14 (2016). Springer
2. Hwang, G.J.: Definition, framework and research issues of smart learning environments—a context-aware ubiquitous learning perspective. Smart Learn. Environ. Open J. **1**(4), (2014). Springer
3. Coccoli, M., Guercio, A., Maresca, P., Stanganelli, L.: Smarter Universities: a vision for the fast changing digital era. J. Vis. Lang. Comput. **25**, 1003–1011, (2014). Elsevier
4. Yonezawa, A.: "Japan". In: Forest, J.F., Altbach, P.G. (eds.) International Hand-book of Higher Education Part, pp. 829–837, Springer (2006)
5. Chu, W.W. (ed.): Data Mining and Knowledge Discovery for Big Data, Methodologies, Challenge and Opportunities. Studies in Big Data, vol. 1. Springer, Heidelberg (2014)

6. Che, D., Safran, M., Peng, Z.: From big data to big data mining: challenges, issues, and opportunities. In: Database Systems for Advanced Applications of the Series. Lecture Notes in Computer Science, vol. 7827, pp 1–15, Springer, Heidelberg (2013)
7. Tan, Y., Shi, Y. (eds.): Data Mining and Big Data. Lecture Notes in Computer Science.In: Proceedings of the First International Conference, DMBD, vol. 9714, June 25–30, Bali, Indonesia. Springer International Publishing (2016)
8. Nomura, S., Yamagishi, T., Kurosawa, Y., Yajima, K., Nakahira, K., Ogawa, N., Irfan, C.M., Handri, S., Fukumura, Y.: Anticipation of the attitude of students: passive or active coping with e-learning materials result in different hemodynamic responses. In: ED-MEDIA, pp. 810–817 (2010)
9. Nomura, S., Yamagishi, T., Kurosawa, Y., Yajima, K., Nakahira, K., Ogawa, N., Irfan, C.M., Handri, S., Ouzzane, K., Fukumura, Y.: Evaluating the attitude of a student in e-learning sessions by physiological signals. In: The IADIS International Conference e-Learning, pp. 323–330 (2010)
10. Handri, S., Nomura, S., Kurosawa, Y., Yajima, K., Ogawa, N., Fukumura, Y.: Study on students' physiological response towards e-learning courses using physiological sensors. In: ED-MEDIA, pp. 4100–4105 (2010)
11. Asanka, D.D., Fukumura, Y., Kanematsu, H., Kobayashi, T., Ogawa, N., Barry, D.M.: Introducing eye blink of a student to the virtual world and evaluating the affection of the eye blinking during the e-learning. In: Procedia Computer Science—18th International Conference on Knowledge-Based and Intelligent Information and Engineering Systems—KES2014, vol. 35, pp. 1229–1238 (2014)
12. Barry, D.M., Ogawa, N., Dharmawansa, A., Kanematsu, H., Fukumura, Y., Shirai, T., Yajima, K., Kobayashi, T.: Evaluation for students' learning manner using eye blinking system in metaverse. Proced. Comput. Sci. 60, 1195–1204 (2015)
13. Kanematsu, H., Kobayashi, T., Barry, D.M., Fukumura, Y., Dharmawansa, A.D., Ogawa, N.: Virtual STEM class for nuclear safety education in metaverse. In: Procedia Computer Science—18th International Conference on Knowledge-Based and Intelligent Information and Engineering Systems—KES2014, vol. 35, pp. 1255–1261 (2014)
14. Kanematsu, H., Kobayashi, T., Ogawa, N., Barry, D.M., Fukumura, Y., Nagai, H.: Eco car project for Japan students as a virtual PBL class. Proced. Comput. Sci. 22, 828–835 (2013)
15. National Institute of Technology, Japan, What is KOSEN (2008). http://www.kosen-k.go.jp/english/what-idx.html
16. Bergmann, J., Sams, A.: Flip your classroom: reach every student in every class every day. Int. Soc. Technol. Educ. (2012). ISBN 1564843157
17. Bonwell, C.C., Eison, J.A.: Active learning: creating excitement in the classroom. School of Education and Human Development, George Washington University. Washington, DC (1991)
18. Renkl, A., Atkinson, R.K., Maier, U.H., Staley, R.: From example study to problem solving: Smooth transitions help learning. J. Exper. Educ. 70(4), 293–315 (2002)
19. Brant, G., Hooper, E., Sugrue, B.: Which comes first: the simulation or the lecture? J. Educ. Comput. Res. 7(4), 469–481 (1991)
20. Kapur, M., Bielaczyc, K.: Designing for productive failure. J. Learn. Sci. 21(1), 45–83 (2012)
21. Westermann, K., Rummel, N.: Delaying instruction: evidence from a study in a university relearning setting. Instr. Sci. 40(4), 673–689 (2012)
22. Hake, R.R.: Interactive-engagement versus traditional methods: a six-thousand-student survey of mechanics test data for introductory physics courses. Am. J. Phys. 66, 64 (1998)
23. Hoellwarth, C., Moelter, M.J.: The implications of a robust curriculum in introductory mechanics. Am. J. Phys. 79, 540 (2011)
24. Michael, J.: Where's the evidence that active learning works? Adv. Phys. Educ. 30(4), 159–167 (2006)
25. Prince, M.: Does active learning work? a review of the research. J. Eng. Educ. 93(3), 223–231 (2004)
26. Khan Academy, "Khan Academy" (2006). https://www.khanacademy.org/

27. Lage, M., Platt, G., Treglia, M.: Inverting the classroom: a gateway to creating an inclusive learning environment. J. Econ. Educ. **31**(1), 30–43 (2000)
28. National Institution for academic degrees and quality enhancement of higher education, role of NIAD-QE (1998). http://www.niad.ac.jp/english/about/role.html
29. Japan accreditation board for engineering education, JABEE and accreditation (1999). http://www.jabee.org/english/jabee_accreditation/

Chapter 6
Building Smart University Using Innovative Technology and Architecture

Attila Adamkó

Abstract The terms Smart University and Smart Campus reflect the various opportunities provided by universities to achieve better experiences for locals to improve campus efficiency and cut down operating expenses. Both directions try to help their users reach their aims much easier. Several new aspects and technologies have appeared in the last few years and gained greater emphasis in areas such as Big Data, IoT (Internet of Things), Future Internet, and Crowdsourcing. A University is a proper place where all of these fields could be examined and applied continuously as a sustainable evolution. In this paper, we investigate software systems and related aspects of this evolution to introduce an open architecture for easily extensible services. These services are responsible to provide their users value-added information. This will increase the smartness level of the campus. We investigated three directions for this reason: IoT to involve sensing, cloud and ubiquitous computing to make our services available everywhere, and to transform our service consumers to content generators or, in other words, data producers and/or developers for new data source connections. The fourth direction, Big Data, is a forthcoming addition to the previous three for enhanced analytical capabilities to Smart Communities.

Keywords Smart university · Smartness feature · Smart campus · Smart community · Cloud · Architecture · Software system · XMPP · IoT · Smart education · Adaptive · Service-oriented · Crowdsourcing

6.1 Introduction

Future Internet research from the last seven years has been focused on establishing cooperation between the industrial and academic communities. The goal is to do research on the Internet, its prospects, and its future. Around 2012, during the

A. Adamkó (✉)
Department of Information Technology, University of Debrecen, Debrecen, Hungary
e-mail: adamko.attila@inf.unideb.hu; adamkoa@inf.unideb.hu

© Springer International Publishing AG 2018
V.L. Uskov et al. (eds.), *Smart Universities*, Smart Innovation,
Systems and Technologies 70, DOI 10.1007/978-3-319-59454-5_6

second wave of this research and experimentation, the technologies such as sensor networks, cloud computing, and service-oriented architectures were in focus. The author of this paper formed a group for the mentioned objectives in the Information Technology Department at the University of Debrecen. We began to investigate these fields and attempted to translate the challenges to our closer work environment. Technological changes continue to influence our everyday life, especially with the use of so-called smart devices and the Internet being so readily available. This means there is an increasing need for systems that could apply these technologies in an effective way.

In our case, we highlight four fields that must be connected for a better experience through better services. The identified areas are as follows:

- ubiquitous and cloud computing;
- Internet of Things (IoT);
- crowd-based solutions; and
- increased amount of data. This does not just mean IoT sensor data, but user generated content and gathered data as well.

The incoming volume of information is significant and is foreseen to expand exponentially. In order to handle the exponential data growth at a scalable and fault-tolerant infrastructure cognitive, smart business intelligence must be applied where information security and privacy are also reliably solved.

Ubiquitous computing, where every device is connected everywhere to the network, is a very promising direction for the future of adaptive services. While borders are blurred between computing devices, it has created a seamless and smooth usage of systems. Obviously, the smartness level should not be measured by numerous different interfaces provided for the devices. More importantly, it should behave as a context-aware and customized environment available through the cloud. While cloud computing architecture, beyond its core benefits, implies a brand new attitude for all of its users as a service based utility model, users need to adopt new ways for smart and connected devices as well.

An important factor is the customization that is supported by the IoT opportunities within the sensors embedded in a user's smart devices. On one hand, this sensor data serves as an endless source that could drive a real-time and/or transactional analytic module to feed services with actual or historical personalized data. On the other hand, this infinite data stream influences the previously mentioned data handling tasks regarding how much data there is.

In order to deal with these factors, our architecture at the data layer involves a unit, referred to as Historical DB, to store the outdated section of the large amount of data. This separation attempts to lower the load caused by queries. Moreover, the historical portion does not need to be in a relational form. We can investigate other data models, such as graph databases, to represent it in a much better form for semantic analysis. In the first step, our vision includes a task for creating trajectories to help users find nearby friends or colleagues based on earlier presence information. The initial precondition for this idea is to have proper location-based data from users, such as GPS coordinates and/or connected Wi-Fi access points. At our university campus, we have a park covered by Wi-Fi signals. This means it could be used for outdoor location data based on the GPS coordinates of people's mobile devices. This is not always available due to the fact that users frequently turn this function off. Furthermore, indoor positioning is not always accurate enough based solely on Wi-Fi signals to manage proper navigational paths inside a building, e.g. for a given room, rest room, canteen, or information points. For this reason, we have started to investigate a BlueTooth Low Energy (BTLE) based solution, in addition to the Wi-Fi access point based one, to increase accuracy.

The last piece in our framework is the role of crowdsourcing. This is a well-known mode of sourcing to divide work between participants in order to achieve a cumulative result. In our case, it is used on one hand for noncommercial work and development of common goods such as the brand new establishment of connections for new data sources based on a common interest. For example, this could be like a new canteen's menu for having a meal or parsing a new newsfeed for recommended events. On the other hand, users are not only consumers as service users, but they are producers as well since they generate data, such as sensed environment by sensors or shared content, to drive the central module and to feed the analytical part for more personalized and customized information.

All of these pieces together help us to resolve architecture for software systems utilizing an open and extensible way to provide services for a better and smarter user experience. The presented work focuses on the University field and discusses solutions and services provided for students and instructors on an extensible architecture. Some of these are supporting the educational line as a smart educational example for enhanced programming experience while others are used in the campus environment, such as smart campuses, to provide a network for communication and collaboration. In a broader scope, these topics are fundamental parts of a newly appeared concept, known as a *Smart University* [1].

6.2 Related Work

Conceptions for campuses with enhanced possibilities are not new ones. The first approach was a joint work between MIT and Microsoft between 1999 and 2006. The established iCampus system tried to alter higher education with the help of the technology. As a base, the iCampus framework was created as a service-oriented architecture to provide the ability to "modularize implementations of educational computing applications, create reusable components, and enable component and resource sharing within the university and across institutions" [2]. Sadly, the website does not appear to have been updated within the last decade; however, the outlined concepts were very straightforward at that time and could serve as a solid foundation for future research.

In 2014, Cisco [4] published a whitepaper about educational innovation fueled by the Internet of Everything (IoE). IoE, in their context, is an extension of IoT that is "allowing people, processes, and things to harness … data to improve decision making for organizations and assist us in our daily lives" [3]. Their approximation is similar to Gartner's recent survey [5] about the number of Internet-connected devices, stating the change will increase from 13 billion to 26 billion by 2020. Their conclusion exposed a lifelong learning process that was brought forth by the rapid technology changes including virtual reality, cyber security, data sciences, and network infrastructure. Another prominent member of the IT sector, IBM, also formed a Smart Education group [6] to develop big data, analytics, and instrumentation-driven techniques to enable personalized education at-scale and improve learning outcomes. They have specified the key concepts as Learner Risk Stratification, Content Analytics, and Personalized Pathways.

6.2.1 Educational Aspects

These directions highlight that more attention has concentrated, not only on the infrastructure required for constructing smart campuses and related applications, but also on learning and teaching aspects extended with the provision of a satisfying personal experience in the college life. These factors altogether are drawing up the outline of a much bigger conception known as Smart Universities.

Following the educational and learning perspective, Ng et al. [7] outlined a new conceptual paradigm as an intelligent campus environment. This was also referred to as an iCampus. It was proposed with the aim of enriching the end-to-end life cycle of learning within a knowledge ecosystem. Cloud based, ubiquitous, and social network attributes also appeared in this field. Research has pointed out problems originating from the distributive nature of the architecture combined with the IoT. These together form a platform. Liu et al. [8] outlined the digital revolution at campuses as an unavoidable process and presented an application intelligence platform using the cloud. Hirsch and Ng [9] have recommended a mobile cloud-based education as a method of ubiquitously providing contextually grounded learning through handheld and smart devices. Nie [10] predicts a campus with smartness as an application of integrating the cloud computing and the Internet of Things. The proposed application framework of smart campuses is a transition from a digital campus to a smart one using IoT and cloud computing to integrate all the affected inner University systems into a unified model. This integration naturally raises issues as well, such as the heterogeneity of the information sources, heavy focus on office management tasks against teaching, and research aspects and the missing data standards for IoT, e.g. RFID card, NFC, or sensor data with their missing unified API for gathering.

Continuing the learning aspects, Atif and Mathew [13] proposed a solution that integrates real-world learning resources within a campus-wide social network. Their model forms a smart campus environment providing its users with context-based personalized learning and feedback opportunities. While smart devices are on a very rapid evolution and became part of our everyday life, users demand content with customization and personalization as well. These directions are under investigation and data mining techniques are involved to fulfill these requirements; however, little evidence is available discussing the potential possibilities. Kuang and Luo [14] constructed a model for an e-learning system to precisely and automatically identify the interests of users from their logs and learning backgrounds. Huayue [15] applied a topic model to develop a model of users' interests, enabling recommendation of resources for users of the system. This direction falls in the Smart Education line and outlines a recommendation model to recommend resources related to the users' interests. Han and Xia [16] found a similar way in their study where they mention a data preprocessing method that was applied to web logs processing. In addition, they made recommendations for systems with good efficiency and reliability based on users' characteristic interests.

6.2.2 Architectural Aspects

Currently, there is not a synthesized solution for developing and deploying a common architecture that applies the above-mentioned IoT and cloud computing inside the academic society to support activities covered by their needs within the academic environment. This includes educational and administrative factors that mention the necessary security and information safety regulations. A good starting point is IoT-A, which is a suggested European architecture establishing the basic layers and components with their interconnectivity for a generic IoT environment. To explain this in detail, nine layers have been developed starting from the topmost application layer as an entry point. Data is passed through the services with management, communication, and security layers before ending up at the devices layer. This IoT-A proposal is an important step in the way of approaching a standard definition and virtualization of the IoT structure [11]. Elyamany and AlKhairi [12] suggested a solution based on the IoT-A architecture that can handle the education management process inside an academic campus with respect to students' activities. This includes the necessity of a safety environment. Their so called IoT-Academia architecture was constructed on the basis of the IoT-A reference model, but was extended with an additional layer to deal with the academia environment requirements expressed as a Service-Level Agreement (SLA) layer. While these layers separate the concerns at a high level, further research is required to translate it into a real service-oriented architecture.

As we identified analytics and data analysis as an important factor for smart campus systems, related research could be found regarding data processing from heterogeneous sources with some focus on semantics and social networking. Very few studies have focused on data mining techniques within campus environments. Boran et al. [17] describe a process that involves including semantic technologies in an application prototype that integrates heterogeneous data. Yu et al. [18] have proposed an architecture based on service-oriented specifications to support social interactions within campus-wide environments using mobile social networking. Xiang et al. [19] introduced a framework based on a chat app for university-based smart campus information dissemination.

Although these results are straightforward for the given domains, further investigation is needed to form more general solutions while their development is still in a primary stage. Moreover, these directions could potentially turn into a new one when we bring up the possibility of using location-based aspects as well. All of these further improvements are necessary in order to achieve a generic model that can facilitate a much smarter campus environment. Based on these works, we could create a list of key elements that are always related to architectural models with service-oriented aspects in mind.

As a side note, another interesting area that should be noted is the field related to the efficient resource allocation from the building's point of view to lower operational expenses. Sensors are used to control different kinds of facility services such as heating or air conditioning in order to achieve a more effective usage. Naturally, these initiatives also have an important impact on the campus life, but we are focusing solely on the previously mentioned directions and concepts while generally accepted solutions do not exist. Our initiative aims to found the establishment of a smart campus architecture and proposes an efficient scheme involving a service centric approach for data gathering, data mining, resource optimization, and crowd-based expansibility.

While formalizing this scheme, we proposed a new concept called *Smart Communities*. This is a community served through smart applications and services by holding the following properties: sensible, connectable, accessible, ubiquitous, sociable, sharable, and augmented. Smart communities are continuously changing and are constantly on the move. Services must follow these changes from time to time, resulting in a continuously evolving system. New data sources could appear, resulting in new information that could be published for the community in order to relay more efficient information to its users. This means that services need to know about the community in order to become smart and share relevant information based on the collected ones. In some cases, the problem originates from the amount of available data. It influences one of our goals, which is to publish relevant information as soon as possible. A centralized and unified way could help us drive these smart services in favor of smart communities.

6.3 Challenges and Objectives for Smart Campus Services

The content and context of this research dates back to the appearance of the Future Internet phenomena. The research objectives include facing the challenges posed by the Internet and carry on investigations covering basic theoretical questions, modelling examinations, the reassessment of architecture, content management, and the creation of an extensible and open application platform.

6.3.1 Smart Campus as a Small Smart City

When we began designing the data management infrastructure, we also started the investigation of *Smart Cities*. This term expresses a trend to integrate multiple ICT (Information and Communication Technology) and Internet of Things (IoT) solutions in a secure manner to manage a city's assets such as local departments' information systems, schools, libraries, transportation systems, and other community services. The overall goal of building a smart city is to improve the quality of life by using urban informatics and technologies, such as sensors for sensing and wireless networks for connectivity, to improve the efficiency of services, including adaptive, shareable, sociable and ubiquitous aspects, in order to meet residents' needs. All of these properties reveal a large number of similarities between the outlined challenges for Smart Campus and Smart City applications. The three main identified characteristics are the following:

- Lots of potential users/consumers: The University of Debrecen has as many students and employees as the number of inhabitants of a medium-sized Hungarian city.
- Large variety of data sources/producers: Information arrives in various formats from multiple sources and should be integrated, e.g., timetable, academic calendar, information on consultation dates and times, energy consumption, etc. Moreover, social networking and sensor-based information, including geolocation, are proper examples of sources that could be common in both domains.
- Need for value-added services/service providers: service providers could give additional value to the crowd-collected raw data using analytics. Basically, we can say this is why someone would desire to join the community. While they are sharing information/data through the system, they are receiving relevant information based on their interests, location, etc., as well. The following use-cases could outline its advantages:

  - get notified when one of my favorite bands will play in a concert venue on campus,
  - get notified when an event's (class, concert, etc.) date and time is being modified,
  - get location data provided by some of my friends and help organizing close get-togethers,
  - get notified if too many students plan to visit the professor's office hour at the same time and assist in reorganizing my schedule based on my task list and the professor's time table.

Naturally, the list of the use-cases is much longer and will be discussed in the following sections. Before this, though, we need to outline one final advantage of the Smart Campus or Smart University field. The users' willingness to use an application and provide valuable feedback is much higher in young people, especially students, than elders. They actively use handheld devices; hence, participatory sensing can successfully be applied as a result.

6.3.2   Challenges

From the Smart Campus' software perspective, one of the most important challenges is to *manage the collection of data from various sources*. Moreover, the list of sources is neither fixed in length, nor the type of the items on it. This can easily change over time. This is a very frustrating factor in the requirements list and this is why we need a highly extensible system designed for change. A Smart University may have lots of various and, most importantly, heterogeneous, data sources like the following ones. This list is just an excerpt from the University where the authors are working:

- an Education Administration System contains information on course enrollments, timetable information of courses, exam dates and times, etc.,
- education offices of the various faculty offering office hours,
- faculty members offering office hours, consultations, etc.,
- the menus of the canteens located at the campus,
- student governments organizing events for students,
- event hosts of actually any events (like public lectures, concerts, exhibitions or whatever users might be interested in), and, which is essential,
- geolocation (e.g., GPS), Wi-Fi or some other sensor data collected by smartphones or similar devices,
- bibliographic databases, such as Google Scholar, DBLP, or Scopus, provide information regarding researchers' published journal articles or conference papers,
- a Library Information System that is able to tell whether a given book is available or not,
- data gathered by environmental and building sensors for temperature, humidity, air pressure, air pollution, etc., and
- the crowd itself with the added value of the capability of generating content that is interesting for a set of people from the community.

Achieving a value-added service typically requires integration of some data originating from several of these sources. Therefore, our main challenge is to create an architecture that allows the access of information from existing sources while also easing the addition of new sources in a seamless manner.

The *data sources could be divided into several not-disjoint categories*: some of these could be gathered in an automated way, similar to sensor data. Some of these may require manual interaction, e.g., canteens' menus or instructors' office hours; some of these data sources offer Application Programming Interfaces (APIs) to access it, similar to social media sites, for instance, while others do not have APIs and some web spiders are required to parse the data; some of the data sources provide built-in notification mechanisms, e.g., an event feed of a social network site using RSS, while others do not support it, such as adding new office hours or changing the daily menu in the canteen, for example.

The last factor is *about the amount of data* that could originate from these sources. We can surely state that a big data management solution is required. Naturally, the frequency of incoming data can significantly differ from source to source. Some might be quite rare, e.g., inserting a new article into a bibliographic database or office hours for the forthcoming semester; others, especially when collecting sensor data such as location, can be frequent and, thus, resulting in a large amount of data to be processed.

6.3.3 Goals

Our primary goal was to create an architectural framework in a university environment that allows community members to create and use value-added services based on the collected data. This data includes information regarding course enrollments, timetable, exam dates, office hours, and various deadlines, as well as data provided by the community. This collected data can be subject to analysis.

Our proposed open architecture should address the challenges described in the previous section and should be as generic as possible in order to integrate a wide variety of data sources regardless of how the data is processed. An architecture for data management and knowledge discovery that fits into the framework based on the Publish-Subscribe model is needed. Extensibility is our key concern. The design should allow the addition of new sources with their processing steps defined in a simple way.

Besides setting our goals, it might be useful to discuss what does not fit into the scope of our goals. On one hand, discussions of the use of XMPP (Extensible Messaging and Presence Protocol) will not be provided as the overall framework was given. On the other hand, the development of value-added services may, and will, require data mining activities including understanding the semantics of data in order to eliminate duplications; however, they are completely out of the scope of this paper as we focus only on the data management platform that can later be complemented with online data analytics solutions currently under development. Data security is an important issue that needs to be addressed in the technology stack. Our opinion, though, is that it must be introduced at a higher level than the data management level.

Based on these project outlines, our solution addresses features like sensibility, adaptation, and inferring. They are centered on sociable communities to fulfill the smartness requirements for applications. It's sensible due to the fact that the sensors sense the environment in order to become aware of events to locate and notify (nearby) users. At a later time, this would include registering infrastructure related processes. It is adaptive and inferring, as recommendations, activities, and personal assessments could be served per user basis. An analytical module could build user profiles to create trajectories as well as follow user interaction and skills. Communities are implemented as groups to make communication available between users and groups in order to become sociable and support security and authorization related elements. The connectable, accessible, and shareable requirements for smart community applications all matter.

6.4 Methods and Results: An Open Architecture as a Base for Intelligent and Adaptive Services

Nowadays, university campuses are constantly changing. Most of the applied tools are extended with online capabilities, new collaboration environments are appearing, and sensors collect data about various things. Following this important shift with semantic aspects, we can state that value-added services can help us achieve a much better and more effective environment at the Universities.

Naturally, this is not just true for this field. Rather, it is true about smart cities and our all-around life as well. It is important to see that current research is heading toward an easier life supported by our environment. It is not just for better and efficient usage of the resources, but for increasing our overall experience.

In our case, at the university level, valuable feedback from the system helps its users to achieve their goals more easily and find relevant information as they need it. For students, this could be classroom information, attendance registration, internal navigation, available workspace lookup, etc.; for the instructors, it could help predict student dropouts, manage attendance, locate students on campus, etc. The possibilities are endless, but IoT could extend it with much more. Not only could internal navigation be developed, but building related services could be added to the system also. For example, the climate control for classrooms could ensure that they are not being heated when lectures are not scheduled. When a small group of students book the classroom for a post-lecture discuss, though, the heat can be turned on for the selected time slot.

Based on these important factors, we have proposed an extensible architecture and implemented some of the aforementioned services to establish a smarter environment at our University. The smartness level could be measured by the value-added services. These need to be able to provide relevant information to its users based on the collected data. As more services have become available, the smartness level increases as an exponential curve. In order to achieve these goals, we need to add some intelligence as an adaption feature to the system by applying data mining techniques. This takes precedence over sensing, which is another distinct feature of smartness.

Naturally, we need to make a distinction between intelligence and cognitive capabilities. At first, only some intelligence can be achieved by the services with the help of adaption. While users feel much better when it comes to personalized information, these are not cognitive abilities from the system. Cognitive infocommunication [20] is a new research direction that should be involved in the future steps. Currently decision making is just a short-term goal for evolving the smartness features in this field. In order to increase adaptation, one should follow the Semantic Web Initiative [21].

6.4.1 Open Architecture for Smart Services

From the Smart Campus or Smart University perspective, we need to design and apply a new form of architectures in order to support services that will be responsible for the smartness level of the system. We need a highly extensible system since we do not know exactly what the sources for the content to be collected will be.

These services should support intelligence, adaption, sensing, and prediction to treat them as a smart service and, therefore, as a result, a Smart University. As we mentioned, these smartness features are not only limited to facility management purposes to increase energy efficiency. Rather, they could support the overall experience of the University locals.

As we mentioned in the Challenges section, data sources can be divided into several categories that are automatically processed through the API provided excluding manually interpreted ones. One of the principal objectives of our system proposal was to capture continuous data streams arriving from diverse sources. We have chosen the Extensible Messaging and Presence Protocol (XMPP) protocol for the base of the underlying communication protocol as it defines a standardized way of event-based messaging and also allows extensibility while applying the publish/subscribe model to fit perfectly into our architecture.

6.4.1.1 Extensibility and the Connectors

Extensibility allows properly preparing input channels for heterogeneous data sources. Our proposed architecture in Fig. 6.1 introduces the term *Connector*, which is a node responsible for collecting data from given sources and sending updates to the Smart Campus Central Intelligence module. Special XMPP messages are used to notify and transfer data to the Smart Campus' Collector Interface. Using this model, the SCCI should decide whether to approve or decline any updates based on the gathered information. The repository update is declined when a Connector sends a redundant piece of information, the Research Connector integrates several sources like DBLP, Scopus, Google Scholar, etc., and the Research Connector could realize that the collected data is the same as what had already been found earlier from another location.

The Smart Campus Central Intelligence (SCCI) component of our architecture provides an interface between the information sources including both the incoming events (XMPP server) and the information stored in the database layer. This is seen in Fig. 6.1.

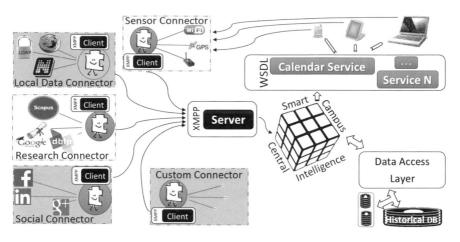

Fig. 6.1 Open architecture developed

The list of sources in this figure does not contain all of the possibilities. Rather it just contains the most prominent ones such as the Educational System and Educational Office information and Library Information System for books and scientific papers or sensor data from smart devices. This could help us demonstrate what kind of diversity in data sources needs to be faced during the design and development. Moreover, it is not just the collection that is a challenging task, but providing value-added services typically requires an integration task as well.

6.4.1.2   Crowdsourcing for Collectors in a Smart Campus

According to our early vision from 2013, we have proposed this extensible architecture that allows the extension of the system with new collectors both at the data producer and at the consumer side. In the previous sections, we have mentioned a small excerpt of the various data sources according to [22].

The Connectors in our proposal are crowd-sourced compound entities without any deep business logic. They contain the necessary parser or API implementation for the proper sources for a given field. This includes a parser for Google Scholar and an API for Scopus. It can be imagined as an open box that can be extended for existing domains or created and attached to the system as a brand new one. This is the power of the architecture because the crowd can do it. The only requirement is to prepare the Connector for the proper XMPP message structure. If the XMPP server has not received the required message type, the extension mechanism of the XMPP comes into the picture. It can be created, inserted, and published to the server.

For example, if a couple of students prefer to have lunch at the small restaurant nearby the campus, they can easily develop a connector that parses the restaurant's web page to provide information regarding the daily menu. The same case can also be true for different newsfeeds. When these sources become available, the potential and interested users can be notified about it. This would increase the smartness level of our solution.

6.4.1.3 Slices of the Service-Oriented View

A service-oriented view of our architecture is shown in Fig. 6.2. The Data Collector layer uses the aforementioned Connectors to connect sources for incoming data. The collected data is then used to populate the databases of the Data Management layer. Applications created based on Web services are at the top of the figure. Inside the SCCI layer, the business rules are defined and validated. This is where the Analytics module resides.

The Data Management Layer should be explained in more detail as it highlights a separate unit referred to as Historical DB. The Data layer's responsibility is to provide quick access to the static personal data and to the "fresh" real-time data for adaption. This means that the frequently changing data, such as sensor data or newsfeeds, can be easily accessed for a limited time, e.g., one, three, or six months. After that, everything is placed in the Historical DB.

This separation could help us to speed up queries targeted for "fresh" data because the data store cannot increase in size; however, the historical portion introduces new challenges as well. On one hand, efficient handling could arouse some problems. On the other hand, the Analytics module would require proper support in order to find new and relevant information from inside the large amounts of data. In this step, we are currently investigating graph databases to achieve a more sophisticated semantic analysis for them. Naturally, the underlying data model for Smart Campus also highlights challenges. We need to apply structured and

Fig. 6.2 Service oriented overview of the system

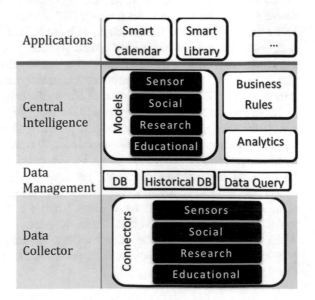

unstructured data as well. Moreover, it should remain an extensible model that leads us to a semi-structured model where we know it does not fit into a uniform, one-size-fits-all, rigid relational schema. Faced with the need to generate ever-greater insight and value-added services, we turn to the aforementioned graph technologies to tackle complexity.

In this context, following the Semantic Web initiative [21], ontology-based annotation helps provide adaptive processes. This means that the incoming content is enriched by semantic data or attached tags. We need to transform this data into triples. The available prefixes are defined in the ontology into which these triples are going to be imported. Traditional SPARQL could be used to make queries, but we have found that graph databases could also be applied for reasoning. When a new, and previously nonexistent, source needs to be attached to the system, the proper edges should be added to the new nodes in the database.

Obviously, at the initial phase, only simple data collection is performed to gather news, timetables, and schedules. In the second step, we have to pay attention to the use of available information and reorganize the information. It is still not an intelligent service, but if we have lots of applications working together and we can share the information in the aforementioned way, the service will become smarter and smarter. In the last step, the recommendation could be based on the learning patterns from the users' behavior. The marked events, places, or friends move the highest level from adaption to prediction. This evolution could be considered the lifecycle of the services.

6.4.1.4 IoT and the Cloud in Smart Universities

Today's two most common buzzwords are the Internet of Things (IoT) and Cloud. While IoT devices collect and submit sensory information flow, the Cloud provides a solution for online services. An ordinary IoT device, such as a smart bracelet, has half a dozen of sensors. These include ambient light, UV, acceleration, GPS, heart rate, etc. Instead of delivering static values, for example, daily average blood-pressure value, these sensors emit a continuous data-flow. This data-flow should be transmitted, stored, evaluated, and transformed. This is the reason that we introduced the Historical DB portion of our architecture.

In our case, IoT is much more related to smart devices that provide location-related data to the services. It is not just for outdoor navigation, but could also be used to search for nearby friends or events provided by the SCCI.

Moreover, indoor navigation could be supported as well. Applying Bluetooth tags around the building, users can get help to find places inside a building. Sometimes it could be useful to get directions to information panels, rest rooms, or even rooms where an event takes place. This is a useful added value from the system at events like conferences where attendees could easily find the location without wandering around the entire building. Naturally, this could be extended to people with disabilities to support voice guidance that can help them reach a given place with ease.

The cloud, in our case, is an online solution that is available anywhere, rather than being restricted only to the campus. When it comes to cloud architecture and cloud computing, the information technology landscape is becoming more and more complex. In addition, the number of different interconnected system components is increasing. Because of this, we are focusing only on the online availability of services. Cloud architecture offers the option to scale up or down as demand varies. The scalability covers storage, memory, computing capacity, and Internet bandwidth as well. While our architecture recommendation is at one step higher in abstraction, we can state that our solution could be applied in this context as well. At the implementation phase, we have chosen a Java EE based solution to address these aspects.

Naturally, several questions could arise around privacy and security for a Cloud based solution. Authentication and privacy are critical aspects of information technology. As data manipulation relates to sensible private information, all the components within the information technology supply chain should consider the suitable and proper authentication and data-protection standards. There are also other techniques for data protection, including anonymization, encoding, and encrypting.

The author of [23] investigated the topics of scalability, sharing, and security in regards to applications in the telemedicine field. The proposed Content Distribution Network (CDN) in Fig. 6.3 could be imagined for the Smart Campus' point of view as a wrapper around the previously mentioned service-oriented architecture. This could deal with scalability and crowd sourced information sharing between Universities in order to drive analytic units for value-added data to its subscribers.

At the center of this, the best architecture to operate a Smart Campus is a self-managed private cloud. A private cloud architecture is able to store a large amount of unstructured data:

- process information with distributed workers,
- dynamically scale the system,
- guarantee information security,
- control accessibility,
- protect sensitive information, and
- monitor the system.

6.4.2 Groups as Small Smart Communities

The last piece in this ecosystem are the semantic tags associated with their groups. These groups could model the subscribers that are the base of our original architectural idea based on the publish/subscribe model. Groups need to be used as high-level entities. Some of them should be automatically created, such as the group for a user's personal calendar or for the newsfeed and all the mentioned cases where the user is involved. As a member of a given group, you will receive the events

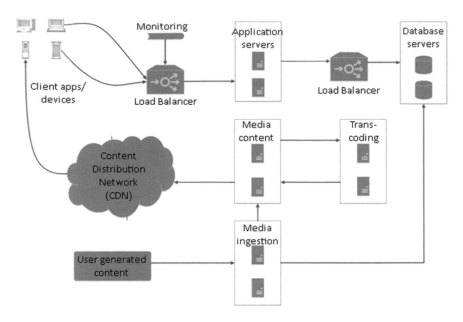

Fig. 6.3 Cloud architecture with CDN

published by that source. Naturally, one user can join as many as he/she wants and one may create their own as well.

The visibility of the groups could also be controlled. There could exist public and private groups, where people can freely join or only invited members can join. A good example of this would be a class where only registered students can be placed. These students are extracted from the Educational System's collector's data.

Another interesting slice of this representation is that the Smart Campus Central Intelligence (SCCI) needs to be represented as a member of communities or groups. The underlying XMPP technology requires 'friendship' between users to allow communication with each other. At our side, it means that we need to represent the SCCI as a user and make it a friend of all users in the system automatically.

6.4.3 Results

An example where our services are created and used by the Smart Campus is the faculty inner portal that is based on Java EE. Its architecture involves three layers: Enterprise layer, Service layer, and Persistence layer. While the Enterprise layer is responsible for fulfilling all of the enterprise needs, including content, workflow, document, user, and management, the Service layer is the core component. It contains the majority of the business logic that performs enterprise needs. We followed the Model Driven Architecture and that made the implementation of a

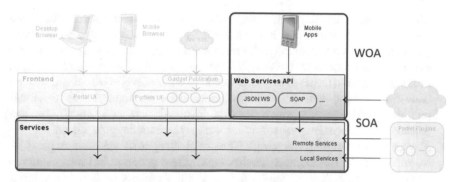

Fig. 6.4 Layered view (source from Liferay documentation)

Service Builder possible. It is used to automate the creation of interfaces and classes for database persistence and a service layer. We applied Web services in the two most common and significant protocols: JSON Web Services and SOAP. These are seen in Fig. 6.4.

6.4.3.1 Smart Campus: Implemented Services

We have successfully implemented services in the portal and created some mobile applications to achieve a smarter environment in our Faculty at the University to emerge as a Smart Campus. The following possibilities are provided:

- **Instructors' office (opening) hours**: The first available service was related to the lecturers' opening hours. A connector is used inside our system's Educational part to periodically check the instructors' published data. When changes occur, the SCCI was notified. Based on the student's actual plan to visit the instructor, all of the affected users are notified when the date or place is about to change for one reason or another. Currently, we are implementing an extension for this portal service to alter it to a fully functional XMPP client. The goal is for the SCCI to provide a direct notification regarding the portal engine changes. It should also request these operations without requiring any inter-mediate crawler to check and gather the information. Moreover, students could see the number of registrations for a given date. If it appears to be full, they can search for another date or request a new consultation time.
- **Online presence and chat**: Another ongoing development is the creation of a notification portal that will be integrated into the Faculty portal where users can see each other's online presence and start conversations without leaving the site. This is the advantage of the XMPP protocol and the usage of LDAP. It is straightforward because we can provide a platform where all of the education-related tasks could be done. Therefore, users can experience the added value of the services. Moreover, messages do not only have to be sent to users. As an extension, it is possible to involve SCCI as a friend of all users.

This gives the possibility of notifying users within the portal system without the need of the execution of any third-party clients. It is similar to push notifications on mobile platforms. Naturally, the mentioned groups are also involved in this extension as smart communities. XMPP supports it as a Personal Eventing Protocol extension (XEP-0163: broadcast state change events associated with an instant messaging and presence account) since it functions as a virtual publish-subscribe service where all of the involved users automatically became friends with the system-generated user who represents the group itself.

- **Nearby events on map (ongoing)**: The portal changes its information provider role to an online collaboration environment where geographical information is applied as well. Figure 6.5 shows a map where a group's events are displayed for its subscribers. We are actually working on an extension to also involve routing between the event's location and the user's actual position.
- **Sensor data collector**: A mobile application was created for recording data provided by various sensors of a mobile device. Users can gain full control over data being recorded. They can switch it on or off whenever they want to. When switched on, the application is running in the background as a service and collects data in the XML format, e.g., gyroscope, temperature, fine-grained or coarse-grained position. This data will automatically be sent to an XMPP server for further processing using data mining techniques. The XML document contains the recorded data with a timestamp along with a header used for identifying the mobile device. Data sampling is 1 ms by default. An Internet connection is not required continuously because the data gathered offline will be sent to the XMPP server foronly online cases. Figure 6.6 shows the activity presented to the user for setting up the application.
- **Indoor positioning (ongoing)**: We introduced an indoor positioning app gathering the information from the building's Wi-Fi access points. The results were good, but it was not accurate enough in all cases. The interference

Fig. 6.5 Map displaying locations for an event's subscribers

Fig. 6.6 Data collector app
(Android only)

sometimes did not allow the ability to precisely determine the actual floor in the building, resulting in confusing outputs. Our new direction is based on BLE (Bluetooth Low Energy). It provides more accurate positioning due to its smaller detection radius. Naturally, it involved some investments because the beacons should have been deployed in the building. Figure 6.7 shows a simulation of the beacons' positioning and interference using RSSI approximation.

Naturally, it is still not a systematic task. While the signal strength is influenced by the building's properties, the signal distribution does not show nominal signs. Good approximations are based on several measurements and are still problematic if a smart device can only see one tag. For better result, at least two are required. Figure 6.8 shows an approximation perceiving two signals based on the blueprint of our building.

The conditions become better if we can receive signals from three or more independent tags. In this case, the location could be determined in a more fine-grained way. The only thing that remained is the proper handling of the floors and the detection of upstairs or downstairs movement.

- **Personal calendar as a service**: One of the first applications in our Smart Campus vision was the development of the personal calendar. The initial version was only a consumer app presenting upcoming lectures to its users. Events such as thesis application deadlines and registration deadlines would show up under the events.

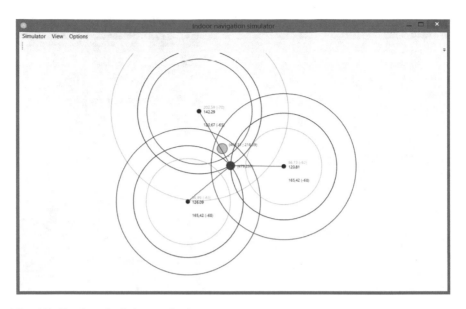

Fig. 6.7 Simulator for indoor navigation

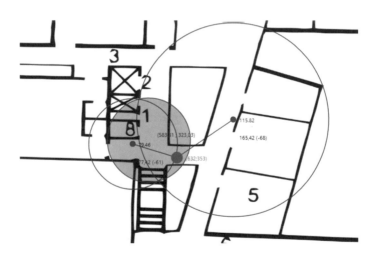

Fig. 6.8 Location approximation based on BT tags

As we mentioned in Sect. 4.1.3, the first step in the lifecycle of an intelligent service is collecting information from various sources and presenting it in a unified way. Obviously, at this stage, it cannot be treated as an adaptive, intelligent, or smart service. Its cycle begins when users can look for and mark events as either relevant or irrelevant. The feedback could serve as a basis for a recommendation system that could display appropriate suggestions for the individual users.

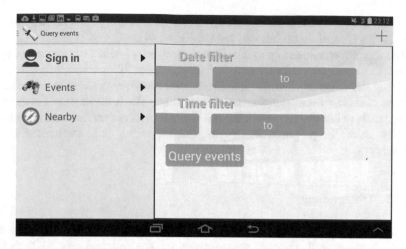

Fig. 6.9 Android app for personal calendar

Moreover, if users could rate those events, the smartness level could evolve because user feedback is always valuable information and could serve as an input for further analysis in the Analytics module. As in an earlier paragraph, we showed that users could see nearby events based on subscriptions for interesting topics. The combination of these two service results in a much smarter service capable of recommending nearby events based on user preferences. A graduated student prepared the working applications. It looks similar to Fig. 6.9.

- **Adaptive meeting planning** (ongoing): For the above-mentioned calendar, we began creating an extension for a meeting-planning module. The original concept is based on the idea that the module will read all of the participants' calendar events and attempts to suggest time slots that could be appropriate for all involved attendees. The adaption was based on properties of the event related to the user as well. It could be mandatory, optional, or reschedulable. Moreover, certain participants have higher priority than others. For example, instructors' calendars receive higher priority than students.

An initial version worked well with Google Calendar, but the connection for the Faculty' official internal Calendar has not been established yet. The provided API currently lacks some necessary functionality.

The recommendation system's evolution has been continued, though, with the help of Artificial Intelligence, whereas some of our colleagues started to design an algorithm based on graphs. According to [24], an Extended State-Space Model was defined. If some of these events are at conflicting times, the calendar service may suggest another schedule by replacing the time intervals of some events with other available intervals. Suppose that the student designates a schedule containing k events $(E_1,…,E_k)$, each with an initial time interval (from $T_{1,1},…,T_{1,N1}$ to $T_{k,1},…, T_{k, Nk}$). The problem, which can be solved with the use of EBFS (Extended

Breadth-First), is to determine a schedule of the same events so that no two intervals conflict. The proper algorithm was published in [24].

The services mentioned up to this point exist exactly inside our system and apply our Smart Community term for groups. The principles followed our primary vision regarding the lifecycle of smart, intelligent, and adaptive services. In addition, we have opened a new direction for Smart Education. The following section contains some details of our ongoing work for this field. It involves services not exclusively residing inside the proposed framework, but supporting it to evolve the overall smartness of it.

6.4.3.2 Observations

The primary motivation behind this Smart Campus project was to make the residents' life easier while they are at the university. The prepared services have been found to be useful for the everyday life as they help their users schedule their valuable time. We have received several feature request and suggestions as well. The most relevant ones are listed as future work in the closing section.

The overall experience shows that it was worth investing more than 2 years in the development of this project. The first steps were the most difficult ones because we could not find a commercial product to fulfill the requirements. Initial investigations showed new results could be achieved when a general architecture is designed first. Most of the problems occurred during the development because application servers and APIs contained several bugs. These needed to be fixed before the real implementation took place.

There currently are no extensive waiting times for lecturer's office hours. Users can also now find each other easily. If they cannot, they can send messages or get notifications regarding events taking place nearby. The calendar module only needs some minor adjustments to attract more attention as students are using messages to schedule meetings.

Students are also motivated to join this ongoing project because they can learn new technologies and collect valued experience on enterprise systems using a service-oriented architecture.

6.4.3.3 Smart Education for Better Programming Experience

The term *Smart Education* is an expression to describe new ways in education. It involves adaptive learning programs and collaborative environments with proper monitoring and reporting capabilities. We can find more details in [25, 26], while a survey could be found in [27].

If we use Smart Education with Smart Communities together in one context, we can reach a much broader sense regarding Smart Universities and Smart Campuses. These terms are all conducting something similar to when the Web 2.0 overtook

Web 1.0. Communities, Communication, Sensing, and Big Data are all related to this topic. They emphasize that users are not just reading the content, but are rather creating it as well.

When we use the smart property for these words, we will create a collection of terms related to adaption, intelligence, and sensing together to reach a better and user-friendly environment where value-added services help users reach their goals much easier. If we translate this concept to the education field, we find applications (collaborative environments) where users get valuable feedback from the system while they are using smart devices to perform different tasks.

In our case, two applications have been developed at the faculty level to help students improve their programming skills. According to [28], our Smart Campus environment used one system to manage programming contests (using ProgCont) and another one to support instructors in offering optional and non-graded exercises or assignments with the help of crowdsourcing.

"The basic goal of the ProgCont system was to support the organization part of programming contests. … web application stores the exercises in a problem cata-logue and collects the submissions including not only the solution source code, but the necessary information about the competitors and contests." [28].

Naturally, this is not a smart service at all, but following our service lifecycle proposal shows that this is simply the first step. The service was not only used to evaluate the submissions, but was used to derive useful information as well. The gathered data could be used to express a user's programming experience through success rate and code metrics like used memory, CPU time, and length of the code for each exercise. Nowadays, the system is also used for classroom tests [28]. The collected data gives the ability to analyze and interpret that from the pedagogy's point of view. This supports instructors and students as well. In summary, three main aspects and services were published:

- *Support for programmers on Smart campus platform*: The aim of the services provided by the application is to help users of the ProgCont system when solving different problems.
- *Hint service*: users can meet diverse difficulties during problem solving. This service tries to provide useful suggestions to achieve goals. The suggestions are based on virtual credits that can be collected by submitting successful problem solutions. More credits mean more hints.
- *Problem suggestion service*: based on the user's previous submissions and experience, the system could suggest problems to solve the next time. The problem suggesting service can contribute to the evolution and gained experi-ence of users. Two types of problem suggestions can be distinguished:

  - problems for maintaining the user's level, and
  - problems for upgrading the user's level.

In the first case, the focus is on practicing while the second case is shifted to the evolution side in order to help the user acquire new knowledge by solving slightly more complicated problems.

The other Web application inside the Smart Education field was created to offer optional exercises or assignments for courses. These are not counted in the final grade. The primary principle was the involvement of the crowd. While in the ProgCont system, Instructors are preparing and analyzing the results. The Web application offers the following major functionalities. It is possible

- to define exercises,
- to submit solutions,
- to assess submissions, and
- to comment and/or rate assessments.

As we mentioned, any registered user (i.e., even students) can define and evaluate exercises. Those who solve an exercise can submit the solution to be evaluated or assessed by someone else. In this manner, practicing students can receive some feedback, even if these answers cannot be trusted.

Naturally, we need to account for the possibility of false positive feedback. To handle this unwanted situation, not only the solutions, but the people giving feedbacks, should also be rated. As a result, the ratings for people who regularly give wrong answers will carry less weight.

Obviously, this should be treated as an early initiative to involve the crowd for educational purposes. The results show that there will be voluntary users whose valuable assessments are good enough to promote them as possible reviewers for given tasks.

6.5 Conclusion and Future Work

Conclusions. The performed research and obtained research findings enabled us to make the following conclusions:

1. The creation of a novel approach for an *open architecture* could be used to provide value-added services for Universities. The original idea [22] was based on a vision and a need for an extensible solution that is able to deal with new trends. These include IoT-, cloud, and crowd-based approaches. During the initial steps, we have identified key factors such as:

   - distributed system architecture,
   - ubiquitous computing,
   - heterogeneous data sources, including internal sites with or without service interfaces,
   - sensors,
   - professional sites,
   - social media,
   - sharing and accessing, and
   - big amount of data.

2. Addressing *smartness levels* in relation to these aspects directly leads to a smart solution that includes (1) sensing, (2) adaptation, (3) inferring, and the (4) social level fulfillment derived from the group and community concept. *Sensible* features are applied through sensors sensing the environment in order to become aware of events necessary to locate and notify nearby users. In the future, the plan is to allow the registration of infrastructure related processes. The *adaptive and inferring levels* could be reached as recommendations, activities, and personal assessments served by user basis. A necessary analytical module will be used to build further detailed user profiles to extend trajectories as well as follow user interaction and skills. Communities achieved *sociable features*. These are implemented as groups to make communication a possibility between users and groups. The connectable, accessible, ubiquitous, and shareable requirements for a smart campus are mandatory and self-explaining.

3. While investigating the available solutions and research tendencies, it has become clear that there is (1) *no unified solution* for Software Systems to manifest a University or campus as a smart solution provider. During the research phase, we have seen (2) several *isolated results* for dedicated usc-cases/problems. This is (3) *without the possibility of integration*, though.

4. At this point, we have concluded a need for an open extensible architecture to support service integration and creation. This makes our proposed architecture better than those available at other universities. The openness has led to an easy integration of any existing service possible. The most useful direction seemed to be to publish/subscribe the model and its XMPP implementation. We have introduced a new term, the *Connector,* that is responsible to collect specific information from different sources. This populates it to the central unit, or the Smart Campus Central Intelligence—SCCI. This makes the necessary decisions based on the business rules for each domain. The openness is provided by the base nature of XMPP as new sources could be connected easily without requiring any modification on the other pieces. When a given type of message does not exist yet to transfer a specific kind of data from a new source to our SCCI, XMPP allows the necessary extension without any pain. Moreover, new sources could be added at any time by anybody because there is no need for any special programming knowledge. This is because the well-defined APIs already exist. This is where we can involve the crowd as, not only data suppliers for the services, but also creators of new services. (1) *Extensibility*, (2) *integration*, and (3) the *crowd as publisher and developer* help make our service more well defined than any other solution. Naturally, the emphasis is not on the provided services because they are not unique; however, there are unique ones as well. These can replace provided way of their integration and (4) *process through the SCCI*.

5. Whereas we find a way to connect heterogeneous sources to our system as data suppliers, we also need to find a solution to deal with the large amount of data. The *Historical* DB was created to do this by showing where the "old" data is

currently stored. Examples of this are sensor data from the past, old events, and outdated information that would only slow down queries if kept in the main storage.

6. To combine all of the concepts discussed, we must use a security principal to give available services to assigned users. We have applied the group philosophy with smart properties and, thus, created the *Smart Community* expression to describe the users of these smart services.

7. Our research includes a *demonstration web site* located under the https:// smartcampus.hu address where authenticated users could use several of the aforementioned services. This helps these users become a member of our Smart Campus' Smart Community. Most of the published services proved to be useful as users are regularly utilizing them. Users can bypass long wait times with the help of the available office hours service. In addition, they can easily locate each other and use the chat feature to communicate. The integrated map becomes useful when looking for nearby events or friends. The first Wi-Fi-based version for indoor navigation was tested during a conference where it was used to locate rooms for participants.

8. Applications related to the *Smart Education line* also proved to be efficient. This is because students are continuously using them to train themselves and organize programming contents; however, the crowd-based assessment evaluation system needs more emphasis as this approach is not widely known and applied by students.

Future work. Naturally, we have several ideas for further steps as well. This includes enhancements over existing services as well as completely new ones such as:

- reminders for upcoming lectures and a possible approximation of arrival when needing to go to off-campus locations,
- enhancing the previous one with a counter to indicate lateness,
- counting missed lectures and display a warning when no more absences are allowed,
- penalty calculator for failed exams or missed lectures,
- nearby events based on recommendations,
- schedule meet-ups based on groups or common interests,
- continuous feedback, and
- from the developer's perspective: push notifications, socket communication, UI redesign and REST-based versions.

The underlying service architecture is in a testing phase currently and several end-user mobile applications have been prepared. Several more must be created, though, to truly call the system smart. The architecture fits well into the more general publish/subscribe based architecture of Smart City and Smart Campus applications. This is because it is extensible with new data sources that provide the capability to integrate heterogeneous data. In the near future, the major goal is the development and improvement of the Analytics module since proper analysis is the key to add more value to the services.

One of the most interesting and exciting new non-education-related services is the introduction of a remote controlled bicycle locking and docking station controller. Users through the mobile app, could read the QR code of the docking place and open or close the lock. This will enhance the system with a completely new direction to make it closer to a Smart University.

Acknowledgements The author would like to thank Lajos Kollár (former colleague) as a founder member of the initial group, Márk Kósa, Tamás Kádek, János Pánovics—the research associates and graduate students of the Department of Information Technology at University of Debrecen—for their valuable contributions into this research project.

References

1. Uskov, V.L., Bakken, J.P., Pandey, A., Singh, U., Yalamanchili, M., Penumatsa, A.: Smart university taxonomy: features, components, systems. In: Smart Education and e-Learning 2016, pp. 3–14. Springer, ISBN 978-3-319-39690-3 (2016)
2. Learning without Barriers/Technology without Borders—Symposium Booklet. MIT iCampus, November 2011. http://icampus.mit.edu/
3. Learning@Cisco DDM14CS4580. The internet of everything: fueling educational innovation. http://cs.co/9001nuAZ
4. What is the internet of everything? Cisco (2014). http://www.cisco.com/web/tomorrow-starts-here/ioe/index.html
5. Forecast: The Internet of Things, Worldwide, Gartner (2013)
6. IBM research—smarter education group, IBM. http://researcher.watson.ibm.com/researcher/view_group.php?id=4977
7. Ng, J.W., Azarmi, N., Leida, M., Saffre, F., Afzal, A., Yoo, P.D.: The intelligent campus (icampus): end-to-end learning lifecycle of a knowledge ecosystem. In: 2010 Sixth International Conference on Intelligent Environments (IE), pp. 332–337 (2010)
8. Liu, Y.L., Zhang, W.H., Dong, P.: Research on the construction of smart campus based on the internet of things and cloud computing. Appl. Mech. Mater. **543**, 3213–3217 (2014)
9. Hirsch, B., Ng, J.W.: Education beyond the cloud: anytime-anywhere learning in a smart campus environment. In: 2011 International Conference for Internet Technology and Secured Transactions (ICITST), pp. 718–723 (2011)
10. Nie, X.: Constructing smart campus based on the cloud computing platform and the internet of things. In: Proceedings of the 2nd International Conference on Computer Science and Electronics Engineering (ICCSEE 2013), Atlantis Press, Paris, France, pp. 1576–1578 (2013)
11. Internet of Things—Architecture IoT-A: Deliverable D1.5—Final Architectural Reference Model for the IoT v3.0. The European Union (2013)
12. Elyamany, H.F., AlKhairi, A.H.: IoT-Academia architecture: a profound approach. In: 16th IEEE/ACIS International Conference on Software Engineering, Artificial Intelligence, Networking and Parallel/Distributed Computing (SNPD), 1–3 June 2015. IEEE (2015)
13. Atif, Y., Mathew, S.: A social web of things approach to a smart campus model. In: Proceedings of the 2013 IEEE International Conference on Green Computing and Communications and IEEE Internet of Things and IEEE Cyber, Physical and Social Computing, ser. GREENCOM-ITHINGS-CPSCOM'13, pp. 349–354. IEEE Computer Society, Washington, DC, USA (2013). doi:10.1109/GreenCom-iThings-CPSCom.2013.77
14. Kuang, W., Luo, N.: User interests mining based on Topic Map. In: Seventh International Conference on Fuzzy Systems and Knowledge Discovery (FSKD), vol. 5, pp. 2399–2402 (2010)

15. Tan, Q., Kinshuk, Jeng, Y.-L., Huang, Y.-M.: A collaborative mobile virtual campus system based on location-based dynamic grouping. In: Proceedings of the 10 IEEE International Conference on Advanced Learning Technologies, pp. 16–18. IEEE Computer Society Press, Los Alamitos (2010)

16. Han, Y., Xia, K.: Data preprocessing method based on user characteristic of interests for Web log mining. In: IEEE Fourth International Conference on Instrumentation and Measurement, Computer, Communication and Control (IMCCC), pp. 867–872 (2014)

17. Boran, A., Bedini, I., Matheus, C.J., Patel-Schneider, P.F., Keeney, J.: A smart campus prototype for demonstrating the semantic integration of heterogeneous data. In: Rudolph, S., Gutierrez, C. (eds.) Proceedings of the 5th international conference on Web reasoning and rule systems (RR'11), pp. 238–243. Springer, Berlin (2011)

18. Yu, Z., Liang, Y., Xu, B., Yang, Y., Guo, B.: Towards a smart campus with mobile social networking. In: Proceedings of the 2011 International Conference on Internet of Things and 4th International Conference on Cyber, Physical and Social Computing (ITHINGSCPSCOM'11), pp. 162–169. IEEE Computer Society, Washington, DC, USA (2011). doi:10.1109/iThings/CPSCom.2011.55

19. Xiang, Y., Chang, D., Chen, B.: A smart university campus information dissemination framework based on WeChat platform. In: Proceedings of 3rd International Conference on Logistics, Informatics and Service Science, pp. 927–932. Springer, Berlin (2015). doi:10.1007/978-3-642-40660-7_138

20. Baranyi, P., Csapo, A.: Cognitive infocommunications: coginfocom. In: 11th International Symposium on Computational Intelligence and Informatics (CINTI), pp. 141–146. IEEE (2010)

21. Shadbolt, N., Berners-Lee, T., Hall, W.: The semantic web revisited. IEEE Intell. Syst. **21**(3), 96–101 (2006)

22. Adamkó, A., Kollár, L.: Extensible data management architecture for smart campus applications—a crowdsourcing based solution. WEBIST **1**, 226–232 (2014)

23. Garai, Á.; Péntek, I.: Adaptive services with cloud architecture for telemedicine. In: 2015 6th IEEE International Conference on Cognitive Infocommunications (CogInfoCom), pp. 369–374. IEEE (2015)

24. Kádek, T., Pánovics, J.: Extended breadth-first search algorithm. Int. J. Comput. Sci. Issues **10** (6), 78–82 (2014)

25. Neves-Silva, R., Tshirintzis, G., Uskov, V., Howlett, R., Lakhmi, J.: Smart digital futures. In: Proceedings of the 2014 International Conference on Smart Digital Futures. IOS Press, Amsterdam, The Netherlands (2014)

26. Uskov, V., Howlet, R., Jain, L. (eds). Smart education and smart e-learning. In: Proceedings of the 2nd International Conference on Smart Education and e-Learning SEEL-2016, June 17–19, Sorrento, Italy. Springer, Berlin, Germany (2015)

27. Global Smart Education Market 2016–2020. Research and Markets (2016). http://www.researchandmarkets.com/research/x5bjhp/global_smart

28. Adamko, A., Kádek, T., Kollár, L., Kosa, M., Tóth, R.: Cluster and discover services in the Smart Campus platform for online programming contests. In: 2015 6th IEEE International Conference on Cognitive Infocommunications (CogInfoCom), pp. 385–389. IEEE (2015)

Part III
Smart Education: Approaches and Best Practices

Chapter 7
Practicing Interprofessional Team Communication and Collaboration in a Smart Virtual University Hospital

Ekaterina Prasolova-Førland, Aslak Steinsbekk, Mikhail Fominykh and Frank Lindseth

Abstract A smart university must utilize different technical solutions to offer its students varied and innovative learning environments optimizing the core learning activities. Contact with patients is at the core of medical and health care education, often taking place at a university hospital. However, students from one profession seldom get the chance to practice in a hospital setting with students from other professions, and they seldom see the whole patient trajectory during clinical practice. Establishing a smart virtual university hospital mirroring a real life hospital can prepare students for direct patient contact such as practice placement and clinical rotation, and thus optimize and sometimes also increase their time on task. Such a virtual arena will support student learning by providing adaptive and flexible solutions for practicing a variety of clinical situations at the students' own pace. We present a framework for a smart virtual university hospital as well as our experiences when it comes to developing and testing solutions for training interprofessional team communication and collaboration. In the main part of the work reported here, medicine and nursing students worked in groups with the clinical scenarios in a virtual hospital using desktop PCs alone and with virtual reality goggles. In the evaluation, it was found that all the students agreed that they had learned about the value of clear communication, which was the main learning outcome. Using virtual

E. Prasolova-Førland (✉)
Department of Education and Lifelong Learning,
NTNU Norwegian University of Science and Technology, Trondheim, Norway
e-mail: ekaterip@ntnu.no

A. Steinsbekk
Department of of Public Health and General Practice,
NTNU Norwegian University of Science and Technology, Trondheim, Norway
e-mail: aslak.steinsbekk@ntnu.no

M. Fominykh
Faculty of Logistics, Molde University College, Molde, Norway
e-mail: mikhail.fominykh@himolde.no

F. Lindseth
Department of Computer and Information Science,
NTNU Norwegian University of Science and Technology, Trondheim, Norway
e-mail: frankl@idi.ntnu.no

© Springer International Publishing AG 2018
V.L. Uskov et al. (eds.), *Smart Universities*, Smart Innovation,
Systems and Technologies 70, DOI 10.1007/978-3-319-59454-5_7

reality goggles, almost all the students reported that they felt more engaged into the situation than using desktop PCs alone. At the same time, most also reported 'cyber sickness'. We conclude that a smart virtual university hospital is a feasible alternative for collaborative interprofessional learning.

Keywords Smart virtual university hospital · Virtual reality · Medical training · Collaborative learning · Virtual operating room · Educational role-play

7.1 Introduction

Patients are the major focus of medical education at all levels. The majority of the student contact with the patient has traditionally been through practice placement and clinical rotation in different health care settings, primarily in hospitals; however, the availability of time for interacting with patients in real life situations is associated with a number of challenges.

First, the increase in the number of students means that each student can interact with the patients less frequently, which is partly ameliorated by increasing the number of patients. Second, there has been a constant drive for many years to reduce the length of each patient's stay, meaning patients spend less and less time in hospitals and other health care institutions. Improvements in hospitals' effectiveness include ideas such as implementing clinical pathways (predefined patient trajectories optimizing the flow of selected patients through the hospital) and the increase in the number of day and outpatient patients. These in combination lead to significantly less time for contact between students and patients. Consequently, students get less time on the task, which is paramount for preparing them for their post-graduate work. Thus, there is a need for smart solutions that give the students more time on tasks or make the time with the patients more effective.

Another aspect relating to interaction with the patient is the increase in team-based work in health care. In a modern hospital, a patient is treated not by a single health-care specialist, but by an interprofessional team with complex collaborative procedures and practices. This means that medical and nursing students need to practice on complex interactions within a team of professionals. Although this is increasingly acknowledged as a central competency by the practice field, it is not implemented in the universities due to practical challenges. Students from different professional health care educations typically have practice placement separately from each other to not overburden those in the clinic. This means that a hospital department can have medical students present at one time and nursing students at another to avoid overcrowding of students in the ward. If there are more students than staff and patients, it would be very difficult to get the everyday work done.

The last aspect concerns the possibilities of a student following a patient over time. When students are on practice placement, they are in one unit at the time. This means that they only see patients while they are in direct contact with the unit. Even if the students have practice placement in other units, they will encounter different patients there. Thus, they are usually not given the opportunity to follow the individual patient trajectory. One consequence is that the students get little experience in understanding the totality of the patient's situation, but also a lack of understanding of how the health care system is organized, the complete process of transition from one unit to another (i.e. seeing it from the perspective of both units involved), and how this organization is experienced by the patients. The transitions are found to be the especially weak points. Students, therefore, need to be skilled in how to ensure the quality of services when the patient moves from one unit to another.

To meet these and similar challenges, there is a need for smart flexible solutions that can utilize collaborative technologies and digital learning resources. These are some of the hallmarks of smart education [1], which is a central concept within Smart universities. By 'Smart university' we mean a university that 'involves a comprehensive modernization of all educational processes' [2]. A smart university exhibits a number of various 'smartness levels' or 'smartness features', such as adaptation, sensing, inferring, self-learning, anticipation, self-organization and restructuring [3].

In the work presented here, we focus primarily on the "Adaptation level" within the smart university framework [3]. The adaptation level concerns issues like the ability to modify teaching and learning strategies to better operate and perform the main business and educational functions. In our case, it is about adapting to the challenge of preparing the students for the changing situations in health care as well as about how we can modify our learning strategies to give students ample time on task with individual patients and preparing them better for team based work along various patient trajectories. In particular, we have focused on the adaptation to the new platforms and technologies, variation in interfaces, and a new style of learning and teaching.

Our solution has been to use Virtual Reality (VR) and to create a smart virtual learning environment. We thus use virtual "technology supported learning environments that make adaptations and provide adequate support" [4]. We have aimed to make a virtual learning environment that will support student learning by providing adaptive and flexible solutions for practicing a variety of clinical situations at the students' own pace, adjusting to the student's level of expertise and scheduling limitations.

Based on information through contacts, web sites, and publications referenced in the next section, it is evident that several world leading universities and hospitals, especially in the US, UK, Australia, and New Zealand, have adopted 3D virtual simulation as a part of their educational programs. Examples include virtual hospitals/medical faculties at the University of South Florida, Imperial College of London, and Auckland University Hospital. Such environments typically include an array of different facilities such as emergency rooms, intensive care units, nursing simulations, and general information for the public. Other examples include the Maternity Ward at Nottingham University and the Emergency Preparedness Training at University of Illinois (secondlife.com).

Our own initiative has been partly inspired by these projects but seeks to achieve a more coherent approach to the development of smart online virtual educational solutions that are embedded in a holistic system to make them more accessible and easier to use. This is done by utilizing an established and integrated framework.

We have worked with the long-term idea of establishing an online Virtual University Hospital (VUH) to create such a holistic system and to be a venue for learning, research, and development. The idea is to make a virtual mirror of the St. Olav's University Hospital (St. Olav), which is integrated with the faculty of Medicine at the Norwegian University of Science and Technology (NTNU) in Trondheim, Norway. St. Olav is newly built and is one of the most modern university hospitals in the world with a state of the art technological platform and modern clinical buildings. Furthermore, the teaching and research facilities are integrated and distributed within the hospital.

As the success of such a smart virtual university hospital is dependent on the activities provided within the framework, we have concentrated our efforts on developing and testing solutions. In short, we have made a virtual part of a hospital in the virtual world Second life (SL) with patient rooms, meeting rooms, operating theatres etc. We first developed mono-professional scenarios for training post-graduate nurses in a surgical environment, anesthesia, and intensive care [5, 6]. Then we moved on to a larger project known as VirSam (virtual communication and collaboration, in Norwegian VIRtuell SAMhandling), where we focused on training for third year nursing and fourth year medical students in interprofessional team communication and collaboration.

The VirSam project is the focus in this paper, but we will describe the activities leading up to this project in some detail as it shows the process we have gone through to adapt our teaching and learning to smarter solutions.

7.2   Background

7.2.1   *Virtual Reality in Health Care Education*

Numerous modes for virtual learning currently exist, e.g. flexible low-cost 3D virtual simulations, 3D virtual environments, and associated infrastructure accessible over the Internet. This technology can benefit the educational process due to low costs and high safety, three-dimensional representation of learners and objects, and interactions in simulated contexts with a sense of presence [7, 8]. It has been suggested that this technology can "considerably augment, if not eventually, revolutionize medical education" [9].

Many studies reported the potential of 3D virtual worlds for educational activities when they have mostly been used on regular desktop computers [10]. Nowadays, these virtual environments can be used in combination with advanced VR technologies, such as motion tracking and head-mounted displays (HMD)/VR goggles to increase the sense of immersion and, therefore, improve the experience. This makes it more believable and transferable to real life. VR goggles are a type of on-body VR devices that are worn on the head and have a display in front of the user's eyes [11, 12]. Most of these devices consist of a display and a tracking system. It allows much greater immersion because the user can control the direction of the view in a virtual world in exactly the same way as in the physical world—by turning the head. The displays of VR goggles have larger fields of view and provide a stereoscopic image, making the experience more believable. One example of this is the Oculus Rift (http://www.oculusvr.com/), which has characteristics making it stand out from its predecessors. At the same time, the device is relatively affordable and, therefore, suitable for educational context [13, 14].

3D virtual environments have been widely used in the healthcare domain, including both desktop-based virtual worlds and other VR applications. Examples include training facilities for nurses and doctors [15–17]. These can be found in palliative care units [18], health information centers, and anatomy education [19, 20]. Such training is, on several occasions, reported to provide a cost-efficient and user-friendly alternative to real-life role-plays and training programs [18]. As demonstrated in several studies, "virtual worlds offer the potential of a new medical education pedagogy to enhance learning outcomes beyond that provided by more traditional online or face-to-face postgraduate professional development activities" [21].

Studies have shown that VR simulations contribute to enhanced procedural skills, with clear transfer of training to clinical practice [22]. A variety of surgical VR simulators have been developed (e.g., dental, laparoscopic and eye surgery) which have shown clear benefits for medical training [23]. The use of VR technologies goes beyond procedural skills. Possibilities for synchronous communication and interaction allow using these technologies, at the moment mostly with the desktop interface, for various collaborative learning approaches [24], as well as facilitate situated learning [25] and project-based learning [26] approaches. In addition, virtual learning environments can also support interprofessional and distributed medical teams working together on complex cases [27].

Desktop-based environments have been augmented with VR elements, such as HMD, for treatment of various neurological and psychiatric disorders including autism, phobias, and post-traumatic stress syndrome, the latter especially in military settings. For example, the Virtual Afghanistan/Iraq system has undergone successful clinical trials in using exposure therapy for treatment of combat-related post-traumatic stress syndrome among veterans [28].

While VR can be used for pain treatment as well [29], there is still little research on smart virtual learning environments in the field of healthcare education. There are, as mentioned above, examples of individual projects, but there is a lack of research-based innovation and research on implementation where such solutions are integrated into existing everyday teaching. For example, Cates [22] and Ruthenbeck [23] have requested more research on the use of VR/virtual environments for procedural training at different levels of medical education and exploration of solutions with greater levels of interactivity and increased realism.

7.2.2   Communication and Team Training

While most of the existing VR simulations focus on training of cognitive and psychomotor skills, there is a lack of solutions that support team and communication training that are essential in modern medical education [17]. For example, Pan et al. suggest further exploration of 'immersive VR' to increase understanding of the social dynamics between doctor and patient [30]. While medical simulation sessions often claim to include team training, they are generally not designed to achieve team training competencies [31]. Furthermore, many medical schools do not have the expertise necessary to implement team-training programs on their own [31].

This emphasizes the need to focus on interprofessional team training as an integral part of the educational program at a smart VUH. Weaver et al. identify identifies the following most common team competencies targeted in medical team training programs: communication, situational awareness, leadership, and situation monitoring. Instructional methods include information-based methods such as lectures, demonstration-based methods such as videos, and practice-based methods such as simulation and role-playing [32]. The latter is a common modality found in 68% of training programs [32, 33].

Baker et al. looked at an array of existing medical training programs, including both simulation-based and classroom-based programs [34]. The former included Anesthesia Crisis Resource Management and Team-Oriented medical simulations and typically take place in a simulated operating room with patient simulator/mannequin, monitoring/video equipment, and pre-post briefings [34]. The classroom-based training programs include MedTeams, Medical Team Management, Dynamic Outcomes Management, and Geriatric Interdisciplinary Team Training. These programs typically include lectures, discussions, and role-plays [34, 35].

Many existing programs focus on team-training within closed environments such as operating rooms and emergency care units [36] while interprofessional team training is important for improving patient safety and breaking down the traditional discipline-based barriers [36]. This motivated the development of the Triad for Optimal Patient Safety program (TOPS) [36].

Another interprofessional training program, originally developed by US Department of Defense and the Agency for Healthcare Research and Quality, is TeamSTEPPS (Team Strategies and Tools to Enhance Performance and Patient Safety) [37]. This framework has gradually reached international acceptance and has served as the main inspiration for our work in this project with developing interprofessional team training simulation. TeamSTEPPS is "an evidence-based framework to optimize team performance across the health care delivery system" [38]. It has several important features, with one of them being the four team competencies: Communication, Leadership, Mutual Support, and Situation Monitoring [38].

It is necessary to develop reliable methodological tools for developing learning content and evaluating learning effectiveness team training in virtual learning arenas [17, 27]. Research is also needed to gain knowledge of what are the good solutions and how they should be adapted to meet students' needs.

7.3 Project Goal and Objectives

The main goal of this project is to increase the students' time on tasks by offering smart online 3D virtual learning environments (with and without VR interface) that provide the learner with practice opportunities, to be completed at their own pace, where patient access is increasingly difficult to achieve. This is done through enhancing the PBL (problem based learning) approach and facilitating online environments for small tutor groups and flexible self-regulated learning, "anytime, anywhere".

7.3.1 Challenges in Medical Education

The medical doctor program at NTNU, where this study was based, has a PBL based curriculum and is based on "spiral learning". This means that the same topics are repeated throughout the study with increasingly advanced level. The system mainly follows the organ system, having one organ (or another topic) in the focus at the time. The teaching is based on learning objectives that are detailed for each semester and there are no fixed set books (students have to find literature themselves based on, e.g. recommendations from teachers). Furthermore, and different from many other programs, there are no separate subjects. This means that when new learning activities are integrated, it is not done on the level of a subject. Instead, the new learning activities are integrated into an ongoing learning activity according to which organ system/topic and the complexity level the activity concerns.

Referring to the McMaster University PBL model as developed by Barrows and Neufeld, which the medical school at NTNU has partly followed, the case-based and explorative methodology consists of three key components: PBL, self-regulated learning, and small group tutorial learning [39]. The method thus acknowledges the complex interplay between the social and individual aspects of learning, as later more fully elaborated and expanded in Nonaka and Takeuchi's SECI (Socialization, Externalization, Combination, Internalization) model for learning and knowledge development [40].

While many medical schools, including NTNU, have adapted the PBL approach for curriculum development and teaching and learning strategies, there is little evidence to suggest that medical schools and postgraduate institutions are sufficiently successful in helping students becoming effective self-regulated learners

[41]. Acknowledging that the medical profession is "in a perpetual state of unrest" [42], developing the self-regulated learner becomes an important aim in itself. The process of becoming an effective self-regulated learner can be greatly supported by technology [9] and offer new possibilities [21].

Solely focusing on the self-regulated learner would obviously be a dead end, though. The community aspect of learning and knowledge development also needs to be included. Drawing upon seminal work of Lave and Wenger [43], Engeström [44], and Brown and Duguid [45], we suggest that the learning process and creation of knowledge is also characterized by narratives, collaboration, and social constructivism. An increasing body of research from different disciplines has suggested that the ability to visualize represents a particular difficulty for many learners. Amongst these disciplines are economics, electrical engineering, the medical disciplines, and biology [46, 47]. The common problem is the ability to perceive unseen forces, patterns, and the transition from 2D (i.e. on paper) to 3D (i.e. "real life").

7.3.2 Smart Virtual University Hospital and VirSam

Based on the context and challenges presented above, we have a vision of a smart approach that integrates educational activities delivered virtually with other educational activities that prepare the students for practice. We have, therefore, worked within the VUH framework introduced in [5, 6]. We identified initial requirements, facilities, and technological solutions for the VUH as an arena for educational activities, for personnel, for students, and for the public [5]. Some examples of these are presented in Table 7.1.

To start the process of filling the VUH with content, some minor projects relating to mono-professional communication and collaboration were started. The first solutions were used for training nurses undergoing specializations in surgery, anesthesia, and emergency care [5, 6]. This gave us valuable experiences in both

Table 7.1 Educational activities in a virtual university hospital

Activity	Content, facilities and technological solutions
Patient simulation	Operating room, patient ward, emergency area, virtual humans, interactive hospital equipment, VR interface
Procedure training	Various hospital departments (operating room, patient ward, emergency area), information using videos and posters, interactive hospital equipment, VR interface
Lectures, e.g. anatomy	Classrooms, lecture halls, 3D interactive models of organs and the human body, posters, videos, VR interface
Role-plays (team training, patient communication)	Operating room, patient ward, emergency area, reception/outpatient clinics, interactive hospital equipment, VR interface

setting up the VR environment and planning a more complex infrastructure and logistics.

We decided to continue with the topic of communication and collaboration. The pressing problem that students from one profession seldom get the chance to practice with students from other professions inside a hospital setting is that they seldom see the entire patient trajectory during clinical rotations. This was the motivation for starting the VirSam project. We received funding from the NTNU's Top Education program for a demonstration project, which is reported on here.

The primary objective of the part of VirSam reported here was: to provide fourth year medical and third year nursing students with the possibility of interprofessional communication training in a smart virtual environment. The secondary objective was to investigate the students' experience with different technological solutions for experiencing VR. This was done by having them use both a desktop version and a set of VR goggles.

7.4 Methods and Technologies for Supporting Interprofessional Communication and Collaboration

7.4.1 Developing Technical Solutions

We chose to use SL as the platform for developing the VR part of our VUH. SL remains one of the most stable, developed, and populated virtual environments, though there are most definitely certain limitations. We developed our virtual hospital on the base of NTNU virtual campus in SL to make it easier for the students to navigate there. Another advantage of using an established platform was the possibility to buy ready-made models in the SL market place.

The virtual environment and avatars for role-playing have been designed in accordance with the learning goals we have set and after consultations with specialists and examinations of the corresponding facilities in the real hospital of St. Olav. We have made the following rooms:

- *Patient room* is the room designed for inpatient visits, which typically contain a hospital bed, some medical devices, patient terminal (TV etc.) as well as a chair, table and washing basin.
- *Waiting area* is an ordinary waiting area that one can find within all ordinary hospital clinics, consisting of a reception desk, sitting chairs for patients and relatives, and a table with magazines and papers.
- *Sluice* is a room that health personnel use for the delivery of patients on their way to a surgery.
- *Operating theatre/room* is a place where surgeries are conducted. The room is usually equipped with operation lamps, different medical equipment, and an operating table for the patient.

- *Emergency room* is a room where emergency patients are received for initial examination. The room bears several similarities to the operating room.
- *Intensive care unit* is a room with several patient beds divided by curtains, with some monitoring equipment, where the patients can be placed for observation after surgery.
- *Meeting room* is where the personnel meet to discuss situations among themselves or where they have, for example, discharge meetings with patients.

The goal was to create a virtual environment realistic enough to give a feeling of being in a real hospital. For example, the virtual operating room was modeled after the one at the Department of Neurosurgery at St. Olav's hospital (Fig. 7.1).

The same design approach was applied to the entire interior. Several equipment artifacts were designed after those found in the real hospital as some of the virtual equipment purchased earlier on the SL marketplace had limited functionality and could mostly be used as an illustration.

An important aspect of using VR is that avatars represented the "players". Thus, to make the experience as real as possible, effort was put into making the avatars and their appearance realistic in term of clothing, gender and general look. The surgical nursing avatars' uniforms were originally pale green and the ward nurses' ones were white. Additional uniforms in other colors for anesthesia, emergency, and intensive care nurses have also been made. In addition, we purchased 'skins' (faces for old people) and other avatar accessories (e.g., a hijab and jewelry, patient gowns) at the SL marketplace.

We made avatars to represent different patient types (children, with foreign background, with serious condition(s), etc.) as well as their relatives. The avatars for the relatives and patients had to match the description in the scenarios, for example, a 'mother', 'pregnant woman' or a 'person with immigrant background' (Fig. 7.2 left). The patient and the staff avatars contained more details and were dressed in accordance with the standards adopted at the Norwegian hospitals.

Apart from avatars operated by the students themselves, we have made virtual patients (programmable agents). The virtual patients could be static or dynamic. The static avatars are placed in the waiting room and hospital corridors. In addition, some beds in the intensive care unit are used to create an illusion of a busy hospital. The dynamic virtual patients are agents programmed to simulate certain diseases for

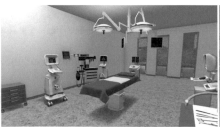

Fig. 7.1 Virtual operating room (*left*) versus a real one (*right*)

Fig. 7.2 A pregnant virtual patient played by a student (*left*) and a dynamic artificial/programmed virtual patient with student avatars around (*right*)

diagnostic training and include a 'boy' avatar with diabetes and an 'old man' avatar with a heart attack. These avatars exhibited certain symptoms (by body language and facial color) and were accompanied with blackboards and pop-up menus where the user can get additional information, order tests, and choose a course of treatment (Fig. 7.2 right). Depending on the chosen treatment course, the virtual patients exhibit pain, recover, or die.

7.4.2 Instructions to the Students

The prerequisite for taking part in this system is that the students must have access to a computer with SL installed. Therefore, we made instructions showing how students could download the SL browser to their laptops, set up a user account, and do some experimentation with the VUH environment beforehand. We also gave the students a presentation of the VirSam project and a one hour long SL tutorial on such topics as camera controls, avatar navigation, familiarizing with the environment, voice communication, and fetching a gurney for patient transport. Later, we developed a short (12 min) video lecture giving the same information in condensed form.

The students log into the VUH with designated avatars. To do the role-play, the students are provided with the description of the clinical situation and role cards for the different roles they can choose, which are described in detail below. Improvisation during the role-play is important in order to mitigate the technical limitations and make the role-play realistic. The technical aspect concerned things that are different to realize in a virtual environment. For example, it is not possible (or difficult) to control facial expressions and perform medical examinations and manipulations. In such situations, the students need to explain what they are doing/planning to do orally. The students are not given instructions on what to say. Thus, they must improvise continuously during the role-play. They are instructed to use their previous experience to play their role, meaning that there has to be some adaption to the role.

Although it is possible to develop solutions as described above with virtual patients (Fig. 7.2 right) with information given on blackboards, this solution is not technically elegant, and it is difficult to incorporate the required flexibility. Therefore, the students were informed that, in one of the clinical scenarios, there would be a game master who had information about the tests and examinations that ought to be done, as well as the results to these. To activate or get access to this information, the students were instructed to clearly say which tests and examinations they would have done. The game master was instructed to only give information on what the students asked for. The game master was to not give out information not asked for.

7.4.3   Learning Outcomes and Clinical Scenarios

Since the purpose of the project was to develop and evaluate smart solutions to give students competencies in interprofessional communication and collaboration, all the work was guided by the following learning outcomes that the students were expected to reach:

1. Practical skills in conducting role-playing in a virtual world,
2. Understanding the importance of:

   (a) clear and structured communication that is comprehensible to all parties,
   (b) clarifying the patient's wishes to set common goals for treatment and follow-up,
   (c) making sure that all roles and responsibilities in a team are properly clarified and distributed,
   (d) asking questions and listening to each other in order to obtain joint understanding of the situation,
   (e) making sure that everybody suggests solutions and contributes to the decision-making process, and

3. Knowledge of collaborative team processes along the patient trajectories at the respective departments and agencies involved.

These outcomes have been mostly inspired by the TeamSTEPPS framework [37, 38] and consultations with subject experts. To achieve these learning outcomes, effort was put into developing clinical scenarios mirroring the real life work in hospital departments. Two clinical cases, geriatric and gynecological, were developed by specialists from the corresponding clinics at St. Olav's hospital, based on a template provided by the authors. The template was inspired by literature review in the previous section and focused on the communicative and collaborative aspects of interprofessional teamwork during different types of patient trajectories.

Furthermore, we wanted to explore the usability of VR in presenting a condensed version of different types of patient trajectories. It was, therefore, chosen to make the geriatric scenario an example of a long-term trajectory lasting several weeks. The gynecological scenario, which was a sub-acute condition, represents a short-term trajectory lasting only a few hours.

In the following, the details of the two clinical scenarios are presented as it was presented to the students. In each scenario, there are several roles, including the role of the patient. A medical student is supposed to play the roles of a medical doctor or a surgeon and, likewise, a nurse student is supposed to play the roles of a nurse. Otherwise, there are no instructions as to who shall play the other roles, i.e. this is left to the students to organize.

In addition to the general description of the scenarios, the students are given one role card describing the role they shall play. For the role as nurses and medical doctors/surgeons, very little information is provided in order for students to use their own experience during the role-play. The motivation is to challenge the students in a safe environment in order to test out their skills and knowledge. For the roles as patients or relatives, some more information is given regarding behavior. For each role, there is information on the corresponding role card, which is not disclosed in the description of the scenarios, such as what the players should get out of a situation.

7.4.4 Geriatric Scenario: Long-Term Patient Trajectory

The geriatric case represents long-term patient trajectory from admission to home care.

Practical information. The role-play may have from 3 to 6 participants. With 6 participants, the roles are patient, doctor at the geriatric department, a nurse at the ward, a relative of the patient, a municipal employee, and a game master. With 5 participants, the role of the relative is omitted. With 4 participants, the municipality employee is also the game master. With 3 participants, there is no municipal employee and the patient is also the game master. The game master acts as an information provider who has information about test results etc., which may be disclosed to other participants upon request.

Clinical situation. The scenario takes place at the inpatient ward at the geriatric department. An elderly patient, 84 years, is hospitalized after a fall at home. The patient spent several hours at the hospital A&E (accident & emergency) department prior to transfering to the geriatric ward in the evening. According to information from the A&E department, a home care employee found the patient on the floor of his home. It is unclear how long the patient has been lying on the floor, but it was probably for several hours. The patient is described as confused and anxious and cannot account for the events himself. The patient has severe pain in the back and is rather thin. It is reported that, according to the home care services, the patient has recently complained about dizziness and feeling shaky. He usually uses a walker. The spouse of the patient died one year ago. He has two children and four grandchildren. He lives in an apartment on the 3rd floor with a lift. The patient is normally fairly self-reliant, but uses home care services for certain tasks such as taking on/off compression stockings. He also gets his medicines in a so-called multi-dose system where the medicines are pre-packed. He also has a remote control safety alarm. His past diseases include hypertension, diabetes, and osteoporosis with previous compression fractures. The drugs he takes include Metoprolol depot, Albyl E, Furix, Metformin, Calcichew-D, Alendronate, Sobril, Imovane, and Codeine as needed.

Role-play. There are 4 scenes to be played. Those who are not participants in the specific scenes are residing in the background or the waiting room in the virtual hospital. Their primary function is to observe.

Scene 1: *The first evaluation.* The scene takes place inside the patient's room in the geriatric department shortly after the arrival of the patient at the ward. Participants include the patient, doctor, nurse, and relative of the patient (Fig. 7.3). The main purpose of this scene is to make an initial assessment of the patient's situation. The doctor and nurse need to clarify who does what and inform the patient and the relative. The scene is concluded with the doctor and the nurse summing up what needs to be done to clarify the situation (e.g., which tests to be done) before leaving the patient room and proceeding to the meeting room.

Scene 2: *Interdisciplinary planning meeting.* The scene takes place inside the meeting room. Such meetings normally are held at a fixed meeting time on the first business day after the patient's admission to the geriatric ward. Participants are the doctor and the nurse (who are often joined by a physiotherapist and an occupational therapist). The purpose is to present the patient (hospitalization cause, findings so far), clarify the treatment objectives and actions, and start planning the discharge of the patient. A designated game master provides information about the test results and findings from examinations that were requested in scene 1. The scene

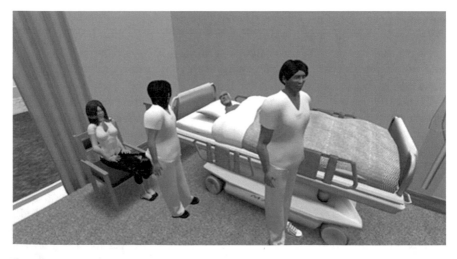

Fig. 7.3 The opening scene in the geriatric case, the nurse and doctor talking to the patient and his daughter in the patient room

concludes with the doctor and nurse summing up a course of action before the doctor leaves the room.

Scene 3: *Phone call to the municipality*. The scene takes place inside the meeting room. The time is a given number of days prior to the planned discharge date (i.e. several weeks after scene 2). Participants are the nurse and a municipal employee. The goal of the call is to start the discharging process and clarify how the municipality might contribute with personalized home care or a temporary admission of the patient to a rehabilitation or nursing home. Ending the telephone call concludes the scene.

Scene 4: *Discharge Meeting*. The scene takes place inside the meeting room a given number of days after scene 3. Participants are the patient, the doctor, the nurse at the ward, the relative(s) of the patient, and a municipal employee. The purpose is to plan the patient's discharge as well as any necessary further action, such as home care. Toward the end of the meeting, the participants sum up the further action plan for the patient.

7.4.5   Gynecological Case: Short-Term Patient Trajectory

The gynecological case represents a short-term sub-acute patient trajectory from admission to surgery preparations.

Practical information. The role-play takes place at the gynecological department and may have from 3 to 6 participants. Depending on the number of available players, the participants are the patient, the doctor of the gynecological department (also surgeon), the nurse on the ward (core participants with 3 players), the surgical nurse (if 4 players), the husband (if 5 players), and the anesthetic nurse (if 6 players). The same person can potentially play the ward nurse and surgical or anesthetic nurse, depending on how many people are participating.

Clinical situation. The patient is a 30-year-old woman with no children (Para 0) who arrives at the gynecological department with suspected ectopic pregnancy (pregnancy outside the uterus). Her husband is also in his 30 s. She became acutely ill at work, coming to the emergency room with severe pain in her lower abdomen. At the emergency room, a pregnancy test was taken that unexpectedly turned out to be positive. Therefore, a tentative diagnosis of ectopic pregnancy has been set. The findings on the examination include tenderness by bimanual palpation of the uterus corresponding to the right adnexa (surroundings of the uterus). A vaginal ultrasound shows no intrauterine pregnancy. Some free fluid in the (ovarian) fossa and a donut-like structure corresponding to the right adnexa of the uterus can be seen on the ultrasound, confirming ectopic pregnancy. It must be removed on the same day with laparoscopic surgery under general anesthesia. Prior to the operation, the surgical team should go through a 'safe surgery' checklist with the presentation of the team and the planned procedure.

The patient's symptoms include severe pain, weak and ongoing vaginal bleeding, and paleness. The labwork showed the u-hCG +, quantitative HCG > 1000 Hb 8 (low).

Role-play. There are three scenes to be played. Those who are not participants in the specific scenes are residing in the background, the waiting room, or outside the operating room in the virtual hospital to observe.

Scene 1: *Assessment and informing the patient.* The scene takes place in the patient's room and the participants are the patient, husband, ward nurse, and doctor. The purpose of this scene is to make an assessment of the patient and explain the situation to her. The patient has just arrived to the gynecological department from the emergency unit and is lying in bed. The husband is sitting in a chair. The nurse enters the room (or is in the room from the beginning), performs some tests and asks questions before explaining that the doctor is coming soon. The doctor enters the patient's room and the nurse presents her findings. Afterwards, the doctor 'performs an ultrasound' and explains the diagnosis, the need for surgery that will follow immediately, and how it will be done. Shortly after, the doctor leaves the room and the scene ends with the nurse also leaving the room to fetch a gurney that will transport the patient to the operating room.

Scene 2: *Transfer and information exchange.* This scene takes place between the patient's room and the sluice. The participants are the patient, husband, ward nurse, surgical nurse (and possibly anesthetic nurse). The goal of the scene is to transfer the patient to the sluice and provide all of the necessary information to the surgical staff. The nurse fetches the gurney and the patient is placed on it before being transported from the patient room in the ward to the sluice. The husband of the patient follows. In the sluice, the surgical nurse (and possibly the anesthetic nurse) is waiting for them. In the sluice, the nurses exchange information relevant to the upcoming surgery. The husband is not allowed to follow the patient further. The scene ends with the surgical and anesthetic nurses rolling the gurney into the operating room.

Scene 3: *Preparing for surgery.* The scene takes place in the operating room and the participants include the patient, doctor/surgeon, surgical nurse, and anesthetic nurse (Fig. 7.4). The patient is 'placed' on the operating table. The doctor and the surgical or anesthetic nurse explain to the patient what they do during the preparation of the patient for surgery. The patient is 'put under general anesthesia' (e.g., being asked to count back from 10). The scene is concluded with the team undergoing the 'safe operation' checklist.

7.4.6 Methods for Testing the Solutions

All the cases and scenarios were tested during the same day and all of the required data was acquired during the test run. The role-play was performed in two phases with a desktop (phase 1) and VR goggles (phase 2). The students were divided into groups and played out both clinical scenarios.

Fig. 7.4 The last scene in the gynecological case with the nurse and doctor making the preparations for sedating the female patient

In Phase 1 (desktop version), the six students in one group were located in six separate rooms with their own PC where SL was installed. They communicated with each other in the group through their avatars and used the voice chat function in SL. The remaining students stayed in the classroom and followed the play either on their laptops through their own avatars or on a big screen connected to a laptop. After the first group had finished their first role-play, the players and observers switched places.

In Phase 2, the students in one group were placed in a lab where every student received one PC set up with SL and VR goggles (Oculus Rift development kit versions DK1 and DK2) (Fig. 7.5). In this setting, the students did not use SL's voice chat for communication due to the proximity to each other as well as the fact that using the voice chat would have created feedback. Instead, they used their own voice to communicate. A microphone was placed on the table to transmit the sound to the classroom where the non-playing group was seated to observe the role-play.

In the beginning of the role-playing session, the students were assigned one of the roles from each scenario and received role cards describing their characters as well as a general case description that included the situation description, preliminary diagnosis, clinical data, and the different scenes in the scenarios (as presented above). The students were told to follow the scenes, but at the same time to improvise their role as best as they could by applying their knowledge and previous experience.

After each role-play, each student group spent 5–10 min reflecting on their experience within the group. After finishing the role-plays, all the students gathered in the classroom and participated in a joint discussion/group interview session.

The data in this study was collected from several sources. The role-play in SL was recorded as a screen capture (with sound). The role-play was observed both in SL and in 'real life' by the authors. The group interviews were recorded with video

Fig. 7.5 Students use VR goggles to participate in role-play. They sit at different angles depending on where they are in the virtual environment

camera and sound capture that was later transcribed verbatim. In addition, 16 of the students filled out a questionnaire consisting of both multiple-choice questions using a Likert scale, 'check-box' questions, and open questions.

7.5 Outcomes

A total of 18 students, including 14 third-year nursing students and 4 fourth-year medical students, participated. The participants were divided into three groups, each group containing one or two medical students. Of these, 16 responded to the questionnaire. Among these, there were three male and ten female nursing students and two male and one female medical student.

Ten students reported that they had little to some experience with virtual environments and 3D games. Only three students did not have any prior experience. When asked to give examples, the majority (10 of the respondents) mentioned Sims. Other games such as World of Warcraft, Minecraft, Destiny, Runescape, and Skyrim, as well as other Oculus demos, were mentioned. When asked to describe their computer expertise, the students almost equally divided between "somewhat good", "good", and "very good".

None of the students had experienced working with students from other health professions before.

7.5.1 Learning Outcomes

There were three questions in the questionnaire asking to what extend the students have achieved the learning outcomes (see details in Sect. 7.4.3). The options given to the students were "achieved", "partly achieved", or "not achieved".

All the participants answered that they achieved the learning outcome we thought were the most important during our work: understanding the importance of a clear and structured communication that is comprehensible to all parties. Some of the other learning outcomes, especially (b) and less so (c) and (d), were not fully achieved (Fig. 7.6).

The majority of the students (14) reported that they achieved the learning outcome of understanding the importance of involving all parts in the decision-making process. As appears from the interviews, the roleplaying session allowed looking back and reflecting on one's own and other's roles and contributions. As one of the students puts it: "...Afterwards all feelings are like that: 'Oh my God we did not think of the blood pressure and this and that ...I should have had a greater role'".

Most of the students (13) answered that they achieved practical skills in conducting role-playing in a virtual world, while only three students replied partly achieved.

More than half of the participants (9) answered that they achieved the knowledge of collaborative/team processes during the treatment and monitoring of patients at the respective departments. Seven students answered "partly achieved".

7.5.2 Experiences with the Second Life Technology

This topic contained five questions, including four five-point Likert scale questions and one open question. Most of the students (14) agreed or strongly agreed that

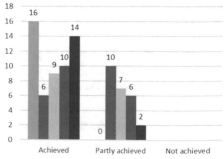

Fig. 7.6 Achieving learning outcomes

familiarizing with SL went quickly enough, while only two were neutral on the subject. Similarly, the majority (9) disagreed that it was difficult to move the avatar, and only four agreed or strongly agreed.

The next two questions received more equally distributed answers. It was easy or very easy for ten participants to observe others in the virtual environment, but five were neutral. Communicating with others was easy or very easy for nine, but difficult for five other participants.

When asked to comment on their positive and negative experiences in an open question, 11 participants provided input focusing on various topics as listed below (the number of participants mentioning them is given in brackets). The positive topics included:

- Rich learning experience (2)
- Good practice for communication/dialog (1)
- Fun experience (1)
- Easy to use technology (1)
- 3D environment resembles real hospital spaces (1)
- Experience causes reflection and learning (1)

The negative topics included the following:

- Difficulties with sound, including identifying who is speaking (4)
- Difficulties with navigation in 3D and view camera (2)
- Annoyance because of delays caused by technical problems (2)
- General difficulty with technology without support personnel (1)
- Difficulties in getting subject information in "game" settings (1)
- Difficult to quickly get into a role (1)

7.5.3 Experiences with Using VR Goggles

All the students participated in the role-play using VR goggles. Two Likert-scale questions were asked to measure participants' perception of presence and physical discomfort when using these (Fig. 7.7). It was clear that using VR goggles lead to increased immersion, but at the same time also physical discomfort was a result, generally accompanied by nausea.

The open question asking to provide comments regarding the use of VR goggles received 14 answers in total. The positive and negative topics are given here, with the number of participants mentioning them given in brackets. Positive feedbacks included:

- Experience is more realistic/easier to immerse in the scenario than on desktop (6)
- Smooth experience overall or in short periods (2)
- Fun experience (1)

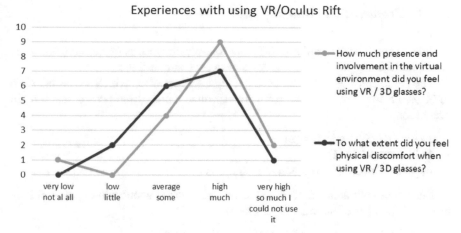

Fig. 7.7 Experience with Oculus Rift: presence versus discomfort

During the interview, the students were asked to compare the VR experience with 'traditional' face-to-face PBL (problem-based learning) sessions. One of the students mentioned that during such sessions "you will need to create those images in your head yourself...It requires more acting skills in a way", while with the VR glasses "you get immersed in (the situation)". Another student noted: "when we had the glasses on, we had really a lot patient contact". The general feedback during the interview was also that it is positive that we (the project team) are early with adopting the VR technology.

Negative feedbacks from the questionnaire included the following:

- Nauseous and dizzy feelings generally, over time, when looking around and when moving in VR (10)
- Technology is unstable and/or crashes (3)
- VR glasses are heavy (1)
- Experience is tiring for eyes (1)
- Difficulty controlling the avatar (1)
- Experience is confusing (1)

Apart from these issues, it was noted during the interviews that the novelty of the technology acted as a disruptive factor: "I think we have been more 'persons' when we used 2D ... Because when we used 3D, there was too much focus on other things".

7.5.4 Preferred Technical Setup

In order to better understand how different role-playing modes functioned, we asked two questions suggesting alternatives and two open questions.

The participants gave the following answers when asked which visual interface they prefer for training interprofessional collaboration (number of answers in brackets):

- Oculus Rift/VR goggles (2)
- PC/Desktop VR (4)
- dependent upon the collaborative situation (10)

Regarding communication, the students were asked about which solution they preferred (number of answers in brackets):

- to be in the same room with other participants (8)
- to sit separately (4)
- dependent upon the collaborative situation (4)

When asked about preferred training interprofessional interaction, the students answered the following (number of answers in brackets):

- in a student group with teacher (7)
- in a student group without a teacher (1)
- dependent upon the collaborative situation (8)

The open questions showed that the students' answers were mainly influenced by technical aspects. Most commented on the discomfort associated with the VR goggles, but that sitting in the same room (which they did when using the VR goggles) was better due to it being easier to hear what the others said. On the other hand, some students commented that sitting apart in separate rooms reduced the amount of small talk. For example, one of the students mentioned during the interviews: "When you sit at separate rooms you know that somebody listens. So you will have to read and think". Another student, preferring sitting in the same room with VR googles on, said, "I got nauseous, but felt to be more present … there was more 'flow'".

When asked about suggestions for technical improvements, especially regarding interprofessional interaction, the students' answers mainly concerned the lack of technical possibilities. Among the things mentioned were (a) making it easier to perform detailed maneuvers such as touching the patient or doing clinical examinations or tests, (b) having a map showing the lay out, (c) automatically getting information like lab results, and (d) more interactive equipment.

7.5.5   Role-Play in the Virtual World

A set of seven Likert-scale questions was asked to collect feedback on the role-playing experience. The first three questions were related to the appearance of different key elements (Fig. 7.8). Most of the students agreed that the inpatient ward gave a good representation of the real one. This was the part of the virtual hospital where most effort had been put into in order to make it as identical as possible to real life in terms of interior, colors, texture etc. The participants were more neutral about the appearance of the operating room. The clothing of all avatars was mostly found appropriate.

Two questions considered the realism of the scenarios. All 16 participants agreed or strongly agreed that the geriatric scenario was realistic, while only five gave such answers to the gynecologic scenario. The rest of the groups were neutral or unsure about the gynecologic scenario. As one of the students stated during the interviews, "Everything I learnt, came out. So for me it was very real". At the same time, the fact that the role-play took place in a virtual environment allowed some experimentation during the patient communication as noted by one of the participants: "In a real situation it would be hard to say (to a patient): "You have fallen before, it is important to prevent falls", but since it was not real, it was possible to say it anyway … You can train on saying things…if you have a patient who is in denial".

The last two questions evaluated how engaged the students felt in the role-play and how fun they thought it was. All the participants agreed or strongly agreed that they felt engaged, while all except one also answered that it was fun to play. One of the factors contributing to the engagement might be the freedom and flexibility the role-play provided, as mentioned during the interview: "I think it was good that we had the time to talk. Because it differs how you are as a nurse and a doctor. Some doctors communicate very little…some nurses chat. So I think it is good that we had some flexibility here. Because it gives more space for how you want it to be".

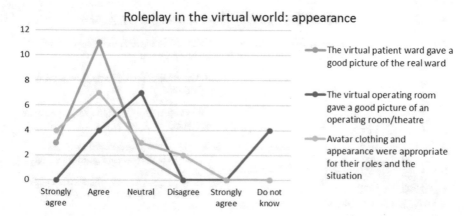

Fig. 7.8   Role-play in the virtual world: appearance

An additional open question was asked to collect suggestions for other similar simulators or scenarios. The suggestions included (a) to generally play more scenarios, (b) play scenarios that are normally not trained in physical reality, (c) play scenarios with difficult patients, and especially (d) play emergency scenarios. Also, during the interviews, some improvement suggestions were given for roleplaying the existing scenarios. For example, it was noted that sometimes knowing one's role was somewhat difficult and it "requires a bit more exercise and warm-up to get into the role". The proposed solutions included "more information to each scene", watching a film about activities at the ward before roleplaying, and spending more time on practicing.

7.5.6   Virtual Simulation and Learning

In a set of eleven Likert-scale questions the participants evaluated how much they learnt playing the scenarios in the simulator.

The majority of the participants agreed or strongly agreed to the general statements related to suitability of Virtual Hospital as a learning platform, with the largest proportion stating that it was relevant to their own study (Fig. 7.9).

The participants showed a similar overall experience when answering more specific subject matter questions. The majority agreed or strongly agreed that the simulation gave them a better understanding of workflow and work distribution in various clinical situations and gave a realistic picture of challenges in interprofessional teams (Fig. 7.9). This is also supported by interview feedbacks where one of the students remarked: "I think it is good to be able to see the whole picture… having done it virtually first and going into practice at the geriatric ward no so long afterwards.…These collaboration meetings…you understand what it is about and you can actually see that this is how this is actually done".

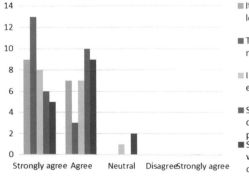

Fig. 7.9 Virtual simulation and learning: general questions

Fig. 7.10 Virtual simulation and learning: increased understanding

The majority of the students answered that they obtained a better understanding of the roles of the other professions involved (Fig. 7.10). In particular, as appears from the interviews, some nursing students felt that the virtual learning environment allowed them to define their roles in relation to the ones of the medical students in a 'safer' way: "People can hide a bit behind their roles and at the same time come with their opinions, be a bit tougher in a way...when you meet medical students, which might feel a bit stressing for us, as they are in a way are our superiors in a hospital...here we meet them a more neutral arena. I think this is a very nice learning base, one can express yourself but also joke a bit". At the same time, most of the participants believed that the virtual simulation could give the students better communication skills among themselves, colleagues, and patients, as well as a better understanding of group processes in interprofessional teams and clinical situations (Fig. 7.10).

Most of the students agreed that virtual simulation should be used in their education. In the interviews, students compared the immersive role-play approach suggested by VirSam and Skills lab with physical mannequins they are already using regularly. One of the students said: "I think this (VirSam) is much more educational/informative than the cases we get at the Skills lab. Where you got a patient that has this and that and you are supposed to roleplay. I think this quickly becomes very unserious, but here I think we managed to keep it pretty well". It was also suggested that VirSam should "target things we cannot do in the Skills lab".

Two final questionnaire questions asked if the simulator should be used in practice. The answers indicate that the majority of the participants agreed and strongly agreed that (a) the simulator can be used as preparation for practice to become familiar with working tasks in advance and (b) it can prepare students for real clinical situations. As one of the students put it during the interview: "It (simulation) kind of 'defuses' the whole thing comparing to going into practice.... You could actually train on this situation...Then going into practice is not going to be as scary as you thought". It was also mentioned that such a preparation would not only be useful for somebody "who has not been so much at the hospital yet", but also "if one has been in this situation before" as roleplaying will allow going through and reflecting on the experience again.

7.6 Discussion

The most prominent finding, given our aim of providing interprofessional communication training in a virtual environment, was that all the involved students agreed that they had learned about the value of clear communication. Communication is one of the core team competencies in medical team training, as highlighted in the TeamSTEPPS framework [37]. It is described as a competency in how to effectively exchange information among team members [38]. As appears from the interviews, the students have reflected on several aspects of communication during the roleplaying sessions, such as own role, communication with the patient [30] and representatives from other health professions. Our work thus supports the notion that VR can be used successfully to make students aware of the need for clear communication but also practice it in a safe environment; however, this study cannot say anything about whether or not the students actually improved their communication skills as a result of the role-play in the VUH. At the same time, most of the students believed that the simulations provided by VUH could improve communication skills and understanding of various interprofessional collaborative processes in clinical settings. Moreover, the majority of students agreed that training provided by the VUH could be successfully integrated in their education and supplement and support existing educational activities. A detailed analysis of the support provided by different components of the VUH is presented below.

7.6.1 Analysis of the Smart Virtual University Hospital Components

As a smart VUH is a part of a smart university and is based on the same major components (such as software systems, technology, hardware, smart curriculum and pedagogy [3]), we will look at how these components support learning in the context of the virtual St. Olav's hospital in the following:

Software systems and technology. The students appreciated the realism of the VUH facilities. Even if there are clear limitations when developing a smart VUH using commercial virtual worlds like SL, the students' evaluation showed that this did not hamper the possibility of making virtual representations that fulfill the expectation of mirroring real life [7, 8]. Due to different standards in the appearance in the virtual hospital (e.g., the inpatient ward which the student found most realistic had been given most resources to be developed) it can be suggested that, although the overall impression is good, it is possible to improve it further. In particular, the major direction for improvement would be more natural interaction with the environment, such as performing examinations on the patient and operating the equipment. Another central aspect is the quality of voice/sound communication in SL as well as other virtual worlds. High quality sound is paramount for team and communication training. This illustrates the need for exploring additional software systems and platforms as well as more optimal ways of integration with hardware components. Also, creating adaptive solutions for presenting students with better and personalized instructions, educational content, scheduling and feedback would alleviate some of the problems we have experienced during the role-play.

Hardware. The most limiting hardware aspect turned out to be the VR goggles [13]. Despite the fact that the majority of the students felt that the use of VR goggles made the experience more realistic and allowed them to be more immersed in the role-play, almost all of them reported 'cyber sickness'. This is a practical problem that must be solved to be able to take full advantage of the higher level of immersion. The 'cyber sickness' is likely to be remedied by using the new consumer versions of the VR goggles, rather than the development kits we used, which includes better tracking solutions. In addition, with the use of more powerful, stationary computers, as opposed to laptops, we might alleviate this. Other types of VR equipment, such as HTC Vive and Oculus Touch, should be tried out in order to investigate to what extent they provide adequate support for capturing hand movements essential for medical manipulations. Different hardware solutions for quality of sound/voice, but also for general communication and collaborative work support in VR, need to be explored. There should also be solutions for adjusting hardware interfaces and different devices/combinations of devices based on personal preferences, working mode (from home/in a lab), and the educational content (such as communication training vs. procedural training).

Smart curricula and pedagogy. The clinical scenarios presented to the students were chosen based on the possibility to explore different patient trajectories. It turned out that it was feasible to condense both longer (weeks) and shorter (hours) patient trajectories. However, there is a clear potential for improvement in how the whole clinical scenario is presented. The two clinical scenarios and the role cards were presented somewhat differently due to discussions on what was likely to function best. The students clearly preferred the geriatric case, and this will therefore be used as the main model for presenting the clinical scenarios.

It was observed that some students did not remember the scenarios themselves and needed help from the other players about what they should do. Even if this mirrors situations that might happen in real life (i.e. that one has to instruct some team members about what to do), part of this confusion was likely due to lack of clarity in preparation and presentation. This is also probably connected to the fact that almost half of the students preferred to perform the training sessions with a teacher from whom they might receive instructions and feedbacks. If the students had been even more systematically exposed to similar training situations, which is what they will encounter in practice, it is likely that this type of problem would not have arisen.

To extend the learning portfolio of VUH, additional cases need to be developed and integrated. These cases will cover different fields of medicine and clinical situations, such as procedural and communication training or both. Therefore, there is a need for a consistent model and template for developing clinical scenarios for specific learning goals. Apart from being medically correct and relevant for the students' teaching plan, the cases should contain some elements of flexibility and adaptability, making it possible for them to be adjusted to the needs of the concrete student groups (such as group composition and number of students representing different professions). Furthermore, there is a need for flexible and personalized solutions in terms of delivering instructions, guidance, and feedbacks before, during, and after the training sessions. For example, this includes student profiles, recommendation systems, and ways of integrating instructions in different interface modes, such as inside VR goggles. Providing tailored and relevant information during all stages of the learning process in VUH requires full integration of smart curricula, software, and hardware elements.

7.6.2 Further Development of the Smart Virtual University Hospital

With respect to the further development of the Smart VUH, we think that the most important lesson has been that it must contain solutions for adaptability and personalization. We build further on the model outlined in Table 7.1 and envision that a future smart VUH should be thought of as a constellation of software and hardware elements, physical spaces, and learners/teachers. It also needs to possess a set of features and characteristics at different smartness levels [3], as outlined in Table 7.2.

7.7 Conclusions and Future Work

Our work shows that a VUH has the possibility to give students more time on task and prepare them for direct patient contact and clinical rotation.

A smart VUH opens several possibilities for health care and medical education. For example, students reported that they found the virtual learning environment to mirror real life and all agreed that they had learned about the necessity of clear and to the point interprofessional communication. Using realistic cases with different team dynamics made it possible to demonstrate the usefulness of a VUH in practicing team communication along different stages of the whole patient trajectory. In addition, using different patient trajectories gave students experience in team communication. Spending relatively few additional resources when it comes to the set up of the virtual hospital facilitated this.

Table 7.2 Important features of a smart virtual university hospital as an arena for smart education and training

Feature	Description	Smartness level
Scenario building	Methodologies for streamlining the process of creating scenarios from medical cases for different learning goals. Intelligent/adaptable or random scenarios to allow for unexpected changes so that playing the same scenario two times will not be the same. Better possibilities and functionalities for user-driven adaptation of scenarios based on their own teams, communication situations, workflow, procedure, skills, etc. including single and multiplayer modes	Adaptation, inferring, self-learning
Personalized learning path	Recommendations of cases and roles to play by using agent technologies, based on the student's study program, preferences and earlier results. Providing individualized information during roleplays and other educational sessions	Adaptation, inferring
Individual feedback	Evaluation facilities in the virtual hospital (e.g., online quizzes). An individual profile/webpage/app with overview of participation, performance, feedback given and received, suggestions for additional training. Personalized feedback during and after simulations	Adaptation, inferring
Personalized interface and interaction	Facilitating using different interfaces, especially VR interfaces, depending on user needs: desktop or VR goggles (ranging from Oculus Rift or HTC Vive to low-cost alternatives on mobile phones). Increasing the realism of interaction with the environment, other characters and equipment, e.g. through motion capturing devices and virtual patients/agents. Facilities for communication and collaborative work, workspace awareness, also in VR	Adaptation, sensing
Virtual agents	Automatic, autonomous and 'intelligent' avatars/virtual patients as well as recommendations systems. Smart virtual medical equipment, e.g. ultrasound images of a virtual patient that are connected to motion capturing devices (such as changing display of images due to change in player's movement)	Adaptation, self-learning

(continued)

Table 7.2 (continued)

Feature	Description	Smartness level
Modeling of virtual and physical environment	Developing virtual hospital environment based on BIM (Building Information Modeling) models of the real hospital when similarity to a physical environment is important, e.g. pre-practice training. Real-time interaction between BIM-database and the virtual environment, i.e. using view-point and direction to extract what you need and not the whole BIM hospital model	Sensing, self-organization
Mobile, ubiquitous and augmented	Linking materials and activities that are accessible through online and virtual system into the physical world, using location services, mobile and wearable devices, augmented reality, and internet of things	Sensing, inferring
General hardware support	Easy plug-and-play support for new hardware devices. Dedicated areas (VUH labs) where hardware is installed and serviced to allow use of more sophisticated hardware equipment	Adaptation
Integration: portal, scheduling and run-time system	Integration of software, hardware and curriculum into a holistic and coherent set of facilities. Web and app-based portal containing all students need to know in order to participate in training, with individual student profiles, fully student/participant driven including scheduling of sessions. Adaptable and scalable run-time system, adjusting to different number of concurrent users, i.e. by allowing several role-plays in the constellation of facilities including virtual environments and physical spaces/labs. Facilitating self-organization of student groups and personalized flexible scheduling of learning sessions	Adaptation, anticipation
Science system	Automatically extracting information regarding what is working and what is not based on the scenarios played and the feedback given afterwards. Covers both technical features such as type of equipment used and subject-oriented issues such as type of scenarios and type of participants	Self-learning, self-organization, anticipation

One of the assets of role-play in a VUH is the possibility for repetitions and the flexibility allowing the learning environment to be used at any time, making it independent of scheduled teaching. The role-playing facilities in the VUH also provide the students with a safe and accessible environment for practicing some of their tasks. In addition, role-play with VR allows a greater degree of immersion into the situation, something that is especially beneficial in a problem-based learning situation. Based on our experiences and literature review, we can outline three major directions for future work: developing *smart educational content, technology, and an overall VUH framework.*

VUH smart educational content. The major challenge of using smart university hospitals for education is in making realistic cases that mirror what the student will encounter in practice. The cases also need to focus on various aspects of medical team training. This is something that is not straightforward given the existing ambiguity in the literature on medical team training [31, 32]. Therefore, there is a need for guidelines and methods for scenario development, including templates for new cases to be efficiently developed by health care professionals, templates for role descriptions, etc. There is also a need for solutions for personalization and adaptation of cases as well as other educational materials for different students and learning contexts.

VUH smart technology. The immediate technological challenges for facilitating communication and collaboration team training in a VUH include physical discomfort caused by VR goggles and the need for better interactivity. The latter particularly applies to capturing hand movement and supporting manipulations on patients and equipment in the virtual learning environment. Apart from supporting training on procedures, such manipulations will create awareness of the learners' activities in a joint workspace. Supporting such awareness and collaborative activities, in general, in a fully immersive virtual environment constitutes yet another challenge. Such solutions are already emerging and, as the VR technology develops further, we believe that these challenges will be fully addressed in the near future by introducing new hardware devices and software platforms. In addition to the VR technologies, other technologies should be taken into account, such as mobile and ubiquitous technologies and augmented reality. This will also open up another dimension of smartness [4]. Additional software elements should include cross-platform solutions linking web, social media, mobile devices, and new software VR platforms, as outlined in Table 7.2.

VUH framework. A fully functioning smart VUH needs to support a wide range of educational activities, as outlined in Table 7.1. This requires integration of smart educational content, curricula, and organizational aspects with smart hardware and software solutions into a holistic and coherent system. An important direction for future work will, therefore, be developing the framework, principles, and methods further for such integration. This is in line with a similar framework for smart universities.

We are currently working on further development of Virtual St. Olav's hospital. We have received an additional grant from the Norwegian Research Council as a part of the FINNUT program (Research and Innovation in Educational Sector). Apart from creating more realistic hospital models, we are exploring new VR interfaces with better support for manual operations (with HTC Vive) and developing a portal (http://virsam.no). Future evaluations will include larger groups of students to evaluate suitability of the VUH for large-scale student training. We plan to extend the portfolio of cases to include additional clinics at St. Olav's hospital, such as oncology, palliative, and especially emergency/intensive care. We will continue developing the Smart VUH of St. Olav from the conceptual framework aspect as well as the technological realization. We plan additional features such as more interactivity with medical equipment and patients (including virtual patients), additional locations (based on the existing plans and models for the physical hospital), and facilities within the virtual hospital to accommodate new cases, anatomical visualizations, and various VR interfaces (including mobile ones).

Acknowledgements The VirSam project has been funded by NTNU Toppundervisning. We would like to thank the students participating in our evaluation, as well as Marion Skallerud Nordberg, Inga Marie Røyset and Cecilie Therese Hagemann who contributed to the case development.

References

1. IBM: Smarter Education: Building the Foundations of Economic Success, pp. 1–4 (2012). ftp://ftp.software.ibm.com/la/documents/gb/mx/Smarter_Education.pdf
2. Tikhomirov, V., Dneprovskaya, N.: Development of strategy for smart University. Paper presented at the Open Education Global international conference, Banff, Canada (2015)
3. Uskov, V.L., Bakken, J.P., Pandey, A., Singh, U., Yalamanchili, M., Penumatsa, A.: Smart University taxonomy: features, components, systems. In: Uskov, L.V., Howlett, J.R., Jain, C. L. (eds.) Smart Education and e-Learning 2016, pp. 3–14. Springer International Publishing, Cham (2016)
4. Hwang, G.-J.: Definition, framework and research issues of smart learning environments—a context-aware ubiquitous learning perspective. Smart Learn. Environ. **1**(1), 1–14 (2014)
5. Kleven, N.F., Prasolova-Førland, E., Fominykh, M., Hansen, A., Rasmussen, G., Sagberg, L. M., Lindseth, F.: Training nurses and educating the public using a virtual operating room with Oculus Rift. In: Thwaites, H., Kenderdine, S., Shaw, J. (eds.) International Conference on Virtual Systems and Multimedia (VSMM), Hong Kong, December 9–12, pp. 206–213. IEEE, New York, NY (2014)
6. Kleven, N.F., Prasolova-Førland, E., Fominykh, M., Hansen, A., Rasmussen, G., Sagberg, L. M., Lindseth, F.: Virtual operating room for collaborative training of surgical nurses. In: Baloian, N., Burstein, F., Ogata, H., Santoro, F., Zurita, G. (eds.) 20th International Conference on Collaboration and Technology (CRIWG), Santiago, Chile, September 7–10. Lecture Notes in Computer Science, pp. 223–238. Springer, Berlin, Heidelberg (2014)
7. Warburton, S.: Second Life in higher education: assessing the potential for and the barriers to deploying virtual worlds in learning and teaching. Br. J. Edu. Technol. **40**(3), 414–426 (2009)
8. Mckerlich, R., Riis, M., Anderson, T., Eastman, B.: Student perceptions of teaching presence, social presence, and cognitive presence in a virtual world. J. Online Learn. Teach. **7**(3), 324–336 (2011)
9. Spooner, N.A., Cregan, P.C., Khadra, M.: Second Life for Medical Education. eLearn Magazine. ACM, New York (2011)
10. de Freitas, S., Rebolledo-Mendez, G., Liarokapis, F., Magoulas, G., Poulovassilis, A.: Developing an evaluation methodology for immersive learning experiences in a virtual world. In: 1st International Conference in Games and Virtual Worlds for Serious Applications (VS-GAMES), Coventry, UK, March 23–24, pp. 43–50. IEEE, New York (2009)
11. Cakmakci, O., Rolland, J.: Head-worn displays: a review. J. Disp. Technol. **2**(3), 199–216 (2006)
12. van Krevelen, D.W.F., Poelman, R.: A survey of augmented reality technologies, applications and limitations. Int. J. Virtual Real. **9**(2), 1–20 (2010)

13. Antonov, M., Mitchell, N., Reisse, A., Cooper, L., LaValle, S., Katsev, M.: Oculus Software Development Kit. Oculus VR Inc., CA, USA (2013)
14. Rubia, E., Diaz-Estrella, A.: ORION: one more step in virtual reality interaction. In: Penichet, V.M.R., Peñalver, A., Gallud, J.A. (eds.) New Trends in Interaction, Virtual Reality and Modeling. Human–Computer Interaction Series, pp. 45–61. Springer, London (2013)
15. Johnson, C.M., Vorderstrasse, A.A., Shaw, R.: Virtual worlds in health care higher education. J. Virtual Worlds Res. **2**(2), 3–12 (2009)
16. Rogers, L.: Simulating clinical experience: Exploring Second Life as a learning tool for nurse education. In: Atkinson, R.J., McBeath, C. (eds.) 26th Annual Ascilite International Conference Same Places, Different Spaces, Auckland, New Zealand, December 6–9, pp. 883–887. Australasian Society for Computers in Learning in Tertiary Education, Auckland, New Zealand (2009)
17. Khanal, P., Gupta, A., Smith, M.: Virtual Worlds in Healthcare. In: Gupta, A., Patel, L.V., Greenes, A.R. (eds.) Advances in Healthcare Informatics and Analytics, pp. 233–248. Springer International Publishing, Cham (2016)
18. Lowes, S., Hamilton, G., Hochstetler, V., Paek, S.: Teaching communication skills to medical students in a virtual world. J. Interact. Technol. Pedag. (3), e1 (2013)
19. Jang, S., Black, J.B., Jyung, R.W.: Embodied cognition and virtual reality in learning to visualize anatomy. In: Ohlsson, S., Catrambone, R. (eds.) 32nd Annual Conference of the Cognitive Science Society, Portland, OR, August 12–14, pp. 2326–2331. Cognitive Science Society, Austin, TX (2010)
20. Huang, H.-M., Liaw, S.-S., Lai, C.-M.: Exploring learner acceptance of the use of virtual reality in medical education: a case study of desktop and projection-based display systems. Interact. Learn. Environ. **24**(1), 3–19 (2016)
21. Wiecha, J., Heyden, R., Sternthal, E., Merialdi, M.: Learning in a virtual world: experience with using second life for medical education. J. Med. Internet Res. **12**(1), e1 (2010)
22. Cates, C.U., Lönn, L., Gallagher, A.G.: Prospective, randomised and blinded comparison of proficiency-based progression full-physics virtual reality simulator training versus invasive vascular experience for learning carotid artery angiography by very experienced operators. BMJ Simul. Technol. Enhanc. Learn., 1–5 (2016)
23. Ruthenbeck, S.G., Reynolds, J.K.: Virtual reality for medical training: the state-of-the-art. J. Simul. **9**(1), 16–26 (2015)
24. Lee, M.J.W.: How can 3d virtual worlds be used to support collaborative learning? an analysis of cases from the literature. Society **5**(1), 149–158 (2009)
25. Hayes, E.R.: Situated learning in virtual worlds: the learning ecology of second life. In: American Educational Research Association Conference, pp. 154–159. AERA, Washington, DC (2006)
26. Jarmon, L., Traphagan, T., Mayrath, M.: Understanding project-based learning in second life with a pedagogy, training, and assessment trio. Educ. Media Int. **45**(3), 157–176 (2008)
27. Alverson, D.C., Caudell, T.P., Goldsmith, T.E.: Creating virtual reality medical simulations: a knowledge-based design and assessment approach. In: Riley RH (ed.) Manual of Simulation in Healthcare (2 ed.). Oxford University Press, Oxford, UK (2015)
28. Rizzo, A.S., Difede, J., Rothbaum, B.O., Reger, G., Spitalnick, J., Cukor, J., McLay, R.: Development and early evaluation of the Virtual Iraq/Afghanistan exposure therapy system for combat-related PTSD. Ann. N. Y. Acad. Sci. **1208**(1), 114–125 (2010)
29. Li, A., Montano, Z., Chen, V.J., Gold, J.I.: Virtual reality and pain management: current trends and future directions. Pain Manag. **1**(2), 147–157 (2011)
30. Pan, X., Slater, M., Beacco, A., Navarro, X., Bellido Rivas, A.I., Swapp, D., Hale, J., Forbes, P.A., Denvir, C., Hamilton, A.F., Delacroix, S.: The responses of medical general practitioners to unreasonable patient demand for antibiotics—a study of medical ethics using immersive virtual reality. PloS one **11**(2), 1–15 (2016)
31. Morrison, G., Goldfarb, S., Lanken, P.N.: Team training of medical students in the 21st century: would Flexner approve? Acad. Med. **85**(2), 254–259 (2010)

32. Weaver, S.J., Dy, S.M., Rosen, M.A.: Team-training in healthcare: a narrative synthesis of the literature. BMJ Quality Saf., 1–14 (2014)
33. Weaver, S.J., Lyons, R., DiazGranados, D., et al.: The anatomy of health care team training and the state of practice: a critical review. Acad. Med. **85**(11), 1746–1760 (2010)
34. Baker, D.P., Gustafson, S., Beaubien, J.M., Salas, E., Barach, P.: Medical team training programs in health care. In: Henriksen, K., Battles, J.B., Marks, E.S., Lewin, D.I. (eds.) Advances in Patient Safety: From Research to Implementation (Vol. 4: Programs, Tools, and Products). Advances in Patient Safety. Rockville (MD) (2005)
35. Flaherty, E., Hyer, K., Kane, R., Wilson, N., Whitelaw, N., Fulmer, T.: Using case studies to evaluate students' ability to develop a geriatric interdisciplinary care plan. Gerontol. Geriatr. Educ. **24**(2), 63–74 (2004)
36. Sehgal, N.L., Fox, M., Vidyarthi, A.R., Sharpe, B.A., Gearhart, S., Bookwalter, T., Barker, J., Alldredge, B.K., Blegen, M.A., Wachter, R.M.: Triad for Optimal Patient Safety P.: a multidisciplinary teamwork training program: the Triad for Optimal Patient Safety (TOPS) experience. J. Gen. Intern. Med. **23**(12), 2053–2057 (2008)
37. King, H.B., Battles, J., Baker, D.P., Alonso, A., Salas, E., Webster, J., Toomey, L., Salisbury, M.: TeamSTEPPS: team strategies and tools to enhance performance and patient safety. In: Henriksen, K., Battles, J.B., Keyes, M.A., Grady, M.L. (eds.) Advances in Patient Safety: New Directions and Alternative Approaches (Vol. 3: Performance and Tools). Advances in Patient Safety, Rockville (MD) (2008)
38. AHRQ: Pocket Guide: TeamSTEPPS. Team strategies and tools to enhance performance and patient safety. Agency for Healthcare Research and Quality, Rockville, MD (2013)
39. Albanese, M.A.: Problem-based learning. In: Understanding Medical Education, pp. 37–52. Wiley-Blackwell (2010)
40. Nonaka, I., Takeuchi, H.: The knowledge-creating company: how Japanese companies create the dynamics of innovation. Oxford University Press, New York (1995)
41. White, C.B., Gruppen, L.D.: Self-regulated learning in medical education. In: Understanding Medical Education, pp. 271–282. Wiley-Blackwell (2010)
42. Cooke, M., Irby, D.M., Sullivan, W., Ludmerer, K.M.: American medical education 100 years after the Flexner report. N. Engl. J. Med. **355**(13), 1339–1344 (2006)
43. Lave, J., Wenger, E.: Situated Learning: Legitimate Peripheral Participation. Cambridge University Press, Cambridge, UK (1991)
44. Engeström, Y.: Activity theory and individual and social transformation. In: Engeström, Y., Miettinen, R., Punamäki, R.-L. (eds.) Perspectives on Activity Theory. Learning in Doing: Social, Cognitive and Computational Perspectives, pp. 19–38. Cambridge University Press, Cambridge, UK (1999)
45. Brown, J.S., Duguid, P.: Organizational learning and communities of practice: towards a unified view of working, learning, and innovation. Organ. Sci. **2**(1), 40–57 (1991)
46. Lawson A.E., Drake, N., Johnson, J., Kwon, Y.-J., Scarpone, C.: How good are students at testing alternative explanations of unseen entities? The American Biology Teacher **62**(4), 249–255 (2000)
47. Korakakis, G., Boudouvis, A., Palyvos, J., Pavlatou, E.A.: The impact of 3D visualization types in instructional multimedia applications for teaching science. Procedia Soc. Behav. Sci. **31**, 145–149 (2012)

Chapter 8
EdLeTS: Towards Smartness in Math Education

Ilya S. Turuntaev

Abstract The emerging field of smart education offers major improvements in education and has the potential to change the very face of a traditional classroom. The application of smart technologies in education has been the primary objective of many research projects over recent years. It is commonly accepted that smart educational technologies are able to boost effectiveness and efficiency of learning. In this paper, the problem of introducing smart technologies in mathematical education is considered. Due to the specificity of the area of math education, some particular requirements should be met by mathematical educational systems. This paper provides a discussion of the problems arising in mathematical educational systems development and presents the project of an educational system for mathematical practical classes support aimed to resolve these problems.

Keywords Smart educational technologies · Mathematical education · Computer algebra systems · Learning spaces · Educational systems

8.1 Introduction

The idea of using information and communication technologies (ICT) in education has a long history. There were a great number of attempts to develop technologically enhanced educational systems. Some of these attempts were quite successful while some were not. Many researchers indicate that nowadays we are systematically moving towards the era of smart technologies. The notion of "smartness" itself is a debatable one. There are continuous arguments regarding what technologies and development projects can be considered smart. Nonetheless, many researchers agree when it comes to what key features should be implemented in smart technologies.

The notion of smart technologies was only recently applied to the field of education. Such concepts as Smart Education, Smart Classroom, and Smart

I.S. Turuntaev (✉)
National Research University Higher School of Economics, Moscow, Russia
e-mail: isturunt@gmail.com

© Springer International Publishing AG 2018
V.L. Uskov et al. (eds.), *Smart Universities*, Smart Innovation,
Systems and Technologies 70, DOI 10.1007/978-3-319-59454-5_8

University started to replace older approaches to electronic education. The new era of educational technologies came with the general idea that a learning environment should not just be filled with various types of ICT. Rather, it should be supported and enhanced with specific smart technologies in order to make education more effective and more efficient. Multiple research projects are dedicated to smart educational technologies as well as other relevant topics. In Sect. 2 of this paper, an overview of some significant publications covering these topics is given.

In this paper, the problem of developing smart educational technologies in the field of exact science, particularly in mathematics, is considered. Due to its specificity, this area presents some unique requirements to educational systems from both technical and conceptual perspectives. This challenge, as well as the growing demand for smart educational technologies, inspired the development project of the interactive educational system EdLeTS presented in this paper. EdLeTS is a web-based educational system focused on supporting practical mathematical classes. The whole project is developed under two general convictions. First, we state that training tasks play an essential role in mathematical education. Through training on specially developed tasks, a student learns to recognize common patterns in various real-word problems. He also learns how to apply his knowledge in order to obtain the final result. It should be noted that the tasks studied during various mathematical courses arise in the future of most students in different technical specialties in one way or another. This is why solving mathematical tasks is a natural activity for the students of technical specialties. Second, we state that technologies cannot and should not try to completely replace teachers or educational techniques that have proved to work well throughout the long history of mathematical education. Rather, they should support the educational process in order to make it more effective and efficient.

A new approach to training task organization is presented in this paper. Namely, the technique of templating training tasks is described. We discuss what benefits are offered by this approach and how it meets some key requirements of Smart Educational Technologies. Further, the use of computer algebra systems (CAS) in mathematical education is considered. Despite a number of advantages that a CAS can offer to the educational process, there are significant drawbacks that prevent many teachers from using such systems in their classrooms. These problems are addressed by the EdLeTS project. Thus, this paper provides an overview of the methods used in EdLeTS, providing a way to avoid the issues of using CAS as an educational technology.

One of the essential features of the presented development project is the ability to automatically check answers. In EdLeTS, due to the two general convictions mentioned above, the traditional form of a task answer is preserved, e.g., all answers are given in a form of mathematical expression or in mathematical notation. Under such conditions, the problem of automatic answer checking becomes formally unsolvable. In Sect. 8.4.4, this problem is discussed and a way to avoid its consequences is presented.

In Sect. 8.4.5, the mathematical theory of Learning Spaces is considered. A relationship between the key concepts of EdLeTS and the theory of Learning Spaces is established and the problem of task structuring is considered. The notion of the precedence relation on training tasks is introduced. This notion is further used to describe the idea of developing interactive learning suggestions and learning path construction. We also refer to interactive learning suggestions as "fill-the-gap suggestions" throughout this paper.

8.2 Literature Review

Many researchers [1–3] indicate there is an extremely fast growth of information flow over the last several decades. The level of development of information and communication technologies (ICT), as well as, particularly, the Internet, offers new facilities for knowledge gaining, creation, sharing, and spreading. Tikhomirov states that, under such circumstances, it is essential for universities to provide students with access to the newest knowledge and technologies as well as to adjust educational techniques in order to meet current demands for education [2]. When applied to education, smart technologies are aimed to respond to this challenge.

According to [4, 5], the term 'smart', referring to various types of technologies, did not appear until recently. Coccoli et al. [4] state that "we have entered the smart-*something* era". This is now replacing the era of "2.0" technologies, which, in turn, have partially replaced the "e-"-technologies. Each of these replacements is associated with significant improvements in the appropriate technologies and/or appearance of new requirements for further development. The prefix 'smart' brings into account the ideas of ease of use, flexibility, mobility, and reliability [4].

The application of smart technologies to education has led to the emergence of concepts such as Smart Education, Smart University (SmU), Smart Classroom, Smart Learning Environment (SLE), etc. This area has attracted much attention from researchers in recent years [3, 5–7]; however, although not surprisingly, there is no strict definition for these terms. R. Koper [7] sees a SLE as a learning environment that is "context-aware and adaptive to the individual learner's behavior". He also indicates that the aim of SLE development is to make learning better and faster, but simply concentrating on technical aspects does not lead to the actual goal. Similarly, Hwang [6] defines a SLE as "the technology-supported learning environments that make adaptations and provide appropriate support... in the right places and at the right time based on individual learners' needs...". To answer the question 'What makes a learning environment smart?', J. Spector comes to a conclusion that SLE should support planning and innovative alternatives, and it "might include features to promote engagement, effectiveness and efficiency" [8]. In addition, Zhu et al. [3], with reference to [9], indicate features of a SLU such as personalization and the ability of students to self-learn. Most of these researchers outline the necessity of ubiquitous learning and the ability of smart learning systems to provide support and guidance at the right place and at the right time.

From this overview, we can conclude that a SLE is based on a learner-centric paradigm. It is generally accepted that SLE should be able to collect and analyze the learners' personal information such as current learning status, learning progress, classes taken, etc., the environmental information, etc., and to adapt its behavior, content, and interface according to the performed analysis in order to provide the best guidance and support for each learner personally and to help teachers by presenting a clear image of their pupils.

8.2.1 Smart Learning

Probably, the most general and complex concept regarding an application of the smart ICT to educational area is the notion of Smart Learning. There are continuous discussions regarding the concept of Smart Learning, but there is still no clear definition. Many researchers [3, 9] agree that Smart Learning can be regarded as a technologically enhanced learning where the focus should be moved from technologies to learners. Kim et al. [9] regard Smart Learning as a paradigm which combines u-Learning (i.e., ubiquitous learning) and Social Learning. Social Learning is characterized by the use of social network services and smart devices for educational purpose. It is stated in [9] as well that Smart Learning is focused on humans and content rather than on technologies and devices. G.-J. Hwang describes Smart Learning as a qualitatively new concept that combines features of context-aware ubiquitous learning and adaptive learning. It also offers some new features that cannot be found in either of the latter two concepts [6]. In particular, these new features include learning tasks adaptation and complex analysis of multiple personal and environmental factors. In various publications, an equivalent concept is considered, but namely the Smart Education. Coccoli et al. define Smart Education as an "education in a smart environment supported by smart technologies, making use of smart tools and smart devices" [4].

The idea of Smart Learning goes much further than the pure application of ICT to educational process even though this is typical for the "e"-era. It is characterized by close attention to learners' personal parameters. Zhu et al. state that technology itself, though essential for Smart Learning, should not be placed in the center of this paradigm [3]. An important remark on the role of ICT in education can be found in [10]: "technologies should not support learning by attempting to instruct the learners, but rather should be used as knowledge construction and representation tools that students learn *with*, not *from*".

There are a number of publications that consider the application of ICT specifically in mathematics education. There are multiple reasons that this area receives such particular attention. First of all, unlike many other disciplines, mathematics is a quite abstract one. Many students suffer from the lack of visual representation of mathematical objects and, consequently, it is difficult for them to build the concepts being studied in their inner knowledge structure [11]. From this perspective, ICT offers a wide range of visualization tools where visual representations can be made dynamic as well as interactive. A. Rubin [11] considered this to be one of several most powerful features that can enhance mathematical education. Second of all, there is a historical factor. For a long time, computers were mostly used by mathematicians and mathematically-equipped researchers. On one hand, this led to the situation where we are "looking at math education from the perspective of the computer" [11]. In the same paper, A. Rubin insists that we should rather "look at computers from the perspective of mathematics education". On the other hand, though, there is a great number of mathematical applications nowadays that provide various features that may help to improve math education.

8.2.2   Computer Algebra Systems in Math Education

Probably the most promising and popular applications that can be successfully used in mathematical education are Computer Algebra Systems (CAS). A CAS is characterized by its main feature. This is the ability to manipulate mathematical expressions given in symbolic form (i.e., to provide symbolic calculations) [12]. Computer algebra systems are widely used among researchers of various specializations; also, they are often used for educational purposes [13, 14]. Drijvers [14] indicates the following benefits of using CAS in math education:

- *Horizontal mathematization.* This refers to the translation of real-world situations into mathematical problems and vice versa. CAS allows the release of students from pure algorithmic operation so they could concentrate on presenting the problems in mathematical notation and interpreting results.
- *Exploration and vertical mathematization* (manipulations with highly abstract mathematical objects). Learners may reinvent properties or theorems by analyzing results of special discovery and classification tasks.
- *Flexible integration of different representations.* Most CAS allow users to easily switch between different representations of an object (e.g., formulae, graph, table, etc.)

In the same paper, Drijvers describes some possible issues of using CAS in a classroom. They are:

- *Top-down tool.* CAS may appear to be too powerful for educational purposes. Unless appropriate measures are taken, it could potentially damage students' motivation for learning.
- *Black box.* CAS acts like a black box because it hides methods and algorithms used to obtain final results. Not only does it give no explanation of the solution, but also the results may have unexpected representation since the inside algorithms are always much more complex and general than the methods students generally use.
- *Idiosyncrasy.* Each CAS uses its own special language. This is always quite complex and requires additional study. Such a language usually differs from the natural language and mathematical notation; a set of rules and constraints has to be kept.

Lavicza reveals some issues, though [13, 15]. He explores the problems of integration of CAS into mathematics curricula. Among other things, he indicates that many math teachers are not going to use CAS in their teaching practice. This is mainly because it requires general course restructuring and many instructors do not think it is worth it. Lavicza also points out that there is a problem of extra CAS-education requirement and the problem of assessment. It is difficult to assess what students really know if they use automated algorithms [15]. Additionally, Rubin [11] reveals risks of inappropriate use: students need to learn to carry out some relatively simple operations by hand before using CAS or any other mathematical application for support.

8.2.3 Learning Spaces

As it was proposed earlier, many researchers indicate the importance of development of learners' automated support based on their current learning status in smart educational systems. There are different approaches when it comes to defining the concept of learning status, recognizing it, and operating it. One interesting approach to this problem is the theory of learning spaces that was originally introduced by Doignon and Falmagne [16, 17].

The Learning Spaces theory suggests a mathematical model for learning processes. Consider a field of knowledge. A set of questions in this field, each having a correct response, is called a 'domain'. The theory operates mathematical structures over such domains. A 'knowledge state' of a student is defined as a subset of questions within the given domain that he masters. In other words, we consider a student to be in a certain knowledge state if he is capable of answering each item in this state correctly. The concept of knowledge state is the key concept of the learning spaces theory. It provides a formal representation for the statuses of the learners in the suggested knowledge field. A set of all possible knowledge states of some domain forms a 'knowledge structure' over this domain. Formally, a knowledge structure is a pair $(Q, \mathcal{K})$ in which Q is a domain and $\mathcal{K}$ is a family of subsets of Q (i.e., a set of knowledge states) such that it contains at least Q and $\emptyset$. A knowledge structure that is closed under a union is called 'knowledge space'.

A knowledge structure is a strongly defined mathematical object that represents a knowledge field and learners' states in this field; however, it appears to be a little bit poor for actual use. Consider a student, say John, is in some certain state $K \neq Q$ of a knowledge structure $(Q, \mathcal{K})$. This means that John can successfully answer each question in K but there are still some questions in Q that he does not master. One day, John picks an item q from Q (an item he hadn't adopted yet) and studies the appropriate material. Finally, he takes some examination to verify that he has actually mastered q. Which knowledge state should be assigned to John? There is absolutely no guarantee that K contains the state $K \cup \{q\}$, i.e. a state which is obtained by adding a new question (which John has just learned) to the original state K, which makes perfect sense in this case. The authors of the Learning Spaces theory introduce two special conditions that improve the concept of knowledge structure.

- *Learning Smoothness.* For any two states $K, L \in \mathcal{K}$, such that $K \subset L$, there is a finite chain of states $K = K_0 \subset K_1 \subset \ldots \subset K_p = L$, such that[1] $|K_{i+1} \backslash K_i| = 1$ for $0 \leq i < p$. In other words, this condition means that if one state (K) is included into another (L) then the more complex one (L) can be obtained from the simplest one by learning one item at a time.
- *Learning Consistency.* For any two states, $K, L \in \mathcal{K}$, such that $K \subset L$, and, for any item $q \in Q$ if $K \cup \{q\} \in \mathcal{K}$, then $L \cup \{q\} \in \mathcal{K}$. Doignon and Falmagne describe this condition as follows: "Learning more does not prevent learning something new" [16].

A knowledge structure satisfying these two conditions is called a 'learning space'. The notion of learning spaces makes more sense from a pedagogical point of view than from a plain knowledge structure. There are several equivalent definitions for this mathematical object. Its properties are well researched and numerous algorithms for building such a structure and uncovering learners' knowledge states are known. These topics are not considered in this paper, but for a more precise overview of the Learning Spaces theory see [16, 17].

8.3   Project Goal and Objectives

In this paper, a development project of the interactive educational system for mathematical classes support, EdLeTS, is presented. The main goal of this ongoing development is to build a smart educational system that meets the requirements of Smart Educational Technologies. EdLeTS is focused on mathematical training tasks generation, delivery and support, and practical mathematical classes, in general, aiming to address various smartness levels of a smart university, namely: adaptation, self-learning, and inferring. EdLeTS is aimed to address these features in order to become a smart educational tool that can be easily integrated into mathematical educational process in universities as a part of smart environment development. The following general features are addressed in this project:

1. *Personalization.* Among the features of smart educational technologies, one of the most significant is personalization; it also tends to be one of the most challenging. Personalization features require a system to adapt to various types of students in order to meet their individual requirements. This means it represents some key aspects of the adaptation level in smart universities. From the mathematical practical classes perspective, there are several directions of

---

[1]$|K|$ denotes the cardinality of set K.

personalization. First of all, there is a problem of personalizing training tasks. It is generally accepted that, for each student, the quality of learning depends on a number of personalized parameters, including both static and dynamic. For example, when working on some technique, each student needs an individual amount of tasks. If extrapolated to the scale of a complex topic, this leads to the personalized training plans requirement. Another important personalization problem is the problem of personalized support. In many cases, it is essential that students receive individual help that takes into consideration their personal errors and provides an explanatory review of the exact tasks in which problems occurred. Training tasks personalization and personalized support are among the key objectives of the proposed project. This requirement follows the general idea of learner-centrism which forms the kernel of the concept of Smart Education.

2. *Adaptiveness*. EdLeTS is intended to help the user find the kernel of his mistakes if there are any. For that purpose, it provides the user with a list of training tasks he might need to work on. This list is based on the analysis of the user's mistakes and the overall structure of training tasks (i.e., learning status). It is updated constantly.

3. *Flexibility and ease of use*. It is important that the proposed educational system meets the following flexibility requirements:

   a   The training task development tool should provide a wide range of instruments in order to meet requirements that teachers could possibly produce.
   b. The training tasks structuring features must be provided.
   c. The system should be malleable enough to the point that it can easily be integrated into any mathematical curriculum with possible cooperation with external educational systems.
   d. Customizable assessment tools must be present.

4. *Mobility, self-learning, and ubiquitous learning*. A primary intention of the performed development project is to provide students with the ability to train anywhere and anytime, using only a laptop, tablet, or smartphone. A possible extension of this idea may include context-awareness and runtime-generated training tasks based on environmental objects. Also, a special self-training mode must be introduced. On one hand, such a mode allows students to explore new techniques while, on the other hand, it allows them to train freely and not be focused on the formal assessment system. The latter can be a serious psychological obstacle.

5. *CAS*. From the perspective of mathematical education, computer algebra systems offer a wide range of algorithms of symbolic and numeric calculations required by educational systems; however, as it was stated in the CAS section of the literature review, there are a number of obstacles that prevent one from using CAS directly. One of the fundamental objectives of the suggested project is to develop an approach that, on one hand, allows the user to utilize the desired capabilities of CAS. On the other hand, it provides reasonable limitations and supplementations that eliminate the weak spots of CAS integration. EdLeTS

aims to be as powerful as CAS within its application to the education area. Therefore, CAS is used here as a computation module. There is no crucial difference between any two computer algebra systems when it comes to meeting some definite requirements. Integration of CAS into EdLeTS is associated with the development of a layer between CAS interface and the EdLeTS, which provides access to CAS features in training task declarations. Particularly, this layer should provide the following features:

a. restriction rules that allow the user to determine a list of forbidden functionality per task;
b. special language for accessing computation facilities;
c. supplemental algorithms that reveal the inside of methods and algorithms used to obtain result

6. *Inferring and the Theory of Learning Spaces.* The theory of learning spaces provides a mathematical formalization for the concepts used in educational processes. This, in turn, helps create strict inferences and introduce algorithms of proven complexity regarding knowledge fields. Specifically, this allows the building of automatic algorithms for students' learning state recognition, providing suggestions and support, etc. based on their responses. A part of this research is dedicated to establishing a relationship between the theory of learning spaces and the concepts used in EdLeTS.

8.4   Methods and Technologies Used in the Development Project

EdLeTS is a web-based educational system for practical classes support in the field of exact science. The idea of such a system was originally introduced by Dr. V. G. Danilov. Within his pedagogical practice, he observed that the process of training tasks generation and assessment could be highly automatized. This idea formed the basis of the concept of interactive educational system, which provides tools for training tasks generation and automatic answer checking. The development of such a system has continued ever since. In this paper, our improved vision of this system implemented in the EdLeTS project is presented.

The key object of mathematics practical classes is a training task. By training on concrete examples, students learn, in the first place, to reveal theoretical concepts in problems and apply their knowledge to real problems. It is essential that students walk all the way through the problem solution up to the very end. This is where they retrieve the final result. The underlying statement of the whole project is that traditional mathematical training tasks have proven well in mathematics education and they must be preserved. Thus, the form of multivariate tests in this area is considered inappropriate in this paper.

8.4.1 Training Task Templates

Traditionally, a training task[2] has a problem formulation, which may or may not include mathematical objects, as well as an answer that, in this paper, is considered to be a mathematical expression.[3] Any training task serves the need of training some technique, or multiple techniques, providing an artificial problem to which these techniques may be applied. It may be observed that a set of training tasks dedicated to the same topic often preserves similar structure. Consider a chain rule of differential calculus usually taught in the early stages of the Mathematical Analysis course. This rule describes the technique of calculating the derivative of a composition of functions. The simplest training tasks for this technique may all have the following structure:

Calculate the derivative of $f(g(x))$,

where f and g are some concrete functions (e.g., $\sin(x)$, $\exp(x)$, $\tan(x)$, etc.). In order to increase the complexity degree of such tasks, one may increase the number of compositions and/or provide more convoluted functions. This is an example of a basic template that describes a general structure for training tasks on the chain rule of differential calculus. Using this example, one may write out a great number of concrete training tasks. Many of these can be found in various schoolbooks on Mathematical Analysis. This case illustrates the idea of *training task templates*.

The concept of a training task template objectifies the proposed idea of the structural similarity of multiple training tasks dedicated to the same topic. The following three main components form the base of a training task template:

- **Problem formulation** describes the problem and the goal of the task
- **Solution steps** are a step-by-step explanation of the procedure of obtaining the result
- **Answer** is a mathematical expression that represents the goal of the problem

Problem formulation, solution steps, and answer are defined within the accuracy of a finite number of dynamic attributes. These dynamic attributes can possess concrete values. Specifying concrete values for all of the dynamic attributes of a training task template defines a concrete training task. Consider the example given above. Variables f and g can be regarded as the dynamic attributes of this task. Now, if someone defines, e.g., f, to be a sine function and g to be a cosine function, then the appropriate concrete task problem formulation will look as follows: "*Calculate the derivative of $\sin(\cos(x))$*". It is easy to see that, given a set of base elementary functions, one can provide a large number of similar concrete training tasks using this template. Take a look at Fig. 8.1.

---

[2]Here and further we consider training tasks on topics involving mathematics or simply mathematical training tasks.

[3]There exist various types of answers; in fact, all other cases can be reduced to the suggested one.

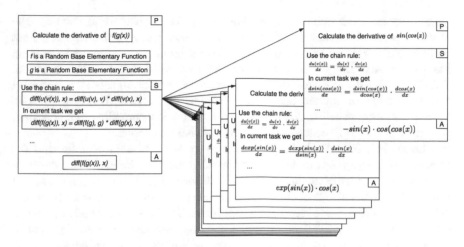

Fig. 8.1 Training task template example

The leftmost rectangle defines a task template for the suggested problem (the problem formulation, solution explanation, and answer areas are denoted by "P", "S", and "A", respectively). Each row, or a directed edge of this graph, represents an independent choice of the values of the dynamic attributes of the task (f and g in this case). It is clear that, given the values of the dynamic attributes, the process of obtaining a concrete task based on some task template is a pure mechanical job. We may also observe that there is no conceptual difference between different implementations of a task template since they all preserve the same topic and the same complexity level. It is important to note that the definition of a complexity level strongly depends on the topic to which these tasks are dedicated to. So, for each dynamic attribute of a task template, there is a set of possible values it can possess. Thus, in order to define a set of training tasks, it is sufficient to define the appropriate problem formulation, solution steps, and answer together with the declaration of the sets of possible values of all dynamic attributes. Since all the concrete tasks in such a set a considered equivalent in a sense that they all implement the same general structure, the process of obtaining a concrete task, given a task template, can be reduced to a pure randomization. The concept of training task templates defines an approach to training tasks development and support. This approach is implemented in EdLeTS.

8.4.2   EdLeTS: System Overview

EdLeTS is a web-based educational system for organizing practical classes on various mathematical disciplines. The main intention of this project is not to replace teachers or change the traditional idea of learning by training on specifically designed problems, but rather, it is intended to improve the educational process by eliminating its weak points and providing some additional useful features. The whole project is being developed with the confidence that a human teacher cannot be completely replaced by a machine and that the approach of learning by training has proven sufficiently well throughout the history of mathematics education. EdLeTS is aimed to provide support for practical classes organization by taking care of dozens of routine and time-consuming operations. Examples include training tasks composition, homework and examination compiling and checking, personalized task review, students' progress statistic gathering, etc.

Classrooms. EdLeTS is built around the concept of training tasks and the principle of training task templating. It provides tools for training tasks development by means of templates and concrete tasks generation. Consider some practical classes. A traditional classroom usually includes a teacher and a number of students. A significant part of teacher-student interaction in a classroom is concerned with training tasks. EdLeTS allows teachers to organize classrooms and incorporate students in a way that would combine people in groups in a social network. Each classroom can be dedicated to one complete course, some specific topic, or virtually any abstract theme. It is up to a teacher to define the level of generality of the classroom. One can run as many classrooms as needed. Building on the same social networks analogy, groups in social networks are focused on sharing media and textual information, but an EdLeTS classroom is focused on training tasks and learning material distribution. The role played by a user in a classroom, whether a teacher or a student, determines the type of interaction with these objects.

Consider a teacher. A classroom teacher in EdLeTS is a user who is capable of the following activities:

- building the task domain of a classroom, i.e. a set of task templates that represent the concepts considered in this classroom;
- providing task assignments, either directly or via homework, quizzes, etc.; and
- getting information on students' classroom progress.

These activities are common for classroom organization in a general sense. At the same time, a teacher traditionally has to perform numerous additional tasks, including creating individual task lists, checking students' works, assessment, and carrying out quizzes. Moreover, it is almost impossible to obtain a satisfactory personalized statistic on each student's progress, or even to provide each student with appropriate attention, in larger classrooms. Contrarily, EdLeTS takes care of these additional jobs, allowing the teacher to focus on analyzing the outcomes and adjusting the curricula.

Human system interaction. As it was described earlier, the idea of training task templates allows teachers to describe a set of concrete training tasks for the given concept by means of a single template definition. In order to define a task domain for a classroom, one needs to define a single training task[4] for every concept

---

[4]Here and further, if it does not lead to a confusion, we will use the phrase 'training task' or simply 'task' to denote training task template. The term 'concrete task' will be used to denote a concrete implementation of a task template.

considered in this classroom. EdLeTS is capable of automatically generating concrete tasks for the given template. The process of training task declaration is shown in Fig. 8.2.

EdLeTS is equipped with a special task designer, or an interactive development environment for training tasks. The task designer provides a user-friendly interface that helps teachers easily develop complex training tasks. Once created, a training task is stored in the task database and can be associated with a classroom. Classroom tasks domain consists of the associated training tasks. These training tasks are available to the classroom students.

From the students' perspective, the interaction with EdLeTS classroom is mostly reduced to tasks solving, whether it is homework, quizzes, or self-training activities). Unlike teachers who mostly deal with task templates, though, the students operate with concrete tasks only. Each time a student requests a task, multiple actions take place, as seen in Fig. 8.3.

The appropriate task template is retrieved from the database and sent to the Task Generator. The latter is capable of compiling a task into a random concrete task. The process of task compilation involves special language blocks recognition and

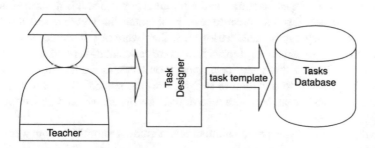

Fig. 8.2 Training task creation process illustration

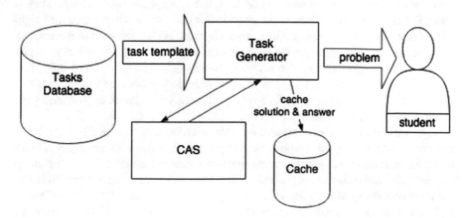

Fig. 8.3 Training task request process

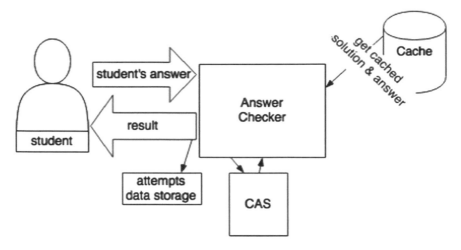

Fig. 8.4 Process of student's answer submission and answer checking

interpreting, symbolic calculations, and randomization tools. It is described in detail later in this paper. Once a concrete task is obtained, its solution and answer are cached so that they can be later retrieved. If they were not, it would be difficult or even impossible to obtain appropriate concrete task solutions and answers in the future due to the fact that every new generation leads to a different concrete task. Finally, the problem formulation is sent back to the student. The appropriate user interface provides the student with a form field for his answer and a math palette to simplify the input.

Having done all appropriate calculations, a student submits his answer back to the system, as seen in Fig. 8.4.

EdLeTS provides the feature of automatic answer checking by comparing the user's answers to the answer calculated during task compilation. This process is described in Sect. 8.4.4. In order to do this, the appropriate concrete task answer and solution are restored from cache. Recall from above that they are cached right after task compilation. The result of answer checking is then returned to the student. In the case that it is negative, the student receives the appropriate solution explanation. A user's attempt to solve a task is considered successful if and only if he submits a correct answer (i.e., the answer checking procedure returns positive result). In all other cases, including wrong answer, cashed task expiry, and page leaving, an attempt is considered failed.

Dynamic attributes randomization. An essential part of task development is the declaration of the sets of possible values of task's dynamic attributes. Recall from the previous sections that a concrete implementation of a task template is obtained via randomization of such sets. Keeping this in mind, we can conclude that the most convenient way to declare a set of possible values of a dynamic attribute is to define it through an appropriate randomization procedure. Consider the example from Sect. 8.4.1. Assume f and g are the dynamic attributes of this task. What sets do they belong to? For example, one may define them to be basic elementary functions.[5] In order to define such a construction, a user would need special randomization tools. Otherwise, he would need to introduce an algorithm of random generation of such functions by himself or reduce the suggested class to some finite dimension and perform random choice on this new finite set defined by hand. Both of these alternatives are unacceptable. This example illustrates the necessity of a wide range of randomization tools vital to meet the possible requirements of task authors. An analysis of training tasks of various directions in Mathematical Analysis was performed in order to obtain a list of randomization tools that the EdLeTS users might require. The obtained list includes, but is not limited to, the following features:

- Integer range randomization
- Float range randomization
- Random choice
- Random trigonometric functions
- Random transcendental functions
- Random algebraic functions
- Random basic elementary functions
- Random elementary functions of the given complexity
- Random polynomials of the given degree

The presented list is not complete and the whole set of randomization tools is being updated constantly.

The proposed technique of training task templating forms the kernel of the EdLeTS. From the teachers' perspective, it allows the user to develop and deliver a huge number of training tasks using one general declaration. The randomization principle implies that, with a relatively high probability, each student receives a unique collection of training tasks. Also, the amount of tasks is unique. In the suggested approach, the task solved does not equal the task revised. A student's attempt is considered successful only if the submitted answer was correct. In case of a failure, he may continue working on this task, but each new request is treated as a new attempt and each new compilation results in a new task because it is randomized. In order to track each student's progress, the result of every solving attempt is stored in the database. Classroom teachers are able review the reports of their students' progress relative to the classroom tasks. In this way, training tasks become personalized for students and the personalized report become available to teachers. It is almost impossible to provide that level of personification in a traditional classroom environment with no ICT involved due to the excessive time requirements.

---

[5]By the set of basic elementary functions we mean a union of the set of algebraic functions and the set transcendental functions.

8.4.3 Task Compilation and Symbolic Calculations

In the general case, the process of training task compilation requires a significant amount of symbolic and numerical calculations. The active use of various randomization tools complicates this process even more due to the fact that the actual values of the dynamic attributes are defined at runtime only. Consider the following simple problem formulation:

Calculate the derivative of $f(x)$ with respect to x,

where f possesses its value in a set of basic elementary functions. An answer for this training task should possess the value of the derivative of f for each f. Thus, on such a task compilation, appropriate symbolic calculations are to be performed in order to determine the actual value of $\frac{df(x)}{dx}$. This example illustrates the need of symbolic calculations, though it is quite simple. In real-world cases, more complicated calculations may be required.

It makes perfect sense to use a CAS to provide symbolic calculations since, by definition, it is meant for this purpose; however, it requires some additional development. There are two general aspects of integrating CAS in the suggested educational system: interface and competence. The interface aspect is associated with CAS-language. The aspect of competence determines if the system functionality meets the needs of EdLeTS as a smart educational system. From these two perspectives, the following propositions can be made:

- CAS native language is too complex. Any CAS is a professional instrument in the first place and such a programming language should meet a number of specific requirements. Many of these requirements, however, do not concur with the requirements of training task declaration. In fact, most of them are excessive. This problem is strongly connected with the idiosyncrasy issue of CAS integration mentioned in the literature review in Sect. 8.2.
- A number of necessary constructions are missing in CAS-languages. Again, the requirements of training task declaration are different from those of a typical programming language. Consequently, the most useful structures and functions are often too complex in a pure CAS. This can also be regarded as a part of the idiosyncrasy issue mentioned in the literature review in Sect. 8.2.
- Most of the CAS hides the methods and algorithms used to obtain results, known as the black-box issue. At the same time, exposing the inside procedure of the expression evaluation can be extremely useful when building a solution explanation.
- Most of the CAS does not allow the ability to hold an expression in pure form with no evaluation. The other requires special constructions that pile the code to do so.

It is clear that a CAS fits well to provide the desired symbolic calculations, yet it cannot be used directly. The solution is to use a CAS as a computational module providing a special interface layer between the end user and this module. For this purpose, a special language, the SmallTask, was developed. The SmallTask language, together with its interpreter, provides tools for training tasks' dynamic attributes management. The particularity of its application area implies specific requirements and restrictions. As a result of the research preceding SmallTask development, the following requirements were established:

- The language should provide functionality that is necessary for training task declaration and be restricted to it
- The language must be as easy and intuitive as possible
- The language should provide a wide range of randomization tools
- The expressions should support fast conversion into representative mathematical notation
- The provided capabilities should be strictly controlled in order exclude the threats, either intentional or not, caused by its inappropriate use
- On-demand evaluation: expressions should not be evaluated until it is required

Some of these requirements may need additional explanation. First of all, there is the on-demand evaluation requirement. A great number of training tasks in various disciplines involving mathematics includes mathematical objects transformation. When transforming a mathematical object, one ends up with an equivalent object in a different form. The process of evaluation of an expression includes a number of subsequent transformation applied to the expression. In order to uncover the process of evaluation, we need to have the ability to preserve expressions in a non-evaluated form. Such a capability is extremely useful when describing a solution of a task. Consider a training task. Assume that, on some solution step, the following expression appears: $\frac{d}{dx}(sin(x) + cos(x))$. If someone wants to explain the process of evaluation of this expression in detail, he would probably write something similar to the following:

$$\frac{d}{dx}(\sin(x) + \cos(x)) = \frac{d \sin(x)}{dx} + \frac{d \cos(x)}{dx} = \cos(x) - \sin(x)$$

Note that, in order to write such an equality, two expressions are to be held in non-evaluated form. The on-demand evaluation feature is essential for training task declaration; it is impossible to provide the desired level of expressiveness without it.

The strict control requirement also needs to be explained. Recall that the language is used for the management of training task dynamic attributes. Since users may use it, the risk of inappropriate use is possible. There are two general cases of misuse. First, unintentional errors can be made. Second, some users may try to damage the system on purpose. In both cases, the system should not crash. This means that it is crucial that all the features provided by SmallTask are threat-safe. Particularly, no direct or latent access to the operating system functionality should be granted. This gives one more reason for a unique language development:

- We have full access to the language interpreter and we can test and improve each individual feature.
- Unless the SmallTask interpreter is published, there is no way for evil-doers to examine it in search of possible vulnerabilities.

SmallTask was developed to meet the proposed requirements and address the problems of CAS integration introduced by Drijvers [14]. It provides an intuitive interface for dynamic attributes management and uses a CAS to perform actual calculations. In SmallTask, each object is preserved in a form of a tree and only minimal simplifications are performed in case it does not break the on-demand evaluation principle. Each SmallTask object supports translation into two languages: the CAS-language, or the native language of the CAS used as a computational module in EdLeTS, and TeX, which allows expressions to be printed in mathematical notation.

It is appropriate to mention that SmallTask differs significantly from traditional programming languages. One may say it includes both programming language and markup language features. On one hand, various types of calculations and transformations on expressions, including variables, are supported. On the other hand, special attention is given to the representation of an object. Specific datatypes are used in SmallTask due to the fact that its standard library is quite atypical. The list of SmallTask datatypes includes mathematical function, derivative, integral, limit, a special subtype of a function known as a polynomial, and plot. The standard library includes a wide range of randomization tools, as described in Sect. 8.4.2, and a number of constructors, which help to construct specific objects. The overall structure of SmallTask is determined by its application area. Inside a system, a training task is represented as an object with several attributes, including problem formulation, solution explanation, and answer. These can be compiled into another object known as a concrete task. From this perspective, both the problem formulation and the solution explanation consist of a markup text with special code insertions written in SmallTask. These insertions can be placed anywhere in text; three main printing options are available: display-none, which does not print the expression; display-pure, which prints the expression as it is with no evaluation performed; and display-evaluated, which evaluates and prints the result. An answer of a task is a mathematical expression written in SmallTask. This expression is always evaluated to its final form.

SmallTask provides an interface layer between the end user and the CAS. Its whole structure is dedicated to training task description. A number of limitations and restriction rules are introduced in order to prevent inappropriate use and to reach the desired level of language simplicity. It provides wide range of randomization tools, supplies CAS with additional functions and structures necessary for training task development, and supports fast conversion in representative mathematical notation and on-demand evaluation. As a result, the SmallTask layer resolves the major drawbacks of CAS use in educational purpose.

8.4.4   Automatic Answer Checking

In previous sections, the concept of training task templating was described in detail and its implementation was exposed. Still, a significant part of EdLeTS training task management remains uncovered. In this section, the problem of automatic answer checking is considered.

The overall concept of EdLeTS prescribes that task authors are only responsible for the appropriate task template declaration and the system takes care of concrete tasks generation and students' answers checking. As it was mentioned above, an answer of a concrete task is a mathematical expression written in SmallTask. Students submit their answers in a plain-math form, which, in fact, is a subset of the SmallTask language. In order to perform answer checking, EdLeTS compares the evaluated answer of a concrete task to the answer submitted by the student. If they match, the student's answer is considered correct. Otherwise. the attempt is considered failed.

Answers equality. It needs to be clarified what we mean by answers equality. Two types of mathematical expression equality are commonly distinguished: *syntactical equality* and *semantical equality*. Two expressions are said to be syntactically equal if they both have the same representation in mathematical notation. On the contrary, two mathematical expressions f_1 and f_2 are said to be semantically equal if they both represent the same mathematical object. This will be written as $f_1 \equiv f_2$. For example, 2^x and $e^{x \cdot \log 2}$ are syntactically different, but at the same time they are semantically equal. Similarly, $\tan(x)$ equals $\frac{\sin x}{\cos x}$ in the sense of semantic equality; however, there is an obvious syntactical difference. It is simple to check syntactic equality since this process can be reduced to a simple string comparison. Unfortunately, the problem of establishing semantic equality is quite difficult. Obviously, in order to perform answer checking, we need to determine whether the two answers are equal or not in the sense of semantic equality. This is done by special comparison algorithms introduced by the CAS being used.

Constant problem. In fact, it appears that the problem of establishing semantic equality is formally unsolvable. The appropriate result was first uncovered by Richardson and is known as the Richardson's Theorem [18]. First, observe that the suggested problem of semantic equality recognition is absolutely equivalent to the problem of determining whether an expression equals zero or not: $f_1 \overset{?}{\equiv} f_2 \Leftrightarrow f \overset{\Delta}{=} f_1 - f_2 \overset{?}{\equiv} 0$. The latter is known as the *constant problem* or the *identity problem*. This states that, given a set of expressions E and a set of functions E^* represented by expressions from E, the identity problem for (E, E^*) is the problem of deciding, given $A \in E$, whether $A(x) \equiv 0$. The Richardson's Theorem states the following [18]:

Let E be a set of expressions representing real, single valued, partially defined functions. And let E* be a set of functions represented by E: if A is an expression in E then A(x) ∈ E* denotes a function represented by A. Assume that E* contains x (the identity function), all rational numbers (as the constant functions) and is closed under addition, subtraction, multiplication and composition.

Then if $x, \log 2, \pi, e^x, \sin x, |x| \in E$ *then the identity problem for* (E, E^*) *is unsolvable.*

Simply put, the theorem states that the identity problem is unsolvable in a common function class; the statement holds for the class of elementary functions, which is the most common class in training task development. From the EdLeTS perspective, the outcome of this theorem says that it is impossible to guarantee the correct automatic answer checking. In order to show that the problem is not only theoretical and far-fetched, the following example is suggested: let $f_1 = \frac{\sec(\mathrm{acsc}(x))}{x^2 \cdot \sqrt{1-1/x^2}} \cdot$ $\tan(\mathrm{acsc}(x))$ and let $f_2 = \frac{-1}{(1-1/x^2)^{3/2} \cdot x^3}$. It can be easily shown that $f_1 \equiv f_2$, both of them representing the derivative of $\frac{1}{\sqrt{1-1/x^2}}$ with respect to x. Multiple CAS have been tested on this example and only one was able to establish this equality.

The form of the final result always depends on the methods used to obtain it. Computer algorithms are often different from those used by humans. A human never acts like a machine and, even among humans, there are different cognitive patterns. The more complex the task is, the more approaches can be applied to solve it. During this research, the problem of identity checking was seldom observed; however, there are still plenty of chances for it to occur and it is almost impossible to predict them. In order to cope with the issue, an additional pointwise checking procedure is suggested.

Additional pointwise checking procedure. Assume that an answer is given by a real-valued elementary function. When the CAS native expressions comparison algorithm turns out to be unable to distinguish whether two expressions are equal or not, neither a *True* (positive result) nor a *False* (negative result) value is returned. Instead, a third value is returned. This means that such an algorithm is reliable in a sense that an unambiguous result (*True* or *False*) is always correct. In the case of ambiguous results, an additional algorithm is needed in order to make the final decision. For this purpose, EdLeTS is equipped with an additional pointwise checking procedure, or a special supplementary algorithm, that takes care of answer comparison if CAS fails to do so. A simplified scheme of this algorithm is shown in Fig. 8.5. The additional pointwise checking procedure considers function $f = f_{system} - f_{user}$, where f_{system} denotes an answer obtained by the system, f_{user} denotes an answer calculated by user, and checks its values in a finite number of points $(x_1, \ldots, x_m)$ one by one. The check points are picked randomly from a uniform distribution on a closed segment. The "Random(A, B)" in Fig. 8.5 denotes a function that retrieves a random number from $U(A, B)$.

The only restriction is that all check points must be different since there is no need to check a function value in some point twice. In order to meet this requirement, a set of visited points is stored and the fresh point is taken into consideration if and only if it does not belong to the set of visited points. As soon as a non-zero value of f is obtained, the procedure returns negative result. Otherwise, if $f(x_i) = 0$ for all $i = 1, \ldots, m$, a positive result is returned.

It is easy to see that the suggested pointwise checking procedure is not precise. An error is possible. The only occurrence of this algorithm failing is when it

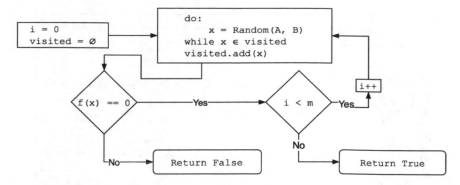

Fig. 8.5 A simplified diagram of the pointwise checking procedure

occasionally picks all the check points from the zero set of f and at the same time $f \not\equiv 0$. This means that the student's answer is incorrect, but the algorithm fails to detect it. It is important that, at the same time, the algorithm won't fail if the answer is correct.

Obviously, the probability of the suggested algorithm failure converges to 0 as the number of check points tends towards infinity. More check points means lower error probability. Still, there are reasonable restrictions on the number of checks produced by evaluation time requirements. Also, we have to consider machine arithmetic and, thus, the discrete probability of error should be considered [19, 20]. The value of the error probability was considered with respect to three attributes: the maximum number of check points (m), the upper estimate for the possible number of zeros of the target function (k), and the lower estimate for the amount of floating-point numbers inside the suggested interval (M). As a result, an expression for this probability as a function of M, m and k was obtained. Figure 8.6 shows the behavior of the suggested probability with respect to k (Fig. 8.6a) and m (Fig. 8.6b).

As we can see, the probability of the algorithm failure falls nearly exponentially with respect to increasing values of m. An analysis of this function shows that this probability becomes quite low for reasonable values of the parameters it depends on. The more precise analysis of the error probability of the additional pointwise checking procedure can be found in [19, 20].

Automatic answer checking is an essential part of EdLeTS. The whole project does not make much sense without it. The identity problem undecidability damages the system's reliability. Fortunately, it appears that the situations in which such a problem may occur are quite rare and, even in cases that it occurs, the use of additional pointwise checking procedure often provides low error probability. In order to lower the risk of failure, EdLeTS presents a precise explanation of the problem of answer checking to task authors and provides them with an ability to adjust task attributes that influence the probability of error.

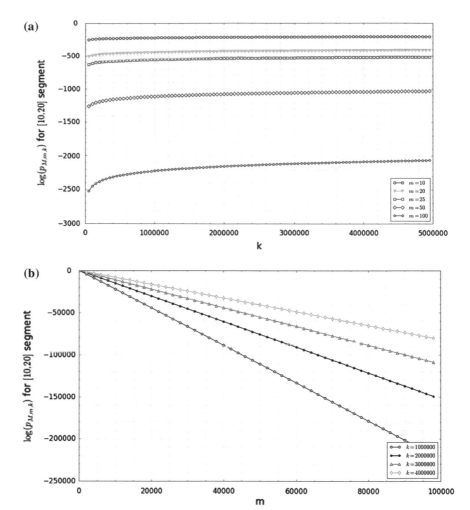

Fig. 8.6 Behavior of the logarithm of the probability of the additional pointwise checking procedure error on a [10, 20] segment: **a** with respect to k for different values of m, **b** with respect to m for different values of k

8.4.5 Task Structuring and the Theory of Learning Spaces

It appears that the theory of learning spaces closely relates to the EdLeTS system. The key concepts used in the suggested approach of EdLeTS can be expressed in terms of this theory. Consider a set of training tasks combined in a single task group. Each training task in this set is defined as a task template and has a number of concrete representations. Such a task group can be regarded as a domain.

Furthermore, a structure on such a domain appears with the idea of training tasks precedence relation.

The EdLeTS framework suggests an ability to structure all tasks inside a group by means of precedence. A training task A is said to precede task B if one needs to master the concept of A in order to be able to solve B. In some cases, such a relation can be obtained in a natural way. For instance, a student would not be able to calculate the derivative of a product of two basic elementary functions unless he is able to differentiate each of these functions separately. Thus, a task on basic elementary function differentiation should precede a task on differentiating a sum of such functions. In some cases, however, such a relation may not be that obvious. Consider the limit theory and differential calculus, for example. From a theoretical point of view, the first one precedes the second one. At the same time, it is possible that some students would successfully learn to calculate derivatives while having difficulty with calculating limits. Should we establish an appropriate precedence relation between the appropriate tasks or should we divide them into two groups and treat them separately? We leave such decisions to task authors. When placing a task in a task group, a teacher is able to specify its direct predecessors or followers. Having this done, he ends up with a structure on a task group that is defined by the precedence relation. This relation is transitive (if A precedes B and B precedes C then A precedes C) and reflexive (in order to master A one has to master A). Thus, it is a preorder, or a quasi order.

It is not surprising that an absolutely equivalent idea can be found in the Learning Spaces theory. In [16] the notion of the surmise relation, or equivalently, the precedence relation, is used. The authors show that there is a one-to-one correspondence between the collection of preorders on some domain and the collection of quasi-ordinal spaces, or knowledge spaces that are closed under intersection, on the same domain. As a corollary, each preorder implies a knowledge space. Since the relation described in the above paragraph is a preorder, then it also defines some knowledge space. In Fig. 8.7, an example of a knowledge space induced by a precedence relation on a domain of training tasks on simple differentiation rules is shown in form of a graph. Each node of this graph represents a state a student might possess. The edges represent the accessibility of these states. We can see, for example, that a student needs to master concept B, which stands for "derivation of base elementary functions", in order to master any other concept: any non-empty state which includes concepts other than B necessarily includes B.

The precedence relation defined on a group of training tasks allows the introduction of personalized suggestions for students. Consider a student who fails to solve some task consistently. At the same time, he didn't perform well on one of the preceding tasks. For instance, he had solved a relatively small amount of such tasks and the overall results were not impressive. It seems reasonable to suggest that the student should work the preceding task over again. On the basis of this idea, a personalized suggestions list of "tasks worth worked off", or "fill-the-gaps suggestions", can be constructed. The system detects possible problems of a user by analyzing his solving progress and provides suggestions based on the precedence relation. In a similar way, a training path can be constructed. If a student has

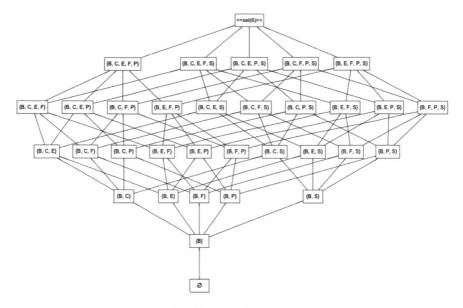

Fig. 8.7 An example of a space induced by a precedence relation on a domain of training tasks on differentiation rules. Each letter represents a task: B—base (derivatives of base elementary functions), S—sum differentiation, P—product differentiation, F—derivative of a fraction, E—power function differentiation

performed well on all of the predecessors of some task, then he is probably ready to move to that task.

The described ideas of "fill-the-gaps suggestions" and learning path construction form a powerful tool that brings adaptiveness to the whole project; however, this is still a work in progress. Some problems arising with the development of the appropriate algorithms are presented in Sect. 8.7. Some algorithms for building learning paths in knowledge spaces are described in [16] as well as other papers dedicated to the theory of learning spaces.

Recall from Sect. 8.2 that a knowledge state is defined by a set of items of the domain that a student has mastered. It is a challenging problem to determine if a student masters a given concept. Consider a student, say Sarah, receives a task on the chain rule of differential calculus and solves it successfully. Does it mean that she has truly mastered the concept? Her success may be due to a lucky guess, or perhaps she recalled the correct answer from her prior experience. There are plenty of possibilities that Sarah had solved the proposed task on occasion and still has not mastered the underlying concept. There are various approaches that can be used to distinguish whether or not a student has mastered a concept. One of them specific to EdLeTS is presented below.

Traditionally, teachers provide students with a fixed amount of tasks so that they can train their understanding of a given concept. This suggests, at least in a latent form, a vision of an average student who is going to master the given concept after training on a given amount of tasks. As it was repeatedly noted in previous sections, such an approach suffers major drawbacks that may be generalized as an absolute absence of personalization. In EdLeTS, a different approach is proposed. Recall that the tasks here are introduced by means of training task templates. From this, a large number of concrete tasks based on a single template can be produced. This implies a new measure for the level of adaption of a concept underlying a task, namely the ratio of successful attempts to the whole number of attempts. This can be referred to as the success ratio. This approach is based on a simple idea that, if someone performs a stable success in solving different tasks on some topic, then he probably masters this topic. Let us return to Sarah from the above paragraph. Now, Sarah is provided with a number of concrete training tasks on the chain rule. She solves them one by one and a track of each attempt is kept. Assume that, after 100 attempts, Sarah reaches the success ratio of 0.9. Now, with a great confidence we could say that Sarah has learned the chain rule well. This new measure implies natural personalization. It shows that each student is going to solve his own amount of tasks in his own way as a way to reach the desired success ratio. EdLeTS allows classroom teachers to define the required success ratio and the lower bound for the total number of attempts for each task assigned to students. The lower bound for the total number of attempts is a prerequisite for ensuring that the success ratio is representative. A success ratio of a total of 1 attempt does not indicate a thing.

In EdLeTS, each task assigned to a student can be supplied with two attributes: the desired success ratio and the lower bound for the number of solving attempts. As it was stated in the above paragraph, this approach is more flexible than the traditional one and it implies a higher level of personalization.

8.5 Testing and Validation of Development Outcomes

EdLeTS is still a work in progress. Despite the fact that the theoretical development, meaning development of concepts and the full system architecture, recognition of tools, and applications to be used, is mostly done and the program realization implements most of the desired features, there is still a lot of work to do. It might be too early to integrate the system in an educational process. This is due to the lack of some features and even some interface gaps. This may lead to disapproval among both teachers and students; however, some significant tests were performed during this research.

8.5.1 Testing of the Task Generation Approach

In this subsection, a test of the general idea of learning by training on multiple examples provided by EdLeTS is described. The main goal of this test was to get insight into how effective the suggested approach really was. The testing was performed on the earlier version of EdLeTS that was provided in a form of an educational computer program with a database of training tasks. This system was able to generate a large amount of different training tasks covering various topics of limits theory, differential calculus, complex analysis, integral calculus, etc. Each of the tasks was supplied with a static solution explanation and the appropriate answer. Altering select predefined attributes through the task generation and automatic answer checking was supported. The system was accorded to a group of 30 students studying at one of the Moscow technical universities. These students were taking classes in mathematical analysis and the theory of functions of a complex variable. The students took the appropriate theoretical classes, i.e., lectures, as prescribed by the curriculum. Some techniques in the practical part of the course were taught with the help of the system. Table 8.1 lists the covered topics.

For each of these topics, a separate test was held. During a test, each student sat in front of a computer with the educational program running. Upon request, a student was provided with a task. Once done, a student submitted his result back to the system. The system then performed answer checking. If the answer was incorrect, the appropriate solution explanation was given. A student then studied this solution and continued on with equivalent tasks. The system kept track of how many solving attempts were done during a session. Each test was held for a limited time, from 20 to 45 min depending on topic. An instructor supervised the whole process. This same instructor recorded the overall results of a test. For each student, the number of tasks it took them to achieve a stable success (3 successes in a row) was recorded (1 if the first three attempts were successful, 2 if attempts 2–4 were successful, etc.). The appropriate data is summarized in Fig. 8.8.

Each pie chart represents the results of the appropriate test. A segment of a chart represents the amount of students that needed a particular number of attempts to achieve stable success. Each segment is labeled with the number of attempts. For instance, Fig. 8.8a shows that 17% of students succeeded starting from the second task. It turned out that, throughout all of the tests, the less enthusiastic students required up to 6 attempts in order to grasp the underlying idea. Starting from the seventh task, they began to solve them correctly. Although the tasks were quite simple and rather algorithmical, the results were above expectations. These results show that, after a given amount of attempts, a student understands what is common throughout all of the tasks, what details they differ in, and how these details influence the problem and its solution. More complex training tasks rely more on a student's background and, thus, the number of attempts required to make progress may grow higher. A student would never differentiate $\sin(\cos(x))$ correctly unless he knows what the derivative of the sine function is. Although the results obtained show that the general idea works fine and if a student faces some serious problems with similar tasks for a long time, it may indicate a significant gap in his background knowledge.

Table 8.1 Topics used for testing the EdLeTS training tasks generation approach

Topic	Description	Problem example
Limits calculation using special limits and the squeeze theorem	By special limits, we denote the two widely known mathematical identities: $\lim\limits_{x\to 0}\frac{\sin(x)}{x}=1$ and $\lim\limits_{x\to\infty}\left(1+\frac{1}{x}\right)^{x}=e$. The squeeze theorem, also known as the *sandwich theorem* or the *two policemen theorem*, states that, if the upper (h) and the lower (g) bounds of a function f both converge to the same limit ($\lim\limits_{x\to a}g(x)=\lim\limits_{x\to a}h(x)=L$), then $\lim\limits_{x\to a}f(x)$ also equals L. The appropriate tasks train the ability 1) to recognize and use special limits; 2) to determine such bounds of a function that the squeeze theorem can be used to calculate its limit	Calculate the value of $\lim\limits_{x\to\infty}\frac{5x+\sin 2x}{2x-1}$
Differentiation rules	These tasks include various examples on the basic differentiation rules such as differentiation of a sum/product/fraction and the chain rule	Calculate the derivative of $2^{\sin(x)}$
Integration by parts	Such tasks are intended to train the technique of integrating by parts using the following formula: $\int u(x)v'(x)dx = u(x)v(x) - \int v(x)du(x)$. In order to solve such a task, a student should be able to recognize a possibility of integrating a part of the original integrand in order to reduce, through integration by parts, the whole expression to the less complex one and calculate it	Find $\int \sec^3 x\,dx$
Complex integration using residues	These tasks are used to train the technique of calculating a line integral of an analytic function over a close curve using the Residue theorem: $\oint\limits_{C} f(z)dz = 2\pi i \sum\limits_{k=1}^{n} Res(f, z_k)$, where C is a positively oriented simple curve, $z_1,\ldots,z_n$ are the isolated singularities of f in a region bounded by C, $Res(f, z_k)$ denotes the residue of f at z_k	Find $\oint\limits_{|z-1|=2} \frac{\sin z}{z}dz$

8.5.2   Teachers and Students' Opinion Poll

In order to find out what students think about the EdLeTS system, a group of 17 first-year students were involved in the test. These students used EdLeTS during their practical classes in mathematical analysis to train the newly learned

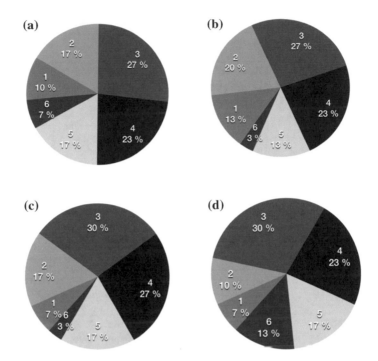

Fig. 8.8 The results of testing EdLeTS training tasks approach on students for the following topics: **a** limits calculation using special limits and the squeeze theorem; **b** differentiation rules; **c** integration by parts; **d** complex integration using residues

techniques. The test was held as a part of each lesson for 3 weeks. At the end, the students were asked to answer the following questions:

(1) How helpful did you find the solution explanation shown in a case of a wrong answer? Choose one of the following:

1. absolutely useless
2. almost useless
3. useful sometimes
4. quite useful
5. very useful

(2) What training approach would you prefer: traditional (traditional schoolbooks, homework, etc.) or the EdLeTS approach? Choose one of the following:

1. I would definitely prefer the traditional approach
2. I think traditional approach is more suitable for me
3. I don't know
4. I think the EdLeTS approach is more suitable for me
5. I would definitely prefer EdLeTS

(3) How convenient/easy-to-use did you find the system in general (user interface, obtaining a task, answer checking, solution overview)?

   1. it was really hard to use it
   2. quite uncomfortable
   3. it could be better
   4. quite comfortable
   5. absolutely comfortable

(4) If you were to decide, would you recommend to use EdLeTS in similar classes?

   1. I would definitely not recommend to use EdLeTS
   2. I likely would not recommend to use EdLeTS
   3. I don't know
   4. I likely would recommend to use EdLeTS
   5. I definitely would recommend to use EdLeTS

The overall results of this poll are presented in Fig. 8.9. It represents students' answers on the proposed questions: Fig. 8.9a shows answers for questions 1 and 2; Fig. 8.9b shows answers for questions 3 and 4. Each group of two bars in Fig. 8.9a and Fig. 8.9b represents one student's answer (e.g., the first pair of bars in Fig. 8.9a together with the first pair in Fig. 8.9b show that the first student's answers were 5, 5, 4, 5 for questions 1–4, respectively). The poll results in Fig. 8.9a show that, on average, students feel good about using EdLeTS in their education. Most of them find detailed solution explanations at least quite useful and all the students said they are likely to choose EdLeTS over the traditional approach. The answers to question 3 are also quite positive. The students did not face any critical problems when using

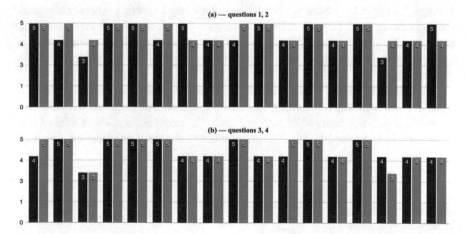

Fig. 8.9 The results of the opinion poll about EdLeTS among students. Each group represents one student's answers: **a**—questions 1 and 2; **b**—questions 3 and 4

the system; however, they did not use the whole functionality. This was due to the fact that the test was held within one particular course and only a limited amount of training tasks was introduced. Finally, question 4 was aimed to reveal the overall level of approval of the system among students. The results show that, in general, the students are most likely to support such an experiment

At the moment, there has not been an opportunity to provide appropriate research for the teachers' opinion regarding EdLeTS due to the small size of the sample of teachers who tried EdLeTS. An investigation in the form of a dialog with various teachers revealed results that are quite similar to the results obtained by Lavicza in [13]. These say that the most experienced and seasoned teachers are often quite sceptic when asked about integration of an educational system into their curricula, while the younger teachers are often more open to such experiments and show interest in EdLeTS.

8.5.3 Pointwise Checking Procedure Testing

A significant part of the EdLeTS project, which requires appropriate testing, is the additional pointwise answer checking procedure. Recall from Sect. 8.4.4 that the procedure is stochastic and errors are possible. Therefore, the first general question here is "How often does this algorithm fail?" Another important characteristic of the algorithm is the number of steps it takes to recognize an incorrect answer. This value varies from run to run.

A number of tests were performed on multiple examples. For each example, a test task was built with a static answer that matched the desired function. For each test task, multiple tests were performed by submitting a function that was somehow close to the answer, but not equal to it. In case the additional pointwise checking procedure was set in motion, the following data was recorded: the overall result of checking and the number of steps it took the algorithm to find out that the answer was incorrect.

The following results were obtained: the algorithm had never failed because, for each test case that required additional checking, the correct response was obtained; most of the runs required only 1 step to obtain correct result; and about 85% of runs were done in less than 10 steps. The distribution of the number of steps is presented in Table 8.2.

Table 8.2 The distribution of the number of additional pointwise checking procedure steps on test runs	Percent of runs (%)	Number of steps
	∼73	1 step
	∼12	2–10 steps
	∼8	11–30 steps

The results of this test show that, in most cases, the suggested additional answer checking procedure is able to detect incorrectness relatively quickly. It is also important to note that no test run lead to a failure.

8.5.4 Comparison with Other Systems of a Kind

The provided research shows that the amount of educational systems considering mathematical disciplines in the way it is done in EdLeTS is next nothing. One project similar has been found so far, namely Maple T.A. EdLeTS and Maple T.A. have much in common, though they have been developed independently. Maple T. A. is a commercial educational web application based on the Maple computer algebra system. It provides tools for training tasks development and organization, affords a user-friendly task designer, and. at the same time, allows the use of Maple language construction in the algorithmic part of a task. The latter feature gives task authors access to the full power of Maple CAS, which, in fact, allows them to implement the same idea of training task templating themselves. This feature is also a drawback, though. The direct use of a CAS leads to most of the issues revealed in Sects. 8.2, 8.4.3, 8.4.4. These issues include the black-box issue, idiosyncrasy, language complexity, etc.

Table 8.3 provides results of the comparative analysis between Maple T.A. and EdLeTS.

Table 8.3 Comparison of Maple T.A. with EdLeTS

Maple is a professional instrument and, thus, its language is too complex for the needs of task development in an educational system	EdLeTS provides a language (SmallTask) that was specially developed for training task declaration and is free from useless constructions (this achieving simplicity)
As a black box tool, Maple hides its inner methods realization. Consequently, it is hard to restore the steps that lead to the final result	EdLeTS provides a number of tools for retrieving a step-by-step solution path
Giving access to the power of CAS may lead to possible vulnerabilities. Consequently, a serious code validation is required	The SmallTask language gives no access to any dangerous areas by construction
Maple T.A. claims that it fully supports automatic answer checking with answer given in mathematical notation. At the same time, it was shown (see Sect. 8.4.4) that this feature has serious limitations. It is possible that this system crashes on unsuitable examples	EdLeTS provides additional algorithm which runs when CAS appears to be unable to solve the identity problem. The probability of its failure is estimated and can be reduced to acceptable values by altering the appropriate attributes of a task

8.6 Discussion

As a result of the performed research, the EdLeTS system was developed. This system implements the key idea of training task templating and resolves various problems described in Sects. 8.2 and 8.3. The idea of training on concrete examples has proven very well throughout the history of mathematical education. Practice shows that, this way, students learn to recognize common patterns in concrete problems and also generalize the solution that finds out how it differs with respect to details. It is always surprising that many students experience serious problems when trying to apply concepts or techniques that they have recently learned to some concrete examples. From a student's perspective, this means that his inner picture of the appropriate concept is incomplete and the only way to fill the gaps is through training on concrete examples. EdLeTS provides students with the ability to train for as long as needed, appropriate solution explanations help to find out the root of mistakes, and, finally, the automatic answer-checking feature makes the whole system usable. EdLeTS takes the idea of learning by training to the next level. Still, some problems remain unsolved and there are features that are going to be implemented within further development. In this section, a discussion of the outcomes obtained and objectives still not achieved is provided.

Smartness in math education. Based on the performed analysis of various concepts in smart learning area and multiple disadvantages of the traditional approach to math education, a vision of Smart Math Education was established. As it was stated earlier in this paper, we believe that a human teacher cannot be replaced. Thus, the general intention of the EdLeTS project is to improve the quality of teacher-student interaction. We find the following characteristics essential for Smart Math Education:

- high level of automatization of training material (automatized training tasks generation and processing, automatic answer checking);
- personalized student support;
- students' personal statistics gathering and analysis;
- adaptive learning plans (e.g., building personalized learning paths based on the current state of a student).

Such features cannot be established within the traditional approach. As it was discussed earlier in this section, it is almost impossible for a teacher to provide the desired level of personalization and adaptiveness in a traditional classroom without specifically designed tools. At the same time, the use of a system such as EdLeTS can improve the situation significantly. From the perspective of hardware equipment, the requirements here can be quite low. It is sufficient that everyone in a classroom is equipped with some kind of a computer, whether a PC, laptop, tablet, even a smart phone. Smart Math Education should also preserve the traditional idea of learning by training on multiple examples and, thus, there may be no specific ideological changes in the curriculum. In practice, the use of smart educational tools would always require a teacher to play the role of classroom developer with tasks including building classroom task domain, inviting students, etc. EdLeTS aims to make this process as easy as possible and also provide a sufficiently low entry threshold level in order to reduce requirements imposed on teachers to minimum.

Compared to traditional university math education, the Smart Math Education should introduce a significantly higher level of personalization and adaptiveness. The traditional educational approach is focused on a relatively large group of students. Contrarily, an individual student is in the focus of Smart Education. The EdLeTS project is aimed at providing the highest level of personalization and adaptiveness in mathematics education while sticking to some traditional techniques that have proven well though its history. In sequent sections, the main features of Smart Math Education addressed in EdLeTS are discussed.

Personalization is one of the most important features of smart education revealed by multiple researchers. In addition, it is one of the key objectives of EdLeTS development. It may be stated, with great confidence, that it is almost impossible to introduce personalized learning plans in a traditional classroom where a group of about thirty students is managed by a single teacher. The suggested approach of training task templating with automatic answer checking and the notion of success ratio together with the notion of tasks precedence relation allow the ability to provide a high level of training personalization:

- the success ratio measure considers each student's personalized progress;
- concretized solution explanations help students easily identify their individual weak spots;
- personal statistic gathering allows teachers to discover both group progress and individual learning status of each student;
- the notion of precedence relation allows to provide automatic personal help, proposing future steps or suggesting task to work on, based on the analysis of his mistakes.

Adaptiveness. The idea of tasks structuring and the precedence relation described in Sect. 8.4.5 forms a basis for intelligent "fill-the-gaps suggestions: and learning path construction. These features make the whole system adaptive to each student's individual requirements. It is often not so easy to find out what obstacles hold a student back from solving a task repeatedly. Specifying precedence relation on a group of tasks allows the ability to build individual suggestion lists based on this relation and current progress of a student. Such a list consists of tasks that a student should pay attention to since he is likely to have problems with some of them. On the other hand, the suggested tasks structuring allows the building of learning paths if a student performs with stable progress. Such paths can be built both horizontally and vertically. If a student succeeds constantly in solving all the tasks that are direct predecessors of task A, then he is most likely to be ready to move to this task A. Some additional theoretical study may be also required. This is the vertical movement. The horizontal movement suggests moving across tasks of the same level. In other words, all predecessors of a task should be worked off before moving to this task. Both of these features, training suggestions and learning path proposal, are still being developed. EdLeTS supports tasks structuring by means of precedence and is able to make simple conjectures on students' problems based on such a structure; however, a user's overall progress has to be taken into consideration. For this purpose, a weighted subtree of the problematic task, implied by the precedence relation, should be walked over in order to build an appropriately ranged suggestions list with the most relevant tasks appearing at the top of the list. The weights of its nodes should represent the possibility of misunderstanding together with the relevance to the original task. Calculating these weights tends to be quite challenging. This is one of the main topics of current research within the EdLeTS development.

Mobility, self-Learning and ubiquitous learning. EdLeTS was developed to easily function on any kind of devices, whether a laptop, tablet, or smartphone. The *self-training* mode allows students to explore new techniques and train freely without the fear of exterior judgement. The overall idea of the project is to allow students to train anywhere and anytime. Particularly, this requires including support for structured theoretical material. This is still a work in progress.

Flexibility and ease of use. In EdLeTS, special attention is given to the system's ease of use and flexibility. It is considered the primary requirement that teachers face no significant problems when integrating the system into the educational process. Classroom structuring and management tools, tasks grouping and structuring, a wide range of instruments for training tasks development, and reports gathering features all serve this purpose. The SmallTask language was developed to be flexible and intuitive, only providing essential features specific for training tasks development, as they are needed. There is still a great need for a graphical user interface, though, as this would allow teachers and task authors to avoid using the SmallTask language as much as possible. Currently, there is an interactive task designer presented in EdLeTS. It allows the ability to build SmallTask constructions using a special keyboard; therefore, there is almost no need to understand or learn the language whatsoever. Meanwhile, a simple investigation showed that many university teachers do not want to see any language constructions at all. Consequently, we must develop a task designer that provides an intuitive graphical representation for each and every construction used in training task development. It must also provide an easy to use interface to operate them.

CAS. EdLeTS introduces the possibility of using the full potential of a CAS while being free from the indicated drawbacks of using CAS in math education. The power of SmallTask is limited to the needs of training task development thus saving students and teachers from excessively complex constructions, consequences of inappropriate use and other possible problems. At the same time, it supplies CAS with a number of methods and algorithms necessary for task development. In such a way, EdLeTS combines advantages of CAS and specifically developed educational system.

8.7 Conclusions and Future Steps

Conclusions. EdLeTS is developed to improve the educational process in the field of exact science. Our vision of smart technologies in mathematical education is inspired by multiple researches in Smart Education, Smart Classroom, Smart Universities, and other relevant topics partially presented in the literature review in Sect. 8.2. According to this vision, the following outcomes were obtained as a result of EdLeTS research and development:

- The concept of training task templating was developed and implemented in the EdLeTS project. This concept embodies the idea of personalization. This is the primary idea of smart technologies and is considered a dominant aim of the EdLeTS project.
- The EdLeTS system can be used anywhere and anytime. It also presents no additional requirements for hardware or environment except for the presence of a computer/laptop/tablet/smartphone. Furthermore, the availability of immediate solution explanations saves students from searching for additional help material.
- Training in self-learning mode allows students to improve their skills without worrying about the possible consequences of failure. These concerns are often rather psychological. The self-learning progress is visible only to the learner. Introducing smart suggestions and adaptive learning paths is generated in this mode so that any student could move forward following his individual path at his own chosen speed.
- An approach to use the power of CAS while keeping off the significant drawbacks of its integration was established. From this perspective, EdLeTS provides an educational interface to the tools provided by a CAS. This interface is customizable by means of restrictions. The CAS itself is supplied with additional learning-specific tools.
- A relationship between the concepts of EdLeTS and the Learning Spaces theory was established. The foundation for intelligent algorithms for learning paths building and "fill-the-gaps suggestions" was obtained. Being appropriately implemented. these features bring adaptiveness to the system.
- The big variety of task development tools enables the ability to expect easy integration in math curriculum.

Future steps. The EdLeTS project is a work in progress. The research and development are going on constantly. New features are introduced as the new requirements are uncovered. Based on the research presented in this paper, the following future steps can be indicated:

- Introducing higher level of adaptation. EdLeTS should provide a deep analysis of learner's progress and indicate possible sources of failures according to the given tasks structure implied by a precedence relation. It should also recognize and report possible inconsistency in a tasks group to classroom teachers. From this perspective, the results obtained in the Learning Spaces theory look promising; however, they still need to be adapted to the needs of EdLeTS.

- Training task designer improvement. Currently, it is clear that the task designer has to support full abstraction from the SmallTask language constructions. It appears to be strictly necessary to provide graphical representation for every object and action used in task development.
- Filling with more tasks. In order to demonstrate the features of EdLeTS to potential users, we need to populate the database with a number of representative tasks. It is considered a good practice because most users prefer to investigate features learning by examples. Also, the process of task development reveals weak spots of the project. One of the primary future goals is to develop a public pull of training tasks which would be observed, used and populated by users.
- Integration is probably the most significant future objective. Once the first production version is obtained, it is going to be integrated into the mathematical analysis course in one of the Moscow technical universities for testing purposes in the first place. This will allow the ability to understand whether or not it works as expected, what the weak spots are, how teachers and students feel about using it, etc. Gathering a precise feedback will determine further steps.
- In the context of using EdLeTS in a Smart University, it is necessary to provide a way to interact with other components of a SmU. From this perspective, a development of an appropriate protocol for learning data export and import is considered.
- Finally, a model for interconnected theoretical learning material is to be developed and implemented in EdLeTS. Currently, the SCORM specification standards are considered as the base for this functionality.

References

1. Khan, B.H.: Developing eLearning strategy in Universities of Bangladesh. In: ICT, p. 78 (2014)
2. Tikhomirov, V.: The Moscow State University of Economics, Statistics and Informatics (MESI) on the way to smart education. In: Proceedings of the 10th International Conference on Intellectual Capital, Knowledge Management and Organisational Learning: ICICKM 2013 (2013)
3. Zhu, Z.-T., Yu, M.-H., Riezebos, P.: A research framework of smart education. Smart Learn. Environ. **3**(1), 1 (2016)
4. Coccoli, M., Guercio, A., Maresca, P., Stanganelli, L.: Smarter universities: A vision for the fast changing digital era. J. Vis. Lang. Comput. **25**(6), 1003–1011 (2014)
5. Uskov, V.L., Bakken, J.P., Pandey, A., Singh, U., Yalamanchili, M., Penumatsa, A.: Smart University Taxonomy: Features, Components, Systems. Smart Educ. e Learn. **2016**, 3–14 (2016)
6. Hwang, G.-J.: Definition, framework and research issues of smart learning environments–a context-aware ubiquitous learning perspective. Smart Learn. Environ. **1**(1), 1–14 (2014)
7. Koper, R.: Conditions for effective smart learning environments. Smart Learn. Environ. **1**(1), 1 (2014)

8. Spector, J.M.: Conceptualizing the emerging field of smart learning environments. Smart Learn. Environ. **1**(1), 1 (2014)
9. Kim, T., Cho, J.Y., Gyou, L.B.: Evolution to smart learning in public education: a case study of Korean public education. In: Open and Social Technologies for Networked Learning, pp. 170-178 (2013)
10. Jonassen, D. H.: Computers as Mindtools for Schools: Engaging Critical Thinking. Prentice Hall, Upper Saddle River (2000)
11. Rubin, A.: Technology Meets Math Education: Envisioning a Practical Future Forum on the Future of Technology in Education, ERIC (1999)
12. Akritas, A.: Elements of Computer Algebra with Applications. Wiley, New York (1989)
13. Lavicza, Z.: A comparative analysis of academic mathematicians' conceptions and professional use of computer algebra systems in university mathematics. Doctoral dissertation, University of Cambridge (2009)
14. Drijvers, P.: Students encountering obstacles using a CAS. Int. J. Comput. Math. Learn. **5**(3), 189–209 (2000)
15. Lavicza, Z.: Factors influencing the integration of computer algebra systems into university-level mathematics education. Int J Technol Math Educ **14**(3), 121 (2007)
16. Falmagne, J.C., Doignon, J.P.: Learning Spaces: Interdisciplinary Applied Mathematics. Springer Science & Business Media, New York (2010)
17. Falmagne, J.C., Cosyn, E., Doignon, J.P., Thiéry, N.: The assessment of knowledge, in theory and in practice. Formal concept analysis, pp. 61–79 (2006)
18. Richardson, D.: Some undecidable problems involving elementary functions of a real variable. J. Symbolic Logic **33**(04), 514–520 (1969)
19. Danilov, V.G., Turuntaev, I.S.: Interactive educational system based on generative approach, and the problem of answer checking. Smart Educ. e-Learn **2016**, 527–537 (2016)
20. Danilov, V.G., Turuntaev, I.S.: Reliability of Checking an Answer Given by a Mathematical Expression in Interactive Learning Systems. arXiv preprint arXiv:1602.00243 (2016)

Chapter 9
A Framework for Designing Smarter Serious Games

Katherine Smith, John Shull, Yuzhong Shen, Anthony Dean
and Patrick Heaney

Abstract As smart technology becomes more ubiquitous, there are increased opportunities to enhance the way that educational content is presented and practiced. There are numerous studies indicating the importance of combining instructional and game design in the process of developing serious games. However, there are no frameworks or processes that have been developed to provide insight on methods for effectively combining these existing design methodologies. To this end, a modular framework has been developed with an accompanying spiral design process that facilitates the design, development, and continued improvement of smarter serious games. The implementation of this framework will be explored through an evolving serious game developed using the framework that is aimed to teach precalculus at the college level.

Keywords Smart serious games · Game design · Instructional design · STEM education

K. Smith (✉) · J. Shull · Y. Shen
Department of Modeling, Simulation and Visualization Engineering,
Old Dominion University, Norfolk, USA
e-mail: k3smith@odu.edu

J. Shull
e-mail: jshull@odu.edu

Y. Shen
e-mail: yshen@odu.edu

K. Smith
Department of Mathematics and Statistics, Old Dominion University, Norfolk, USA

A. Dean
Engineering Fundamentals Division, Old Dominion University, Norfolk, USA
e-mail: adean@odu.edu

P. Heaney
Department of Mechanical and Aerospace Engineering, Old Dominion University,
Norfolk, USA
e-mail: phean001@odu.edu

© Springer International Publishing AG 2018
V.L. Uskov et al. (eds.), *Smart Universities*, Smart Innovation,
Systems and Technologies 70, DOI 10.1007/978-3-319-59454-5_9

9.1 Introduction

As of 2015, 86% of 18–29 year olds owned smartphones while 78% owned traditional computers [1]. As today's students are constantly engaging with technology, the ability to further engage them in education on that same technology through gaming is an incredible asset. However, many educators and instructional designers have cited lack of content relevance as a major deterrent to the adoption of serious games in education [2, 3]. There are studies showing serious games are highly effective at conveying content from training for Navy recruits [4] to medical instruction and training for both patients and health care providers [5]. While some efficacy studies of serious games have yielded mixed results, this result is not surprising. Just as there are effective and ineffective teachers, there will be effective and ineffective serious games [6]. The challenge then becomes how to develop serious games that are not only effective, but smart. There are many papers focusing on characteristics of serious games that have been developed [7–9], but there has been no overall methodology put forth to allow game developers to design serious games that get smarter as the player model is developed. In particular, due to budget and scheduling constraints it is often necessary for developers to choose a subset of these characteristics to incorporate into an initial version of the game. From there, it will be important that the overall framework is flexible enough to allow developers to remove characteristics that are not effective for their target population and add elements that may increase learning and engagement. In this process, it is vital for developers to focus on three main aspects. The first is the target player and how they may be different from the general population. The second is the content as they carefully determine what content can be effectively delivered through a serious game and the optimum way to present this content to enhance learning and maintain engagement. The final focus is the game itself and using effective game design techniques to build a game that is engaging as well as instructive. The approach highlighted aims to build upon approaches to instructional design [10] and game design [11] to present a clear method for combining them to create a framework for smart serious games. In addition, this framework supports the enhancement of smartness features including adaptation, sensing, and inference. This is demonstrated by showing how serious games developed under this model can be considered smart learning environments and enhance the overall smartness level of a university.

In this chapter, the discussion will focus on the development of a framework which combines game and instructional design and can be applied to produce an evolving smart serious game that becomes smarter as data collection informs the developers. The end result is iMPOS^{2}inG: A Model and Process for creating Smarter Serious Games which is an adaptable framework that contains each of the modules needed for a smart learning environment [12] while ensuring that those modules are able to adapt and change as more information is revealed about the individual users. A spiral process involving (Re)Definition, Development, and Enhancement of player, instructional and game design characteristics is carried out repeatedly to create a final product that is smart, effective, and entertaining. At each stage, the player, content, and game are emphasized to ensure that all three considerations are properly balanced in the final product.

Beginning with assumptions based on research into the target population, game and instructional design principles are combined to create a serious game that targets general deficiencies in the student population. Through data collection from in-game action, each component or module of the game can be updated individually to take into account learned features about individual students. Finally, an intelligent tutoring layer is added to redirect students through the game modules based on the user model learned from initial trials. This customization is facilitated by having all information about the player's path associated with the player model rather than with individual objects within the game.

The remainder of this chapter will be organized as follows. Section 9.2 provides a review of significant literature related to serious games, instructional design, and smart learning environments. Section 9.3 defines the overall goals and objectives that will be addressed in this chapter. Section 9.4 discusses the overall framework and process along with methods used in their development. Additionally, Sect. 9.4 indicates how this framework can be used to develop smarter serious games that enhance particular smartness features at a university. Section 9.5 discusses applications and implementation of the spiral process and framework for designing smarter serious games. This is demonstrated with examples from a serious game following the framework that has been developed by the authors. Section 9.6 provides discussion of results and lessons learned. Section 9.7 addresses conclusions and future work.

9.2 Literature Review

Smart learning environments. Hwang [12] presents criteria for smart learning environments as well as a set of modules that comprise a smart learning environment. According to Hwang [12], smart learning environments are aware of the learner's context and environment. This can include their physical location and surroundings as well as their behaviors. The smart learning environment is then able to adapt and provide support to the learner based on their specific context. Additionally, smart learning environments are aware and able to respond to unique learner requirements based on information gathered from the user throughout their interaction with the environment. Finally, smart learning environments have user interfaces that are able to change to present content in different ways based on the learner's context and needs. In addition to providing criteria for smart learning environments, Hwang proposes a set of modules that are required to implement a smart learning environment and presents a framework that links content and learner profiles to the learner through a user interface.

Reviews of serious game effectiveness. Connolly et al. [6] provide a highly detailed survey of the literature regarding positive outcomes from studies of serious games. They discover that while negative outcomes from serious games are more publicized, there are actually a greater number of studies reporting positive learner effects. They also note that there is great variability in the methodology used to collect these results which they attribute to the multidisciplinary nature of serious game development. In response, they recommend more rigorous methods for evaluating serious games.

Study of educational game designers. Ruggiero and Watson [9] focus on how game designers approach praxis in their games by interviewing twenty-two educational game designers. Based on the compiled results, they find that many experienced game designers describe the entire process of game design as an action-reflection cycle that is focused first on the content they are trying to convey to the learners. In addition, they find that it is important to understand and reflect on any design and project constraints as soon as possible. The authors recommend using an action-reflection cycle in the educational game design process.

Game characteristics in game and instructional design. Charsky [7] provides an overview of not only how the implementation of game characteristics has evolved over time, but also the importance of collaboration between game and instructional design in implementing these characteristics. The author focuses on competition and goals, rules, choice, challenges, and fantasy and how incorporating these characteristics requires knowledge of topics that span instructional design, game design, and computer programming. He also points out that there is no set procedure for balancing these considerations. More research is required to understand how these various components can be integrated in successful serious games.

Need for a serious game framework and process. Bellotti et al. [13] highlight opportunities for research in the design of serious games particularly identifying a need to study different methodologies and architectures for developing serious games that promote the development of serious games that are effective for a variety of learners across a range of content areas. In our work, we aim to provide a methodology for development as well as an overall game architecture that can be adapted to serve the needs of many content areas and user groups.

Smart features. Uskov et al. [14] identify key features that indicate the level of development of a smart university beyond a traditional university. These features include adaptation, sensing, inferring, self-learning, anticipation, and self-organization. From our perspective on investigating a model for the development of smarter serious games as smart learning environments, we will highlight how smarter serious games can address adaptation, sensing, and inferring. Uskov et al. [14] define these as follows:

- Adaptation describes a smart university's capacity to alter the way it approaches functions, such as teaching and learning.
- Sensing describes a smart university's ability to collect data regarding changes that may affect its interests.
- Inferring describes a smart university's ability to use collected data to make decisions that change the way the university functions or help students.

Smart serious gaming. Uskov and Sekar [15] identify a set of trends which they expect will be followed as serious games progress into smart serious games. A few of these trends are:

- Serious games will evolve by incorporating and enhancing the same smartness features from [14].
- User engagement will tend to increase.
- Collaboration and integration between diverse platforms will be enhanced.

9.3 Research Project Goal and Objectives

This framework for smart serious games has been developed as part of a smart learning environment for military veterans that aims to help universities become smarter by anticipating the needs of a target student population and providing adaptive programs to serve those needs. While this particular project focused on student veterans and STEM courses specifically, the framework and lessons learned can be transferred to support the development of smarter serious games for a range of student populations across a broad array of content areas. This chapter will focus mainly on the development of the framework and components of the program that directly impact this aspect of the project. First, a modular smart learning environment with components that continually evolve to become smarter is presented. In addition, an overall design process that accounts for player, instructional, and game design considerations is proposed. The overall goal is to develop and demonstrate a framework that can be used to develop smart serious games.

9.3.1 Goals and Objectives for Smart Serious Games

The goal for a smarter serious gaming framework is to target many of the deficiencies that traditionally plague serious games while simultaneously serving the target population. In order to do this, the framework needs to be content focused. In addition, the framework needs to focus particularly on challenging content that can be enhanced by interactivity, compelling visualizations, and dynamic rapid feedback which are all characteristics of any good game. Furthermore, providing an adaptable framework that allows small teams to develop games that become smarter as more information about the player is revealed was of great importance. Designing an adaptive learning environment requires a detailed model of the player population which can usually only be obtained by gathering data as the game is played. This requires the development team to use limited knowledge to develop a smart player model that can adapt to become smarter. Going beyond the traditional idea of an adaptive learning environment which allows players to proceed through a custom path, this framework needed to be adaptable to allow incorporation of additional game play enhancements such as score boards to promote competition and integration of external resources for just-in-time (JIT) assistance.

9.3.2 Stern2STEM Program Goals and Objectives

This smart serious game approach has been developed as part of *From Stern to STEM* which is a pilot program designed to develop and investigate techniques to assist military veterans in attaining STEM degrees [16]. This program is designed to recruit driven, capable military veterans with technical STEM experience into engineering and engineering technology degree programs. Once the student veterans are enrolled, the program aims to support the veterans throughout their degree program. First, veterans are aided in preparing for their first college classes and entrance exams by providing STEM leveling assistance and tailored advising. Many of these students have been out of the classroom for an extended period of time and require a refresher on introductory precalculus, calculus, physics, and chemistry courses to allow them to start the program at the correct level. Students are provided with tailored support through tutoring, mentoring, advising, online resources, and interactive gaming to allow them to prepare to begin successfully. The tools provided through this program are designed to combine proven pedagogical practices with smart technology such as interactive gaming to provide the veterans with an experience that is tailored to their individual needs. After graduation, the program aims to provide graduates with career placement resources and provide workplace development throughout the career of Navy STEM professionals.

9.4 Development of IMPOS²inG

While many serious games have been designed and analyzed [6], there has been no adaptable framework presented to guide game developers through the complex process of designing smarter serious games. Serious game design requires understanding of both game and instructional design in order to combine game characteristics and content in a way that leads to a motivating game focused on learning [7]. In order to resolve this lapse, a framework that incorporates game design, instructional design, and player consideration has been developed. As recommended by a survey of experienced game developers [9], this framework involves a spiral process that promotes activity and reflection on that activity throughout the development process. After reviewing a model for instructional design and a model for game design, the discussion will focus on combining the two into a unique process as well as the resulting model for a smarter serious game. Finally, the ramifications of this development on smartness features and smart universities will be discussed.

9.4.1 Overview of the Successive Approximation Model for Instructional Design

The Successive Approximation Model (SAM) was introduced by Allen and Sites in 2012 as a replacement for the more traditional Analysis, Design, Development, Implementation, and Evaluation (ADDIE) model that took advantage of iterative design processes [10]. While an extensive discussion of the model is not necessary, the features and components of the model are summarized here for readers not intimately familiar with instructional design models.

The SAM process starts with information gathering to collect background information needed for a successful project start. The process then moves to a "SAVVY Start" phase which is a short brainstorming session meant to collect ideas from key team members and kick off the iterative design phase. In the iterative design phase, the team follows a design, prototype, review cycle in order to solidify a design that will be ready to move into the iterative development phase. During the iterative development phase, developers follow a cycle where they develop, implement, and evaluate to develop and enhance a product that is ready for market. Once a product is released, the iterative design and development phases continue in order to continually improve the product and incorporate feedback.

The entire process is focused on developing a series of effective learning events that are meaningful, motivational, and memorable. These characteristics are often easily incorporated in games as many of the key game characteristics directly support them. In addition to these characteristics, learning events are comprised of four components. The first component is context which provides the background and setting for the learner's task. The next component is challenge which defines the problem the user must solve or the adversity they must overcome. The third component is activity which is the actual set of actions the user can take in their attempt to complete the challenge. The final component is feedback which informs the user about their performance and potentially generates a new learning event based on the outcome of this current event.

The iMPOS^{2}inG model incorporates aspects of the SAM model to support player completion of effective learning activity throughout game play.

9.4.2 Overview of a Game Design Model

The U.S. computer and video game market had a revenue of 22.41 billion dollars in 2014 [17]. As the market has grown steadily over the last decade, the emergence of large studios and distribution companies has steadily followed suit. With so many commercial developers working to release games, it would seem intuitive that there would be many models for game design. However, as market-leading games can take a great deal of time and capital to produce, these studios are hesitant to turn over their models. Thankfully within the past decade, casual game design models

Fig. 9.1 The MDA model by
Hunicke et al. (*source* [18],
p. 2)

Designer   Player

have emerged. In general, these models share many aspects and some would argue
that their differences lay in more semantics than functionality. For our intentions we
will identify three models that have been highly referenced and reviewed. These
models are the Mechanics, Dynamics, and Aesthetics model (MDA) by Hunicke,
et al. [18], a model of lenses by Schell [19], and a playcentric approach by Fullerton
[11].

The MDA model was developed using an iterative process through a game
design workshop run by Robin Hunicke [18]. A graphical overview of this model
from [18] is shown in Fig. 9.1. Within the MDA approach, the three design elements are broken down into specific components based on the interaction between
the rules and the system and focusing on the fact that the result of this interaction
should be enjoyable gameplay [18]. This framework is linear and viewed by
designers and players from different ends of the process. Designers approach the
game starting with Mechanics, while players approach it starting with aesthetics.
From the developer perspective, mechanics would be addressed first as this is
related to the overall structure of the game and how algorithms are developed to
implement the game rules. Next, dynamics focuses on how the elements of the
game and player interact as the game is played and time passes. Finally, Aesthetics
is described by Hunicke as "…desirable emotional response evoked in the player"
[18]. These experiences and responses are a result of the player's interaction with
the game. Good design would have the goal of evoking these positive responses
from the player. Overall, the three design components are considered as a lens
through which the developer and player view the game. The idea behind viewing
the game through the lens of MDA is to view the game as separate components that
are casually linked.

The lens approach is expanded greatly by Jesse Schell who developed the
second game design model as a model of lenses [19]. Schell lays out a list of
hundreds of lenses that build upon the MDA concept. These lenses present the
design element as a series of questions that are intertwined with what Schell calls
the four basic elements, *Mechanics, Aesthetics, Technology, and Story* [19]. Schell
has these four elements linked as shown in Fig. 9.2. In this model, the definition of
mechanics is very similar to the definition from the previous model as it states that
the mechanics components consist of the rules and procedures for the game to
function. In addition, Schell's model highlights that these rules and procedures
should also describe the goal of the game. In this model, a separate story component
is identified where the story is the driving engine that helps to define what
mechanics will be needed. The Aesthetics are similar to the MDA model in that
aesthetics are described as the medium in which the player perceives the game

Fig. 9.2 The elemental tetrad by Jesse Schell (*source* [19], p. 51)

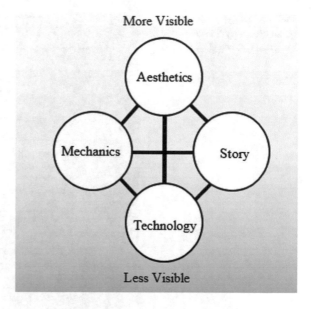

mechanics and the story. This naturally includes any sensory perception as well as emotional response from the player. Schell includes technology as a fourth aspect and particularly emphasizes that the technology used to deliver the game experience deserves consideration in its own right. The definition of technology is loosely stated to encompass anything that can be used to deliver the game experience including pen and paper, cards, computers, and mobile devices. He states that the "... technology you choose for your game enables it to do certain things and prohibits it from doing other things" [19]. These four elements are built upon to show how each plays a role in the overall process of game design. Similar to Hunicke's approach of considering the player perspective, Schell includes an interpretation of which elements are more versus less visible to the player.

Tracy Fullerton leads the USC game design lab and has developed an approach similar to Schell's and the MDA framework. Fullerton's approach is composed of three main elements, *Formal, Dramatic, and Dynamic* [11]. These three elements share commonalities with MDA as well as Schell's four elements. Fullerton wraps up mechanics, logic, rules and procedures into the 'formal' element as she agrees with Schell that these elements are what separates games from other media avenues [11]. Fullerton combines the aesthetics and story from Schell's model into her dramatic element. She expands greatly in areas of what makes a game challenging and how to incorporate the nature of play into the process. Fullerton's dynamic element is similar to MDA, but expands into a description of how defining simple rules and logic in the formal element can lead to a changing dynamic environment in which the player interacts. Fullerton presents her model as an iterative game design model which can be seen in Fig. 9.3.

Fig. 9.3 Model for iterative
game design by Tracy
Fullerton (*source* [11], p. 272)

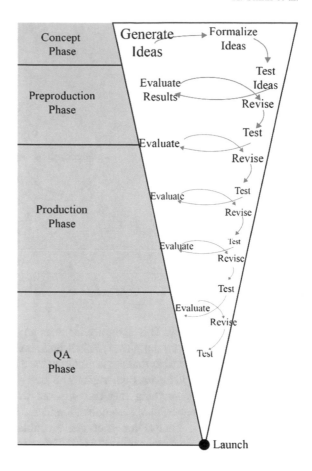

Each model uses similar constructs but takes a slightly different approach. The MDA framework focuses on the different ways in which game designers and players approach the game. A key point from this model is that when designing games one should truly consider how the player will experience the game. Schell's model [19] is a detailed breakdown from a developer's point of view, where each of his four elements can be assigned as tasks to various team members. Fullerton's approach defines three components that can be reworked in an iterative process. When considering these three models and factoring in the importance of incorporating instructional design elements, Fullerton's approach was selected as a starting point to build the new framework as it was the most adaptable and allowed the instructional design elements to flow freely.

9.4.3 A Model for a Smarter Serious Game

In his discussion of smart learning environments, Hwang [12] discussed seven modules that comprise a smart learning environment including a learning status detecting module, a learning performance evaluation module, an adaptive learning task module, an adaptive learning content module, a personal learning support module, a set of databases, and an inference engine. The iMPOS^{2}inG model for smarter serious games (Fig. 9.4) incorporates each of these modules. Each module will be discussed in detail below. These modules are not only meant to provide conceptual modularity, but also modularity in software design. By designing the code in a modular way, individual modules can be modified much more easily without greatly affecting other modules.

Capturing in-game actions and events. Every smart learning environment needs a learning status detecting module. In smarter series games, this is the portion that allows for monitoring of player actions and detection of in-game events. Depending on the type of game, this process can be very simple or very complex. For example, in augmented reality games, the player's physical movement through space, geographic location, surrounding objects, and gesture interactions all need to be tracked. In a physical fitness training game, wearables can be used to monitor the player's heart rate, movement, and location. In contrast, for a simple turn-based game where the player interacts with the mouse, in-game actions would consist of a player taking their turn or interacting with menus.

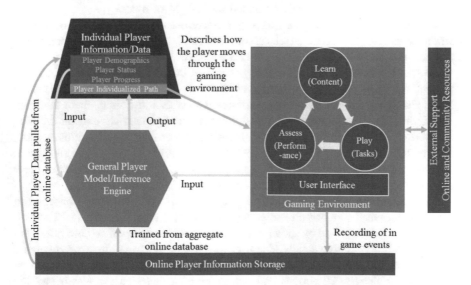

Fig. 9.4 Model for smarter serious games incorporating modules needed for smart learning environment

Storing User Data. In order to record progress and inform a player model, player actions and in-game events need to be stored. The simplest way to do this is using local storage on the user's device. It is relatively straightforward to store data objects in a serialized format that can be accessed throughout game play. Local data storage also has the advantage of allowing users to play without an internet connection. However, storing data locally does not allow players to maintain progress between devices or provide a backup in case the local file is corrupted or lost. In addition, locally stored data is inaccessible to game developers. By using one of the many available online storage systems, developers have access to player data to monitor player progress, improve game play by analyzing player results, aid in identifying user reported problems during troubleshooting, and inform a player model to support an adaptive game.

Player model and inference engine. The gold standard for smart educational tools is to provide tools that adapt to the learner to support their learning based on learner characteristics, learning style, as well as past results. This requires a refined player model that serves as an inference engine where the inputs are information about the player and player actions and the outputs are in-game events such as levels being unlocked or additional content and resources being provided.

Individual player information and data. Individual player information needs to be accessed and stored throughout the game. By centralizing all of this information as a single entity that is persistent throughout the game, all other elements within the game have a single point of reference for the player's progress, current status, and progression through the game. Other details unique to a certain serious game implementation can easily be added to this player model.

Learn, play, assess in the gaming environment. The *Learn, Play, Assess* structure mimics a traditional classroom structure where material is presented, students practice the material, and finally are assessed on what they have learned. In some smart serious games, these three activities could be carried out seamlessly throughout game play so that the player is not aware of a transition between the three. However, the game developer still needs to consciously implement all three. The *Learn, Play, and Assess* modes provide the adaptive learning content module, adaptive learning task module, and learning performance evaluation module, respectively. Even though in reality these three modes of play may have significant overlap in keeping with a system dynamics approach, we will discuss each of these aspects as a mode in which the user is engaged during game play.

User interface. An adaptive and responsive interface is critically important in serious games as it supports all components of effective learning activities. The user interface provides context by showing the setting and background and providing the player with any narrative details. The user interface needs to accurately convey the challenge to the player and potentially provide clues on how the player can overcome the challenge. All actions and communication between the player and the game during the activity phase are provided by the user interface. Finally, feedback on performance is provided to the user through the user interface.

External and internal learning support. Throughout game play, content resources can be made available to players to facilitate their learning. In the

Stern2STEM program, this involves directing users to other program components including tutoring and advising. However, in general, there is a vast array of resources available for most topics in education. Simply redirecting student to a short video example at an appropriate time can be considered incorporating external resources.

9.4.4 The Cycle for Smart Serious Game Development

Now that a model for a smarter serious game has been presented, a process that will allow for the development of models within that framework is required. A process incorporating game design, instructional design, and player considerations is shown in Fig. 9.5.

Environment. The first step in the process is similar to the collection of background information from SAM. The process begins by identifying and collecting information from subject matter experts (SMEs), target student population, curriculum and content requirements, and any existing resources including existing resources that are smart. All this information will be combined to define the environment within which the smarter serious game will be developed.

(Re)Define. The outset of the project will begin in a Define phase. As we will cycle through this phase many times, it will become a Redefine phase that will allow for updates to the existing definitions from feedback information. During this stage, we are defining the target audience needs and characteristics. Initially, this may be a general set of characteristics that will eventually evolve into a player model that can be used in combination with an inference engine to allow the game to become smarter.

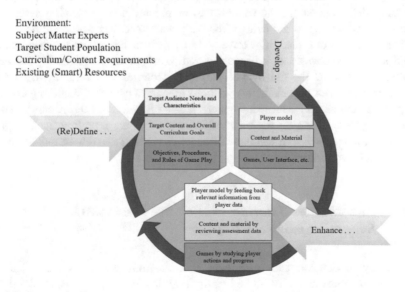

Fig. 9.5 Process for developing smarter serious games that fits within an adaptive framework

The content and overall curriculum goals are also under development during this stage. Initially, this involves selecting a group of topics at a high level. As the process continues, determinations are made about which of those topics contain learning objectives that can be effectively conveyed through game play. Further in the process, learner feedback will be reviewed to determine if there are topics that are not being conveyed effectively or additional topics that can be added to further assist learners. Finally, as part of the game development, the overall game objectives, procedures, and rules are being defined from both a player and developer perspective.

Develop. Proceeding into the develop phase, development begins on the software that will eventually become the completed game. The player model will start very simply. As more information is collected, the player model will become more sophisticated with the goal of eventually having a player model that supports an adaptive learning experience. In addition, content and material development begins by creating both in-game and external resources as necessary. At this stage, the structure for the games and user interface is cultivated. As more cycles of the process are completed, these will become more mature. It is best to follow a modular design so that items in the user interface can be swapped out and replaced as needed. In addition, designing games that are reusable for a number of different topics not only shortens development time, but also decreases the burden on the user to learn a large number of different modes of play.

Enhance. The key benefit of rapidly developing a playable model is that feedback can be easily gained from demonstrations and players. During demonstrations, developers can collect feedback on how easy the controls are to use, if there is enough direction and help, and what may be frustrating or preventing players from completing certain tasks. Once there are a significant number of players, the data collected during game play can be used to inform and redefine the player model. In addition, the content can be enhanced by reviewing assessment data to see which topics are being effectively addressed and which are not.

This process is designed to be traversed many times throughout the development spiral so that feedback can be incorporated to improve the game. By having a modular framework for the overall game, it becomes easier to modify certain aspects that may not be satisfactory without having to change the entire game. For example, if the current data management system is not working, it can be easily replaced by only changing the parts of the code that are reporting back in-game events. A good code design will have these items isolated to a few scripts so that the modifications are minimal.

9.4.5   Focusing on Smartness: How the Framework Supports Smartness Features

The ability to develop and implement smart serious games as smart learning environments enhances a smart university's ability to adapt, sense, and infer to provide an enhanced teaching and learning experience. These are three of the smartness features based on the Taxonomy of Smart Universities developed by Uskov et al. [14]. Each of these smartness features is discussed as it is highlighted in the iMPOS^{2}inG model as well as an example game that was developed using this system. The game highlighted is called MAVEN and was developed to help students obtain a better understanding of precalculus topics.

Adaptation. The very purpose of developing a modular framework for smarter serious game development was to enhance the developer's ability to modify existing elements to allow them to adapt to player needs. In fact, adaptation is highlighted twice in this approach. By using modularity and well-defined interfaces in the software design, the model itself is adaptable by providing the ability to interchange various components to better meet the needs of the learners. In addition, the model provides the opportunity to collect data to inform an adaptive player model and inference engine that will then be implemented to allow the game itself to adapt to the needs of individual learners.

In MAVEN, the adaptive model has been used to allow design changes driven by initial feedback and testing. For example, the original in-game menu user interface was separated and positioned at corners of the screen. Figure 9.6 shows an older implementation of the interface with the menu in the top left, help in the top right, and a back button to return to the main navigation menu in the bottom left. In addition to this physical separation which required users to interact with different parts of the screen to conduct menu interactions, the control of these areas was separated. Changing the menu user interface to one that was more compact and centrally controlled was made easier due to the modularity of the overall design. The only components that had to be modified were the menu items and their interfaces which had been kept simple to support the modular design. The new compact design is shown in Fig. 9.7.

Fig. 9.6 A scene showing a former implementation of the menu user interface

Fig. 9.7 New menu user interface where menu controls and player status are consolidated and available in one location

Additionally, the player model itself supports adaptation. Currently in MAVEN, the player model adapts to the player's performance by only allowing the player to progress when they have successfully completed previous tasks. During play, the game is constantly collecting feedback and player data that will allow the game designers to adapt the player model to be more intelligent as time goes on. Eventually, it is feasible for the framework to support a player model that adapts the path of the user through the game based on previous performance as well as external factors.

Sensing. Since the game is constantly collecting player data, it is continuously monitoring the actions of the players. It collects data about how and when learners are using the game. This can contribute to awareness of unique ways in which students are engaging with the content. For example, a variety of information is collected as individuals play MAVEN. Initially, players are required to create an account and answer a series of demographic and educational background questions. This will allow developers to assess whether or not external factors contribute to the overall effectiveness of the game. Additionally, each gameplay action in MAVEN is mapped to a question that the user gets either right or wrong. For example, one of the games in MAVEN requires players to match the graphs of functions with the correct category of parent function as shown in Fig. 9.8. The player is presented with a set of missiles, each displaying a flag that shows the graph of a function. There is also a set of planes that each display the name of a type of functions.

The goal is to load the missiles on the planes that correspond to the correct parent function type for the graph. The player controls a cart that they must use to pick up missiles and transport them to the correct planes. In this game, each time the player loads a missile they are actually categorizing a function by type. The player's action is recorded as the response to a question and scored as right or wrong. Information about the problem, including the graph of the function, and the player's answer, and whether it was correct or incorrect, is stored in a time stamped online database of player information. This information can be used by developers to review player performance. It also provides training data for an intelligent player model and will provide inputs to that model for future players. For example, given that a player has missed a certain number of questions of a certain type, the model may recommend they play a different game or provide information on more formal remediation.

Inferring. In addition to the player inference model which makes inferences to inform the player's path through the game, the model can be used in a more general

Fig. 9.8 An example of a game from MAVEN where players practice their knowledge of library functions by matching functions with their parent type

sense. For example, by collecting data and pairing it with student data already maintained by the university, recommendations can be made in order to alert educators of students that are not performing well in particular areas. These early alerts could allow interventions that would help the student to get back on track before it was too late.

Currently, inference is implemented on a smaller scale. The Stern2STEM program provides tutoring to on-campus student veterans as well as providing education games for their use. While students are being tutored, the tutors will often encourage them to download the game and play it either in the tutoring center or at home. The tutors keep a list of the usernames of the students who are being tutored and monitor student progress through the game to help inform them about the student's progress.

9.5   The Model and Process in Action: Discussion and Supporting Examples

Now that the iMPOS^{2}inG model has been presented and defined, it is important to demonstrate how it can enhance the development of serious games and help them to become smarter. Following a discussion of how the framework can be applied, this will be demonstrated by specific examples using a series of games that were developed to help military veterans enhance their understanding of precalculus content through the Stern2STEM program [16].

9.5.1 Refining the Player Model

At the beginning of a project, game developers are unlikely to have a pre-developed, sophisticated player model. However, in the early stages of development, developers can research to find general information about the target audience as well as information about the content that can inform a simple player model. For example, developers may interview potential players to gather information or talk to a content expert or instructional designer to determine the best method or order in which to present content. The simple model developed may be a linear model that describes a common path that all players take through the game, progressing only when they master the previous content. As user data is collected and analyzed, this simple model can be replaced with a higher fidelity model that allows the game to adapt to the user's needs based on performance. In fact, multiple models could be developed and compared to see which best support the learning goals of the game.

As an example, the Stern2STEM project performed a literature review and relied on subject matter experts in both the veteran population and academic community to devise a descriptive set of student characteristics [16]. This set of student characteristics in addition to information from STEM educational content subject matter experts helped game developers design a primitive player model at the onset of development of MAVEN. This primitive player model was used to develop a linear gameplay progression. An illustrative example of this gameplay is described below. An image showing one of the content submenus from MAVEN is shown in Fig. 9.9.

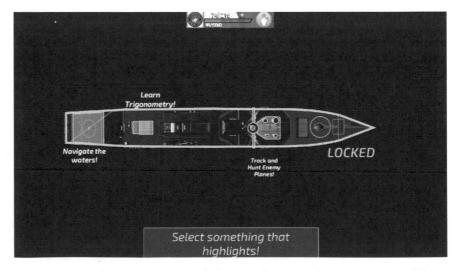

Fig. 9.9 An menu scene from MAVEN which is a serious game developed to help military veterans obtain a better understanding of precalculus content

This menu shows that the player has limited access to game areas. This access is updated based on completion of specific tasks. For example, when the player first gains access to the Trigonometry Destroyer shown, they can only access the "Learn Trigonometry!" area. Once they have gone through some content introduction in this area, they will gain access to the first play area "Track and Hunt Enemy Planes!" where they will practice the skills they have just learned. They will continue to play until it is determined that they have mastered the previous content and are ready to continue learning. This process of unlocking will continue until the player has unlocked and mastered all available play areas. We can see in Fig. 9.9 that this particular player has one play area left to unlock. Once all play areas are mastered, the player is assessed to determine their level of understanding of the content presented in the learn and play areas on this ship.

While this is certainly a primitive model that makes many assumptions about how the target population learns and employs a simple adaptive approach that is responsive only to player mastery, the modular design of Fig. 9.4 makes it easy to interchange a more advanced player model with this one. This simple model allows the development of the game to move through a full cycle, so that player data can be collected to inform a more sophisticated model. In addition, it is important to note that the player's individual path through the game is associated with the player profile rather than then overall player model. This is essential as it allows the individual player's path to be an output of the player model so that it can be easily individualized for more sophisticated models.

In addition, the player inference engine has inputs both from the player's individual data including demographics, player progress, and player status as well as from the game itself. This makes it possible to develop a model that adapts the game play in a way that takes into account in-game actions, learning styles and behaviors.

9.5.2 Learn, Play, Assess: Bringing Together Game and Instructional Design

In the Learn mode, players are presented with content. Technology provides a variety of ways to present different content including interactive visualizations, videos interspersed with checks for understanding, and interactive pop ups that remind players of key principles during game play.

In Play mode, players engage in game play focused on practicing the content knowledge they are learning. As discussed above, in the Successive Approximation Module (SAM) for instructional design, effective learning events involve context, challenge, activity and feedback [10]. Each learning event should be structured to include these four components while also adhering to good game design. In Fullerton's game design model, there are *Formal, Dynamic, and Dramatic* components that must be implemented throughout the game [11].

9.5.2.1 Structuring Learning Events by Interweaving Instructional and Game Design

Since the goal of a serious game is to help the player learn, the first focus is on the four components of an effective learning event and the second is on ensuring that each of the components needed for game design are incorporated into each component of the learning event. This process is described below.

Context. The context is the way that the game is presented to the player. The Formal elements incorporated in context are the rules and the instructions for the player to follow. The Dynamic elements of context relate to how the context changes as the player takes actions within the game as well as the strategies that are promoted to the player through cues from the context. The Dramatic elements of context involve the background story and setting for the game. The context is usually dominated by the Formal and Dramatic elements.

Challenge. The challenge is the problem that is presented for the player to solve. The Formal element of the challenge is the actual problem itself. The Dynamic elements of challenge are how the player devises a strategy to overcome the challenge. The Dramatic elements of the challenge are the visual features and feedback that help the student recognize the problem they need to solve. The challenge is usually dominated by the Formal element.

Activity. The activity is the set of actions that the player needs to complete in their attempt to solve the challenge. The Formal aspect of the activity is related to the boundaries on the gameplay and the resources the player has available to them. The Dynamic element of the activity is the set of actions that the player actually takes as they attempt to solve the challenge. This includes player strategy and behavior within the game. The Dramatic element includes the visuals that change as the game state changes. This may be due to a player action or simply the passage of time within the game. The activity is dominated by the Dynamic elements of game play.

Feedback. Feedback is provided to the player to inform them of changes in the game due to player action or the passage of time. For example, if they solve a challenge successfully, they would be rewarded with positive feedback. If there is a time limitation, they would be made aware when the time to complete the challenge was running out. The Formal element of feedback involves the rules for awarding player score and resources based on how they behave in the game. The Dynamic aspect of feedback focuses on the relationship between game and the player. For example, if a player fails to solve a challenge, they receive an indication that encourages them to modify their strategy. The Dramatic element of feedback includes visual effects, sound effects, and text or narrative feedback that result from a player's action during activity. The feedback component is dominated by the Dramatic elements of gameplay.

9.5.2.2   An Example Learning Event

Following this general overview of how instructional design and game design are interwoven throughout learning events, it is helpful to go through an example of learning events in a game that follows this model. For this example, the focus will be on the game shown in Fig. 9.10 which shows a serious game designed to help players practice their manipulation of trigonometric identities.

Context. The player is presented with a set of missile expressions. Each missile type has a different trigonometric expression and limited ammunition. In addition, there are ships positioned on a grid. Each ship has between two and four targets, where each target is associated with a trigonometric expression. The Formal elements are the defined rules of gameplay that stipulate that the player is able to cycle through their missile expressions and search through the target expressions. In addition, the restricted number of missiles is also a Formal element in that it places a resource restriction on the user. The Dynamic elements are involved with the ability of the player to cycle through the missile expressions and search through the ships to find target expressions. The player has a choice between keeping a set missile expression and selecting different ships, keeping a set ship and cycling through missile expressions, or some combination of the two. Each of these strategies may result in varied degrees of success. The Dramatic elements are the background story and setting. In this game, the player is attacking enemy ships in an attempt to protect their own ship. They can see their battleship in the foreground while hunting through the open ocean in the background.

Challenge. The player needs to fire missiles at targets that match the missile type. A match is determined by whether or not the trigonometric expression

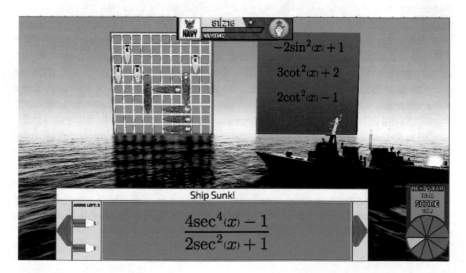

Fig. 9.10 An example of Play mode in a serious game designed to help players practice their knowledge of trigonometric identities

associated with the selected missile and target are mathematically equivalent. This defines the Formal element of the challenge by formally identifying the problem. The Dynamic element focuses on how the player changes their strategy to address this challenge. For example, they may at first try to only cycle through the missile expressions while focusing on a single ship, but find that this takes a long time because there are a small number of target equations on each ship and most of the missile expressions will not match. They could adapt their strategy to better address the challenge. The Dramatic element of the challenge is how the challenge is interwoven into the narrative. In this game, the player needs to sink ships.

Activity. The player can cycle through missiles and search through various ships to find a match. They must use algebraic manipulation and trigonometric identities to decide if they have found a match. Once they have found a match, they indicate their answer by clicking on the target expression. The Formal aspect is the set of affordances and limitations that are placed on the user's activities. They can use the arrows to cycle through expressions and click to select different ships. However, their actions are limited to the number of ships on the game board and the number of missiles remaining for each missile expression. The Dramatic element is revealed in how the player interacts with the game. Their activities during game play and behaviors are both Dramatic elements. The Dramatic element is how the setting and user interface change as game play moves forward. In this case, the expressions are different on each ship and the player can see which targets they have already hit.

Feedback. Once the player fires a missile, feedback is immediately provided in a number of ways. The Formal aspect defines the rules by which a hit or a miss is awarded to the player. In addition, the rules for the number of points awarded for each hit and how ammunition is removed are Formal elements. The Dynamic element involves how the player may modify their strategy based on the feedback. For example, if a player got a hit, they may look to see if there are any other target expressions that are the same as the one they just hit. The Dramatic element is revealed in the way the sound and visual effects are used to indicate the result of the player's activity. In the event of a hit, an explosive sound effect is triggered, the target expression is replaced with "Hit!", and the player is rewarded with an increase in score. In the event of a miss, the player hears the missile splash into the water near the target and "Miss!" is displayed in the text feedback area directly above the missile expressions. In addition, feedback on overall progress is displayed at the bottom right of the screen showing a green segment for ships that have been sunk and a red segment for ships that are no longer able to be sunk due to a lack of ammunition.

Assess mode evaluates the player's content retention over a range of previous topics. The assessment could be built into game play, take the form of a more traditional test, or be presented sporadically throughout game play for players to earn bonus points.

9.5.3   Topic Selection: Focusing on Topics Appropriate for Gaming

While topic selection may at first seem like a purely instructional or top-level project consideration, game developers quickly realize that their input in topic selection can have a large effect on the overall success of a serious game. Because of the time and expense required to develop a serious game, it is first important to narrow down the list of topics to those that students struggle with that are not appropriately addressed by existing methods. If there is no demand for a new way to teach a certain topic, developing a game to do so will have little chance of being successful. Once the list of topics has been narrowed to a subset, the game developers need to consider how effectively the content can be conveyed within the confines of a game. Sometimes this takes more than a little imagination; but making sure that the game play is well designed and based solidly in the content is important [20]. In a traditional classroom, getting students to do their homework is a constant struggle; however, one of the main assets favoring gamifying education is that when students are engaged in game play, they are effectively practicing skills. Whether they are practicing skills through homework or game play, the outcome of practicing and engaging with the material is the same [21].

The list below defines a set of criteria that are important in selecting content areas which can be successfully incorporated into games. While it is not necessary for each topic to meet all the criteria, identifying topics that meet most of the criteria helps to narrow the range of topics to those that are more easily adaptable to being presented through game play.

- The topic is not satisfactorily addressed by an existing resource.
- There is a known deficiency in the understanding of the topic in the target population.
- The topic would be enhanced by interactive visualizations.
- The topic involves some process of cause and effect that is not easily visualized by traditional methods.
- The topic involves a learning process that can be broken down into manageable steps that can be turned into effective learning events.

It is important to note that some topics may not be covered completely by serious games. However, taking the most common stumbling block in a process and creating a game around that portion can help students gain the skills and confidence they need to learn the rest of the topic by more traditional methods. Game designers should not be afraid to mix serious games with external resources to present the student with the best possible learning environment. In fact, one of the key features of a smart learning environment is that it provides timely access to external resources to promote student understanding.

Example topic selection. Students in precalculus traditionally struggle with graphing using transformations. This topic is particularly important as they move on in mathematics as more advanced courses are often dependent on student

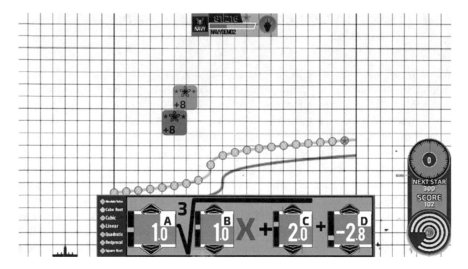

Fig. 9.11 A serious game that helps students learn about graphing by transformations

understanding of how functions change. The existing method to teach this in the classroom is to process the function transformations one at a time and move the individual function points. This leads to difficulty for the students as they attempt to visualize the overall function transformations. This topic meets all five of the conditions above.

In order to address this topic, an interactive game has been developed. This game is shown in Fig. 9.11. This game will be presented in terms of the components of an effective learning event to give another example of how to apply the iMPOS^{2}inG framework. After giving this overview, emphasis will be placed on the components that make this game a good match for the selected topic.

Context. The player is presented with two graphs that are identified as paths for their ship to take through the ocean. The darker grey path is the desired path while the lighter gray path is the route that the player is charting. The equation for the lighter gray graph is shown at the bottom of the screen.

Challenge. The player needs to choose the correct parent function from the grey area and the correct equation coefficients to match the lighter gray graph to the darker grey graph. There is limited time as the ship is already traveling along the path.

Activity. The player can use the toggle menu to the left to select the correct parent function. Then, the player can use the up and down arrows to increase or decrease the values of the four coefficients in the equation to modify the graph to match the target graph.

Feedback. Every second, the player is awarded points based on how closely their graph matches the target graph. These points are shown using score pop-ups. This rewards players for working more quickly. In addition, players can see how close their graph is by viewing the bull's-eye pattern at the bottom right that indicates a percent match between the two curves.

This game overcomes deficiencies in the traditional method of instruction by allowing the player to manipulate the equations themselves and immediately see how their changes affect the graph. This is a quicker way for them to understand the function behavior than graphing by hand. Once they have a firm understanding of the effects of changing the different coefficients, they will be better equipped to move to the next learning objective which would be sketching the graphs by hand.

9.6   The Model and Process in Reflection: Discussion and Lessons Learned

9.6.1   Broader Applications

Though the game used as an example in this work was developed to help learners master precalculus skills, the overall framework and model developed can be used to develop games for a variety of disciplines. Particularly, it would be easy to extend the model to cover courses that are problem solving focused since the model depends on mapping in-game actions to questions that can be scored as either correct or incorrect. Examples of subjects that meet this description include calculus, physics, chemistry, statics and dynamics. Students in all of these subjects traditionally have difficulty visualizing problems and situations that arise, and a virtual interactive environment would therefore support learning.

In fact, if a series of interconnected games were to be developed, instructors could use a similar environment across a range of classes while receiving feedback and early warnings when students were struggling with particular sections of content. Stretching this idea further, virtual and augmented laboratories could be incorporated as well to bring students a common virtual experience that would allow them to engage with a number of smart systems as they completed their coursework.

One thing that is interesting about the overall process is that it is general enough to be applied to a variety of courses at a smart university. Though some of the games developed for MAVEN are specifically tailored towards precalculus or mathematics content, other games could easily be modified to include material from a variety of subjects. For example, the destroyer game shown in Fig. 9.10 is used to help students learn how to solve trigonometric identity problems. However, the root game play action is matching. Therefore, any class which poses questions that can be expressed in a matching context can be employed in this game. This includes a variety of subjects including science, history, and business.

9.6.2 Comparison to Existing Products

At this time, there are no other serious game in mathematics for higher education that truly incorporate mathematics into gaming. There are other excellent interactive mathematics resources, but they are mainly sources of content with interactive questions interspersed. Khan Academy [22] is one excellent example of an interactive educational tool that is available on a wide variety of devices and presents a broad range of topics including precalculus and calculus. Khan Academy presents content mainly as videos with some typed notes interspersed. Students are invited to practice what they have learned by completing questions between videos. There are a variety of question types, including multiple choice, text entry, and a graphing utility. Even though Khan Academy provides some aspects of gamification including badges and achievements, it is not technically a game. Serious games, such as the one developed here, complement the existing content resources by encouraging students to engage in practice in the form of gameplay.

9.6.3 Lessons Learned

Throughout the development of this process and framework, there have been quite a few bumps in the road. By being fastidious about employing modularity to both the conceptual and software design, the flexibility required to adapt to changing requirements and incorporate feedback along the way has been maintained. In this section, a discussion of several issues that were encountered and how these issues were surmounted will be provided.

9.6.3.1 Developing and Integrating External Tools for Equation Display

The first issue was the display and manipulation of symbolic expressions and graphs. The solution was to develop a custom toolkit to support the development of the games [23]. Once this toolkit was implemented, it underwent many revisions as more content was continually added to the game. By having a modular design where the toolkit was only accessed through a single problem generating script, the ability to make changes easily was preserved. In fact, even switching to another tool, if required, would not have been difficult. In addition, there were instances where real-time player interaction with functions was required. Because the toolkit was calling Python scripts from the game, the implementation was not fast enough to provide the desired real-time performance. In these instances, custom solutions were developed while only modifying the interface between the problem generating code and the specific game under development.

9.6.3.2 Using Modularity and Communication to Support Collaboration

Another issue encountered was with version control. A common problem in game development is the handling of asset files by version control software. The solution to this turned out to be a combination of modularity and communication. Before the framework presented in this chapter was being utilized in the project, an unsuccessful attempt was made to implement version control. This led to a great deal of copying files, sending updates back and forth through email, and a significant amount of repeated work. Since the framework has been implemented and project components have been carefully set up to be independent, different developers are able to work on different aspects of the project simultaneously. As long as there is good communication in place, there are very few issues merging after modifying asset files in binary format. One of the keys to this success has been the use of Bitbucket [24] as a cloud hosting service for the entire project. Using Bitbucket along with the git bash terminal rather than a git tool that had a graphical user interface, provided a high level of control when merging files which prevented many issues.

9.6.3.3 User Interface Redesign and Implementation

Another issue faced involved the user interface. Initially, the assumption was made that the user interface would be essentially the same in every scene, so that it would be smart to have it be one consolidated piece where individual components could be turned on and off. This design seemed modular at first because the components of the user interface were all together. However, over time the user interface became unwieldy to work with and modify and was a bottleneck in the work flow. Looking back at the overall framework, there was not enough separation in this original model. While the user interface can be viewed as one component, different pieces of the user interface are intimately involved with different modules in the framework. After this realization, it made much more sense to break the user interface into functional pieces that were associated with different framework modules. This current implementation has a player menu, feedback menu, and user input menu. Each of these menus is highly customizable and closely tied to the module it is associated with rather than being tied to other user interface elements.

9.6.3.4 Integrating an External Player Data Management System

Data management also became a problem early on. Initially, all of the player data was kept locally on the user's machine, but this limited the model use case and jeopardized the player's experience by prohibiting developer access to player data and not allowing players to maintain progress between devices. As the development of MAVEN continued, this concept was partially reworked as the decision was made to host player data on a server so that the players could access their own data from different devices and the developers had access to player data to allow for analysis that would lead to game improvements.

This solution was implemented quickly by taking advantage of the existing project web hosting services and using a Wordpress plugin within the development engine. It turned out that this setup is limited and should only be used for small amounts of player profile data including email, player avatar name, and other non-duplicate data structures. The approach does not work well for storing data that is generated from in-game actions and events because of the sheer quantity of data generated during play. Since a great deal of data needed to be collected in-game to improve the game and eventually inform an adaptive player model, alternative solutions were considered. A better approach is to use a system that stores data in a relational database system. This allows not only for the storage of data, but the eventual retrieving and processing of data to inform developers based on user progress and actions. While there are many existing services that would meet the project needs, one example of a suite of packages that addresses data management problems is Amazon Web Services (AWS) [25]. A major benefit of these services is their ability to be integrated with existing social media sites where users may already have accounts. This can be vitally important to retaining users and being able to compare various user player data models. The combination of offloading data management to a system like AWS and using a complementary social media platform allows more interaction among players and allows us to take advantage of leading cloud storage systems. The cloud model is an industry standard in software development as it provides the developers with an enhanced package of tools, takes advantage of economies of scale, and automatically adjusts to handle various user capacities. Players will be able to log in from a wide range of platforms and experience the same result.

In particular, cloud based storage systems offer the opportunity to use their services to handle data management, employ industry standard encryption for user data, integrate social media platforms, broaden the user base, and importantly build a complex system of player data models. These player models can take advantage of leading technology trends in machine learning, adaptive learning, and smart learning models. Using a cloud based relational storage system in combination with an accurate player model can provide developers and educators with valuable information and allow that information to inform complex player models that will provide players with the smartest player model. This player model can dynamically adapt to the player's current level of knowledge. Figure 9.12 provides a visual representation of how these updated services would connect into the existing

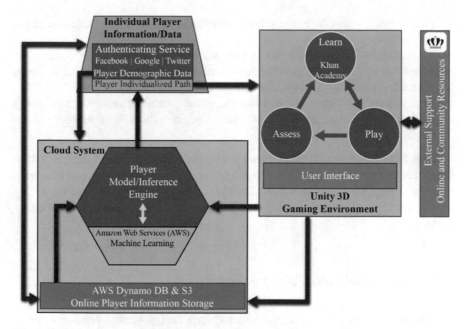

Fig. 9.12 Using the model for smarter serious games incorporating AWS service modules needed for an adaptive, smart learning environment

model. The overall conceptual model does not change, but the modular components are replaced to provide a smarter learning environment.

9.6.3.5 Game Manager Implementation

The final lesson learned was about overall game control. Since the design is modular, it makes sense to have a state machine that can transition the game between a set of known states. This allows for events that are common for all modules to be centralized while individualized events can be handled by their respective modules. These states are easy to implement from a single script and it is relatively simple to add another state should the need arise. Currently, MAVEN has the following six states:

- Initialize: Allows for initial variables to be set at the beginning of a menu or game. The online data functions that need to be called at the beginning of every game are centralized.
- Warm-up: Allows for a variable length pause between player entry into a scene and the start of a game.

- Demo: Incorporates the use of guided prompts to instruct a user in how to play a certain game.
- Play: Regular mode of play where the user is completing learning events.
- Menu/Pause: Allows the game to pause and open the menu at any point. Allows for central handling of pause and menu functions from any scene.
- End: Allows for a common end to each game where a round over menu with feedback on the level just completed and navigation buttons is presented to the player.

Centralizing the transition of these events to a single game manager which interfaces with individual game controllers in every game has proved to be a successful way to handle transitions between states as well as assisted in getting new games up and running quickly by handling the common events that occur in every game.

9.7   Conclusion and Future Work

In this chapter, instructional design and software design models have been combined into a novel modular framework and accompanying development process aimed to facilitate the design and development of smarter serious games. Under this framework, each smart serious game can be viewed as a smart learning environment that enhances the smartness of an overall university by contributing to the development of individual smartness features. In addition, this framework and process have been demonstrated through their application to the development of an evolving serious game designed to assist military veterans in enhancing their understanding of precalculus topics as they return to pursue engineering degrees. This process demonstrates how to maintain a focus on the target player population, instructional design principles, and game design principles in order to develop a serious game that has interchangeable components to facilitate continued enhancement and development over time.

In the future, additional data will be collected from this game and others developed using this framework. It will be particularly interesting to perform an efficacy study and then incorporate the results back into the games that have been developed. In addition, future work will include incorporating additional elements that are known to engage users in game play, such as high score boards to facilitate competition, additional rewards including unlocking special items for successful game play, and interactive in-game assistance to provide just-in-time learning during game play. Additionally, it is of interest to further modularize the user controls in order to modify controls by player preference for a variety of devices and to address accessibility concerns. Finally, the work will be continued by developing additional games within STEM fields including calculus, physics, and chemistry.

Acknowledgements Many of the games that have been provided as examples to elucidate game and instructional design principles were developed as part of MAVEN: Mathematics for Veteran Education, which is part of the Stern2STEM program made possible through the Office of Naval Research STEM under ONR GRANT11899718.

References

1. Anderson, M.: Technology Device Ownership (2015)
2. Rubin, A.: Technology Meets Math Education: Envisioning a Practical Future Forum on the Future of Technology in Education (1999)
3. Williams, D.L., Boone, R., Kingsley, K.V.: Teacher beliefs about educational software: A delphi study. J. Res. Technol. Educ. **36**(3), 213–229 (2004)
4. Hussain, TS., Roberts, B., Menaker, ES., Coleman, SL., Centreville, V., Pounds, K., Bowers, C., Cannon-Bowers, JA., Koenig, A., Wainess, R.: Designing and developing effective training games for the US Navy. M&S J., 27 (2012)
5. Kato, PM.: Video games in health care: Closing the gap. Rev. Gen. Psychol. **14** (2): 113 (2010)
6. Connolly, T.M., Boyle, E.A., MacArthur, E., Hainey, T., Boyle, J.M.: A systematic literature review of empirical evidence on computer games and serious games. Comput. Educ. **59**(2), 661–686 (2012)
7. Charsky, D.: From edutainment to serious games: A change in the use of game characteristics. Games Cult. (2010)
8. Murphy, C.: Why games work and the science of learning. In: Interservice, Interagency Training, Simulations, and Education Conference, Citeseer, pp. 260–272 (2011)
9. Ruggiero, D., Watson, W.R.: Engagement through praxis in educational game design common threads. Simul. Gaming **45**(4–5), 471–490 (2014)
10. Allen, M., Sites, R.: Leaving ADDIE for SAM: An agile model for developing the best learning experiences. Am. Soc. Training Develop. (2012)
11. Fullerton, T.: Game Design Workshop: A Playcentric Approach to Creating Innovative Games. CRC press, Boca Raton (2014)
12. Hwang, G-J.: Definition, framework and research issues of smart learning environments-a context-aware ubiquitous learning perspective. Smart Learn. Environ. **1**(1):1(2014)
13. Bellotti, F., Berta, R., De Gloria, A.: Designing effective serious games: Opportunities and challenges for research. iJET **5**(SI3): 22–35(2010)
14. Uskov, VL., Bakken, JP., Pandey, A., Singh, U., Yalamanchili, M., Penumatsa, A.: Smart university taxonomy: features, components, systems. In: 2016 Smart Education and e-Learning, pp 3–14. Springer, Cham (2016)
15. Uskov, A., Sekar, B.: Smart gamification and smart serious games. In: Fusion of Smart, Multimedia and Computer Gaming Technologies, pp 7–36. Springer, Cham (2015)
16. Dean, AW.: A Pilot Program for the Recruitment and Education of Navy Veterans Based on System-level Technical Expertise and Leadership Maturation Developed during Service
17. Entertainment Software Association: Essential facts about the computer and video game industry: 2015 sales, demographic and usaged data. Entertainment Softw. Assoc., Washington, DC (2015)
18. Hunicke, R., LeBlanc, M., Zubek, R.: MDA: a formal approach to game design and game research. In: Proceedings of the AAAI Workshop on Challenges in Game AI, p. 1 (2004)
19. Schell, J.: The Art of Game Design: A Book of Lenses. CRC Press, Boca Raton (2014)
20. Squire, K.D., Jan, M.: Mad city mystery: developing scientific argumentation skills with a place-based augmented reality game on handheld computers. J. Sci. Educ. Technol. **16**(1), 5–29 (2007)

21. Bavelier, D., Green, C.S.: The brain-boosting power of video games. Sci. Am. **315**(1), 26–31 (2016)
22. Khan Academy (2016). https://www.khanacademy.org/. Accessed 18 Nov 2016
23. Smith, K., Shull, J., Dean, A., Shen, Y., Michaeli, J.: SiGMA: A software framework for integrating advanced mathematical capabilities in serious game development. Adv. Eng. Softw. **100**, 319–325 (2016)
24. Atlassian Bitbucket (2016). https://bitbucket.org/. Accessed 27 Sept 2016
25. Amazon Web Services Inc. (2016). https://aws.amazon.com/. Accessed 22 Sept 2016

Chapter 10
Using a Programming Exercise Support System as a Smart Educational Technology

Toshiyasu Kato, Yasushi Kambayashi and Yasushi Kodama

Abstract During the completion of programming exercises at higher educational institutions, students typically must complete the assigned exercise problems individually. While there are certain students who can easily solve the problems independently, many students require too much time to do so. For this reason, most institutions use teaching assistants, or TAs, to help teach programming classes. In this chapter, we propose support functions to assess the learning conditions of a programming practicum. The aim of these functions is to reduce the burden on instructors by supporting the assessment of learning conditions in order to improve the quality of instruction. In the current study, we have designed support functions to assess learning conditions for a support system. We also conducted experiments in actual classes to assess the results. We propose three functions for a programming exercise support system. The first function is to support teachers with the analysis of students with common problems. The second function is to support teachers with the analysis of students who are having difficulties. The third function is to provide TAs with the features of students' programming behaviors. We have developed smart educational environments through these three functions of our programming exercise support system. The system has successfully supported instructors and TAs in their provision of smart pedagogy for students.

Keywords Programming exercises · Learning conditions · Instructor assistance · Face-to-face class · Smart education · Learning analytics

T. Kato (✉) · Y. Kambayashi
Nippon Institute of Technology, Saitama, Japan
e-mail: katoto@nit.ac.jp

Y. Kambayashi
e-mail: yasushi@nit.ac.jp

Y. Kodama
Hosei University, Tokyo, Japan
e-mail: yass@hosei.ac.jp

© Springer International Publishing AG 2018
V.L. Uskov et al. (eds.), *Smart Universities*, Smart Innovation,
Systems and Technologies 70, DOI 10.1007/978-3-319-59454-5_10

10.1 Introduction

Higher education institutions use various learning management systems for laboratory-style lessons [1, 2]. In addition, researchers have conducted studies to analyze the learning histories stored in such learning management systems [3, 4]. When doing programming exercises, students independently complete the assigned problems. While there are certain students who can easily solve the problems independently, there are many students who require much time to solve the same problems [5]. In order for an instructor to effectively support students' problem solving, the instructor is expected to not only answer the questions from the students, but also understand which particular students really need help [6]. The instructor needs to understand which students need assistance, what kind of assistance is needed, and in what situations it will be most effective [7].

As for the current state of programming exercises for entry-level classes, the number of instructors and the number of TAs are limited [8]. The instructor can recognize the learning situations of students from observing their computer screens; however, accurately assessing the work progress of the class as a whole as well as each student's individual learning needs are two completely different tasks. Assessing student's individual learning needs is difficult to achieve. Instructors can encourage students to raise their hands to indicate their need for help or check student submissions of the assigned problems. This method is inefficient, though, because it takes a lot of time when the number of students in the class is large.

This chapter presents the learning situation awareness functions of the Web-based learning management system for the implemented programming exercises. This function enables instructors to assess the learning situation of each student. This cannot be understood by simply viewing a student's computer screen. The proposed functions present the information necessary for the instructors to guide each student individually. The instructors and TAs can assess the learning needs of each student and give appropriate guidance according to the information that the function provides.

In order to implement the given function, we performed a requirements analysis of the programming exercise designated for learning situation awareness. Next, we executed the design and implementation of the function on the basis of previous investigations. The assessment of student learning situations is generally performed in a face-to-face manner. Therefore, the present study extracted the functions from the requirements analysis that assesses the student learning situations during a regular, face-to-face class. In addition, the authors clarified that the problem of the present study is the techniques used in previous studies. The present study requires the use of a learning management system that can collect student learning history data in real time.

10.2 Literature Review

Kurasawa [9] developed a support system for assessing learning conditions that provides instructors with information regarding students with common problems. It collects the compiler history and, from the analysis of trends in past compiler errors, it estimates and aggregates the locations and causes of errors. By supplying the instructor with error messages, the number of errors, their causes, and error syntax, the instructor can receive information on students' common problems. This, though, requires analyzing and preparing error cause candidates beforehand and, as the instructor must analyze error causes, it adds to the workload of the instructor. Moreover, it is limited to whole-class instruction and does not provide information regarding individual students. Individual learning is fundamental in programming practicums and information necessary to individual instruction, such as student logs and repeated mistakes, is indispensable. In the present study's proposed method, error analysis takes place automatically. There is no need for the professor to do any work. Furthermore, it assesses and displays assessments of common student problems as well as individual solution histories to allow for both class and individual instruction.

Higher education institutions could harness the predictive power of CMS data to develop reporting tools that identify at-risk students and allow for more timely pedagogical interventions, as well as which student online activities accurately predict academic achievement [1].

Colthorpe [10] investigated the relation between the presence of self-reflection based on material access and the report date of Learning Management System (LMS) submission. Students that reported reviewing lectures as a learning strategy were more likely to access the online lecture recordings, but higher access was actually associated with poorer academic performance. Cluster analysis of all available data showed high academic performance was positively associated with early submission of intra-semester assessment tasks but negatively associated with both the use of, and the reported of use of, lecture recordings by students. Therefore, using an online test enables more realistic feedback.

Research on the assessment of programming behaviors includes debugging training, programming tutorials, and error analysis. Ryan [11] studied how to improve student debugging skills. He constructed a model from debugging and development logs. He also found that students could improve their debugging skills using the model. The present study also analyzes debugging logs and other programming records.

Alex [12] proposed employing tutors to support students in developing programs step by step. The tutors examined students' processing and monitored whether they were on the right track. Students received advice when their programs were incorrect. For example, the tutor provided hints to students on how to refactor their programs. The present study builds upon this work by using TAs to provide students with problem-solving techniques to solve their own problems.

Serral presented a synthesis and update of a long-term project that addressed this challenge in the context of conceptual modeling by developing SAiLE@CoMo, a smart and adaptive learning environment [13]. By crafting innovative process analytics techniques and expert knowledge on feedback automation, SAiLE@CoMo automatically provides personalized and immediate feedback to learners. This research will deal with the problem of feedback deficits reported by students and will significantly alleviate teacher effort. This leaves the instructor more time for in-depth discussions with students.

Truong's paper described a "fill in the gap" programming analysis framework that tests students' solutions and gives feedback on their answers, and detects logic errors and provides hints as to how to fix them [14]. The framework makes use of client-server communication architecture. This is where the execution of students' programs takes place on their own machines while the evaluation is carried out on the server. With this framework, teachers can immediately confirm changes to program sources.

In light of the abovementioned research, we are able to assess student work progress from the start of the exercise to the submission of their work. The purpose of this is to find students with ongoing errors at an early stage. In addition, we will be able to grasp the compilation errors occurring throughout the entire class without requiring faculty error analysis. The reason for this is that the teacher must analyze all of the errors if a single student generates multiple errors at once. Furthermore, we are able to notice the students who are behind in their work in relation to the whole class. This can help find and instruct students who are not progressing even if no error has occurred.

Smart education is rapidly gaining popularity among the world's best universities because modern, sophisticated smart technologies, smart systems, and smart devices create unique and unprecedented opportunities for academic and training organizations to improve their educational standards [15].

The smartness level here is the discovery of students experiencing difficulties with a focus on assessing individual learning situations in detail.

10.3   Research Design and Research Objectives

A teacher can more accurately assess student progress on learning tasks and provide feedback when using smart educational technology. This creates a smart learning environment [16] for students.

10.3.1   The Problems of Programming Exercise Support

In programming exercises, each student must individually tackle the assigned problems. The progress of each student is very different and depends on their skill

levels [5]. An inefficient programming mode is generally the main reason a particular student cannot progress. This student occasionally comes to a standstill and
does not know what to do next. Such students can often become unmotivated [7]. In
programming exercises, it is important to identify these students during the early
stages of the exercise.

In the programming exercise, the instructor and TAs go around to students'
consoles. Even in this situation, it is not easy for the instructor and TAs to determine which students need extra assistance [17]. The instructor and TAs check
students' progress through submitted programs and the students' screens; however,
the number of TAs is limited. They are always too busy to check simple errors. We
have to rely on the instructor's teaching experience and intuition to solve the
problems of individual students [7]. The instructor and TAs answer students'
questions as needed. The instructor teaches the class based on the information they
have collected from individual student cases. The instructor must advise students
when it comes to challenge topics and how to understand the learning method [17].

The objective of the learning management system for programming exercises is
to reduce the workloads of the instructor and TAs [18, 19]. The function of the
system is composed of the functions of the instructor and students. Figure 10.1
shows the basic functions of the programming exercise support system.

- Functions of the instructor:

  - Exercise making: the question contents and the date of setting questions are
    input and registered.
  - Exercise demonstration: the specified exercise at the date of setting questions
    is presented.

- Functions of the students

  - Exercise receipt: the exercise is selected and the development of the answer
    begins.
  - Program edit: the answer program to the exercise is edited.

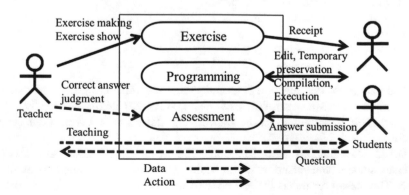

Fig. 10.1 Functions of the programming exercise support system

- Temporary preservation: the answer program is temporarily preserved.
- Compilation: the answer program is compiled and it presents the results.
- Execution: the answer program is executed and presents the results.
- Answer submission: the answer program is submitted and the exercise is complete.

10.3.2   Request Analysis for Smart Classroom Realization

In order to utilize a smart classroom for programming exercises, the following three principles of instruction should be followed [20, 21]:

(1) Assess each individual student's progress toward the learning target. The objective is to consider subsequent lesson plans for each student based on his or her current problems.
(2) Assess each individual student's understanding of the learning content. The objective is for the instructor to recognize the exact area where the student is having difficulty so that he or she can arrange the guidance contents.
(3) Assess each individual student's difficulties. The objective is for the instructor to recognize where each student is having problems so that appropriate instruction can be provided.

As for the proposal of the present study, we set three functions listed below on the basis of the principles of pedagogy.

Work progress sum function. The work progress sum function presents the achievement situation of the learning target as work progresses in the exercise. The learning target of the programming exercise is to solve the exercise. The achievement situation refers to the advancement towards execution and compilation of the exercise. The work of the programming exercises is as follows:

1. The answer begins
2. Input and compilation of the program
3. Confirmation of compile errors
4. Execution
5. Confirmation of execution results
6. Submission of exercise

The reason for this function is to understand the work progress of students on exercises in class. Moreover, the function should be suitable for the intention of the exercise. This function checks the answers and detects mistakes in the execution results.

Error classification sum function. The error classification sum function assesses student understanding of the learning contents as an error status of the class. The reason for this is that the error of understanding of the shortage and the programming is the same. The goal of this function is to identify any common errors occurring in the class.

Work delay detecting function. The work delay detecting function shows delays in the completion of students' work. The purpose of this function is to inform instructors of which students are slow or late in completing the exercise.

10.4   The Function of Assessing the Learning Situation of the Class

This function is a work progress sum function and an error classification sum function. We first designed the function, then defined the algorithm, and then completed the mounting.

Work Progress Sum Function

Function design

The objective of the work progress sum function is to present the number of students that have completed the exercise at any time. The previous work is specified for one work [12]. Therefore, the technique was not applicable to the entire exercise. The present study was designed to assess student progress at each step of the exercise. The proposal function presents the sum result at each work completion time. The instructor modifies the instructions based on the display contents. Therefore, the contents include work completion time, matriculation number, name, and seat number. This information is input at the login of the learning management system.

Work completion with correct answers is defined by the presence of the correct answer according to the correct answer judgment. The correct answer judges the student's execution result, the instructor's execution result, and the keyword. This is done according to the timing of the students' execution of the program. The function then presents the presence of the correct answer or the incorrect answer and the keyword. The work progress sum function calculates the number of incorrect answers and the number of correct answers. The instructor only prepares the example answer program and the keyword. As shown in the following example of correct answer judgment, the correct answer is "for" in the answer program and 55 of the execution result. "The total from 1 to 10 is output by the use of the 'for' sentence." This correct answer judgment can be applied to the exercises that obtain the output result and ask for the grammar.

Algorithm

1. The function initiates the processing by the instructor's access.
2. Each work completion time of the class set is input.
3. The number at each work completion time is totaled.
4. The total number of each work is output.

User Interface

Learning situation screen (displayed in Fig. 10.2): This shows data for each student's work at the beginning, compilation, execution, correct answer, and answer submission stages of the exercise. When the number of presented items is clicked, it shifts to the learning situation screen according to the classification.

Learning situation screen according to classification (displayed in Fig. 10.3): This shows students' matriculation numbers, names, seat numbers, and work completion times that come under each item of the learning situation screen.

Error Classification Sum Function.

Function design

The objective of the error-classification sum function is to present the result of the error classification from the student's program and the error message of the compilation. The previous work will prepare the error factor and the error pattern [9, 11]. Therefore, the present study was designed to identify student compilation errors without requiring these analyses. The proposal function presents the sum result of the error classification. The error classification presumes the place of a common compile error. The number at the head of the error corresponds to the line number of the example answer program (hereafter, correspondence line number). Because the first error leads to other error factors, the error of the head is targeted. The object language of the error classification uses the same Java language as the lesson of the assessment experiment. The analysis object of the error classification is a compile error of the student who does not arrive at execution. This is because it acquires the error that occurs when performing the function. The reason to assume the analysis object to be a compile error is that there is a necessity for the guidance of the instructor in the programming exercise [7].

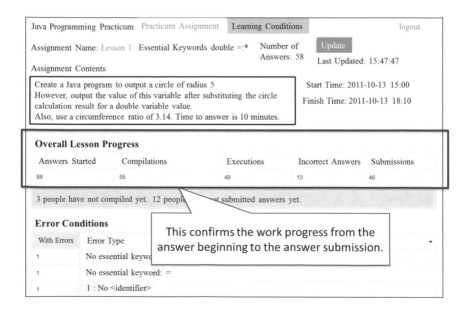

Fig. 10.2 Work progress displayed on the learning situation screen

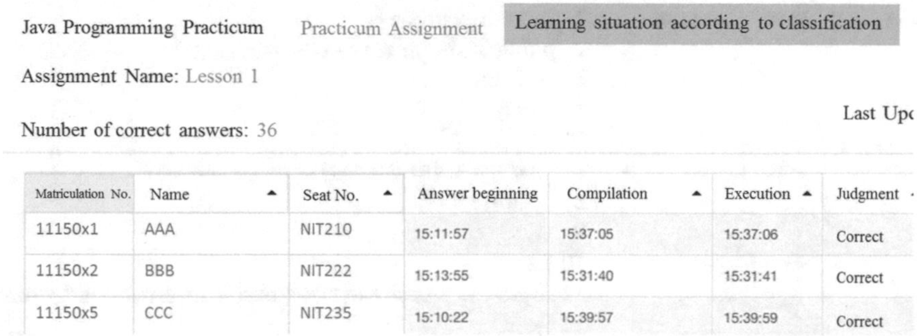

Java Programming Practicum	Practicum Assignment	Learning situation according to classification

Assignment Name: Lesson 1

Number of correct answers: 36

Last Upc

Matriculation No.	Name	▲	Seat No.	▲	Answer beginning	Compilation	▲	Execution	▲	Judgment
11150x1	AAA		NIT210		15:11:57	15:37:05		15:37:06		Correct
11150x2	BBB		NIT222		15:13:55	15:31:40		15:31:41		Correct
11150x5	CCC		NIT235		15:10:22	15:39:57		15:39:59		Correct

Fig. 10.3 Learning situation screen according to classification

The error classification method uses the difference of the text by diff [22] in UNIX. The line can correspond. "Diff-b answer program example answer program" of the command in the diff is space and a tab by one and counted optional-b. This method can be classified as an error to which the error of a different line is common in two or more programs. Moreover, the correspondence line number is "Line of error line c example answer program of the answer program" in the result of diff. Therefore, the proposal function can identify the error in the line number and types of classes regardless of the way the answer program is written. This error classification can be done according to the line number and the error type in the error message. Therefore, this can also be applied to C language, C ++ language, etc.

Algorithm

1. When the instructor accesses the function, it initiates the processing.
2. Input of answer program, error message, and example answer program.
3. The first error line number and the error kind are extracted from the error message.
4. The correspondence line number of the example answer program corresponding to the error line is extracted.
5. If the correspondence line number does not exist, the correspondence line number is assumed to be unclear.
6. The correspondence line number and the error type of pair are output.

An example of executing the error classification follows. Figures 10.4, 10.5, and 10.6 show examples of the answer program, the error message, and the example answer program, respectively.

1. The instructor accesses the error classification sum function.
2. Input of answer program, error message, and example answer program.
3. The "11" of the error line number and error kind of "';' expected" are extracted from the error message.
4. Correspondence line number "7" is extracted from the answer program and the example answer program.

Fig. 10.4 Answer program

```
1 class ForExample {
2  public static void main (String[] args) {
3   int sum = 0;
4   int i;
5
6   for (i = 1; i <= 10; i++) {
8     sum += i;
9   }
10
11  System.out.println("The total from 1 to 10 is " + sum)
12  }
13 }
```

Fig. 10.5 Error message

```
ForExample.java:11:error: ';' expected
    System.out.println("The total from 1 to 10 is " + sum)
                                                          ^
1 error
```

Fig. 10.6 Example answer
program

```
1 class ForExample {
2  public static void main (String[] args) {
3   int sum = 0;
4   for (int i = 1; i <= 10; i++) {
5     sum += i;
6   }
7   System.out.println("The total from 1 to 10 is " + sum);
8  }
9 }
```

5. Because the correspondence line number exists, nothing is done.
6. Correspondence line number and error kind of "7 ';' expected" are output.

The line numbers of Figs. 10.4 and 10.6 provide additional explanation. This technique does not target an irrelevant character string in the error line. Therefore, this can display the tendency for compile errors.

User Interface

The learning situation screen (shown in Fig. 10.7): The correspondence line number and the kind of error pair are displayed in the order of the sum number. When the number of presented items is clicked, it shifts to the learning situation screen according to the classification. Refer to Fig. 10.3.

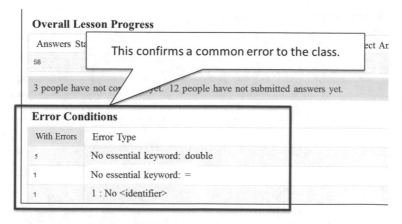

Fig. 10.7 Error classification sum on learning situation screen

10.5 Function to Assess Slow or Late Students

The method of achieving the function designs the function of the work delay detecting function. Afterward, the algorithm is defined and mounted.

Work Delay Detecting Function

Function design

The objective of the work delay detecting function is to detect the students who are behind. The pattern of the work delay is prepared during previous work [8]. Therefore, the present study detects the student who submitted their work late without requiring the pattern of the work delay. The proposal function detects students who have not finished working when the time expires. The function performed statistically labels these as outliers for the distribution of the class at the work completion time. The reason at a time now is that there is no work time data for the students who have not worked yet. Therefore, the student for whom work is late is shown at the time of the outlier. The detection by the outlier function also used the threshold method. This method is problematic, though, as the delay of continuous work is detected without fail according to the work time. The instructor should instruct only the students who are working slowly [23].

Analysis methods of the outlier use the Smirnov-Grubbs test to consider the delay of work to be an outlier by the elapsed time of work [24]. The Smirnov-Grubbs test is a technique for giving official approval of the maximum or minimum value. Figure 10.8 shows an actual class distribution of the answer beginning, compilation, execution, and submission during the laboratory class. The present study assumed a normal distribution of added time at the completion time of work, as shown in Fig. 10.8. The outlier at a time now can define delayed students who have not yet begun working. This detection can present students who are behind in their work progress because it changes the amount of detection due to an increase in the problem presenter.

Fig. 10.8 Distribution of
work time

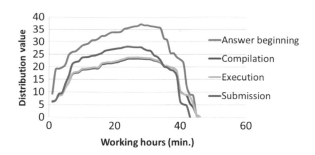

Algorithm

1. When the instructor accesses the function, it initiates the processing.
2. Input of work time set and significance level and time now.
3. The time now is added to the work time set and the outlier is analyzed.
4. The coming off standard value is calculated from the number of the Smirnov dismissal authorization table and work time set.
5. The test statistic of each work time is calculated.
6. The maximum value of the test statistic is extracted.
7. The maximum value of the test statistic comes off and, if it is larger than the standard value, the work time is assumed to be an outlier.
8. This algorithm is ended if the outlier is not time now.
9. Information on students that the work time does not exist in the work time set is output.

An example of executing the work delay detection is as follows. The submission time set is shown in Table 10.1 as an example of input data.

1. The instructor accesses the work delay detecting function.
2. It is assumed "17:38:25" of time now and the work time set. The submission time set and the significance level "5%" are input.
3. The time now is added to the work time set.

Table 10.1 Submission time set

Students	Submission time
A	17:09:00
B	17:12:00
C	17:12:40
D	17:15:10
E	17:16:13
F	17:21:40
G	17:25:10
H	–
I	–
J	–

Table 10.2 Test statistic of work time set

Students	Test statistics
A	1.029
B	0.714
C	0.643
D	0.381
E	0.270
F	0.303
G	0.671
Time now	2.064

4. The coming off standard value is from a Smirnov dismissal authorization table to "2.032" when the number of work time sets is 8.
5. The test statistic of each work time is requested. Table 10.2 shows the test statistic of the work time set.
6. The maximum value of the test statistic is "2.064" at the time now.
7. The maximum value of the test statistic comes off and the time now is assumed to be an outlier because it is larger than the standard value.
8. The outlier is next at time now.
9. The matriculation number and the name of student "H, I, J," who does not have the work time, are output.

Tables 10.1 and 10.2 arrange the time in ascending order for explanation. Students who have not submitted their work receive a "-". Moreover, expression (10.1) shows the test statistic using the Smirnov-Grubbs test. t of the coming off the standard value of Smirnov-Grubbs test is the number n of specimens, significance level α, and $\alpha/n \times 100$ of t distribution of the degree of freedom n-2.

$$\tau = \frac{(n-1)t}{\sqrt{(n(n-2)+nt^2)}} \tag{10.1}$$

User interface

Learning situation screen (shown in Fig. 10.9): This shows the students who are late completing their work, their working names, and how many there are. When the number of people is clicked, it shifts to the learning situation screen according to the classification. Refer to Fig. 10.3.

10.6 Evaluation of the Smart Classroom by Instructors

10.6.1 Objective

The objective of the experiment was to evaluate the utility and effectiveness, in an actual lesson, of the achieved learning situation assessment function (hereafter, the

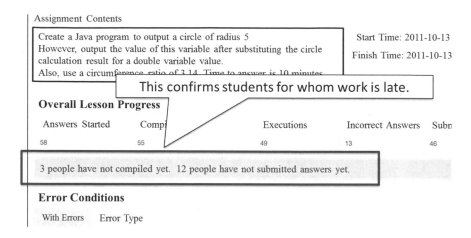

Assignment Contents

Create a Java program to output a circle of radius 5
However, output the value of this variable after substituting the circle
calculation result for a double variable value.
Also, use a circumference ratio of 3.14. Time to answer is 10 minutes.

Start Time: 2011-10-13

Finish Time: 2011-10-13

This confirms students for whom work is late.

Overall Lesson Progress

Answers Started   Comp   Executions   Incorrect Answers   Subn

58   55   49   13   46

3 people have not compiled yet. 12 people have not submitted answers yet.

Error Conditions

With Errors   Error Type

Fig. 10.9 Display of students whose work is late on the learning situation screen

achievement function). The method of evaluating the utility measures was to use the achievement function in an actual lesson. The method of evaluating the effectiveness measures the presence of guidance by the achievement function. Moreover, we questioned the lesson instructor.

10.6.2   Method

Outline of the lesson. The course used was "Basic Programming and Exercises" based on the console application in Java. Table 10.3 shows the outline of the experiment subjects. This subject has two classes, each of one instructor, and TA of two people and four people. The classes have 38 people and 73 people respectively.

Experiment system. Figure 10.10 shows the composition of the experiment system. This experiment system is a client-server method of the Web-base that can,

Table 10.3 Outline of subjects in the experiment

Learning contents	Number of exercises	Performance target
Arithmetic operator and expression	3	Do the learning of something as the expression
Condition branching	4	Do the learning of the method of switching processing on the condition
Boolean	4	Do the learning of expressible of the combination of two or more conditions by the use of Boolean
Repetition	4	When the loop construct is used, do the learning of can the description of the repetition instruction of the same processing

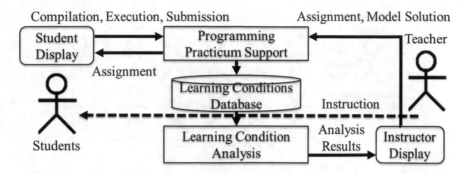

Fig. 10.10 Composition of the experiment system

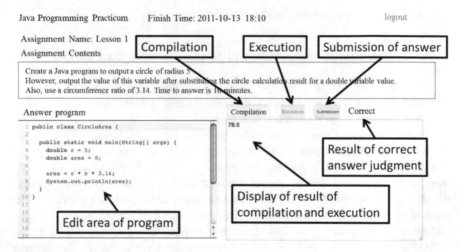

Fig. 10.11 Student screen

in real time, collect students' learning history data in the programming exercises. The student screen of the client requests preservation, compilation, execution, temporary submission, etc. from the server for the input answer program. Figure 10.11 shows the student screen. The instructor screen shows the exercise and the example answer program. These are transmitted to the server and the result is received. The server does the compilation and execution of the received program and returns the result to the client afterward. The programming languages used in the experiment were PHP, HTML, Ajax, JavaScript, and MySQL.

Experimental conditions. Two experiments were conducted. In the first, the achievement function was used twice. In the second, the achievement function was not used twice. This looks at the user's experience through the achievement function and presence of the effect. The environment of the experiment is composed

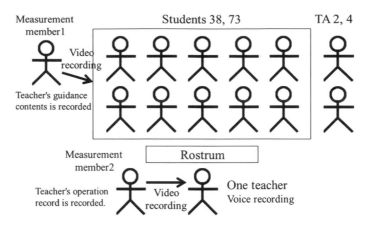

Fig. 10.12 Measurement environment of two classes

of the learning management system that can be accessed from all PCs on the network.

The significance level in the work delay detecting function was fixed at 25%. This setting was an average value used to detect outliers. Not using the work delay detecting function to evaluate the examination.

Measurement method. Measurement items for usefulness shall be the guidance frequency that is the cause of guidance on the presentation screen of realized functions in order to measure whether or not the realization function is used for instructors. Moreover, the items that measured effectiveness were the guidance frequency for each realized function based on the guidance content provided. The method of measuring these included a voice recording of the instructor, taking a picture of the lesson scenery, and taking a picture of the operation record on the instructor screen. Figure 10.12 shows the measurement environment. The learning situation screen, instructors and TAs walking around, and students raising their hands all lead to the assessment that a student requires assistance. The result of guidance was a pair of the entire guidance or an individual counseling and specific guidance content. We recorded the guidance that the instructor directed the TA to provide, but we did not record the guidance given by the TA to the student. Guidance from the TA is described in Chap. 7 of the following description. Moreover, we did not record the contents of student questions following the instructions. Finally, we interviewed the instructors at the end of the experiment.

10.6.3 Results

There were 30 instances of guidance based on the screen of the achievement function. Table 10.4 shows the guidance contents shown on the screen of the

Table 10.4 Outline of the experiment

Result	Classification	Class	Individual
	The main guidance contents	The demand of the support is pressed from the entire work progress to late students[a] 1 (5) Use the Boolean[a] 2 (2) Confirm the method of connecting character strings[a] 2 (2)	Student with a long interval time of the compilation is urged[a] 3 (2)
Guidance frequency		10	20
Total		30	

where: [a] 1, 2, and 3 show guidance given unconventionally
() the frequency of common guidance contents is shown

Table 10.5 Guidance frequency by instructor rounds and students raising hands

Achievement function		None	Used
Case	**Rounds**	42	23
	Raising hands	22	10
Total		64	33

achievement function. Guidance was given five times regarding the work progress of the class based on the work progress sum function ([a]1 in Table 10.4). Guidance was given 4 times to correct common errors of the class using the error classification sum function ([a]2 in Table 10.4). Guidance was given 2 times to slow students using the work delay detecting function ([a]3 in Table 10.4).

The numerical value in () in Table 10.4 is the frequency of common guidance contents. Moreover, guidance was given 64 times by the instructors and TAs walking around and from the students raising their hands. The achievement function was not used in these cases. The use case was 33 times. Table 10.5 shows the guidance frequency by the instructors and TAs walking around and students raising their hands.

The results of the interview with the instructor are seen below. The first question was "Please tell us a good point and a bad point about the proposal function." Their answers to this question were as follows:

Answer of Instructor A:

Good points:

- Everything from the beginning of the exercises to submission is automated as a system. Therefore, the instructional workload is decreased during the exercise and the confirmation of the problem submission.
- The error status when compiling can be understood. Therefore, I can understand the students' standstill situations in detail. This can be used as prior information before guidance is provided.

Bad points:

- Nothing in particular.

Answers of Instructor B:
Good points:

- It is possible to provide comments about compile errors to many students at once.
- The learning situation can be assessed remotely from the rostrum. Therefore, I can issue instructions to the TA based on the contents.
- Moreover, I can understand the causes for which student work is late based on the information of the function.

Request:

- Please let me know (send me the alert message) about those students who need guidance.

The second question was as follows: "Please tell us about the effect of the display contents of the learning situation assessment function on guidance." The instructors' answers were as follows:
Answers of instructor A:

- Information on student work progress can provide understanding of the learning situation and needs of the class.
- Information on error classification sum can be provided for the entire class or individual students for guidance.

Answer of instructor B:

- Information on students for whom work is late can identify students who are slower than others and need guidance.

10.6.4 Consideration of the Results

Guidance based on the proposal function was given 30 out of 63 times. Additionally, the instructor interviews provided useful feedback regarding the contents of the achievement function. The achievement function presents useful information regarding the areas in which students require guidance. This is effective for the assessment of student learning situations during programming exercises.

The guidance given when only the achievement function was used is detailed in [a]1, [a]2, and [a]3 in Table 10.4. Additionally, the interview results provided answers that confirmed that the contents led to guidance. Moreover, the ratio of guidance from the instructor was high in the lessons that used the achievement function. Table 10.6 shows the ratio of guidance from the instructor. The numerical value of

Table 10.6 Ratio of active guidance by the instructor

Achievement function		None		Used
Case	**Observing screen and making rounds**	The instructor's active guidance	65.6% (42)	84.1% (53)
	Raising hand	The student requests guidance	34.4% (22)	15.9% (10)
Total			64	33

where the guidance frequency to the cause of guidance is shown in ()

() in Table 10.6 is the frequency of guidance for each cause of guidance. The ratio of active guidance increased because the instructor was able to understand the learning situation from the screen display of the achievement function. The ratio of guidance given at the students' request decreased because the instructor was able to identify the question from the display screen before the question came from the students. The ratio of this guidance is intentionally high at a significance level of 1% by the chi-square test. The instructor was able to assess and instruct a difficult learning situation. Moreover, the instructor can give guidance before help is asked for by the student. Therefore, the achievement function is effective for guidance during programming exercises.

According to the results of the interview, the instructor should return to the rostrum during the achievement function. Instructors can not check their own computer while teaching at student's desk. Instructors can direct and instruct more students without returning to the rostrum if they use the accomplishment function. A method to overcome this problem is the use of the tablet terminal. Figure 10.13 shows instruction using a tablet terminal in the lesson in the following year of the assessment experiment.

Students can request instructor guidance remotely through the use of tablets. Therefore, tablet use is more effective than the instructor making rounds when time efficiency must be taken into consideration.

Fig. 10.13 Assessment of learning situation using a tablet terminal

When the achievement function is developed with other languages and courses, it raises the issue of what should be how much time can be given to each learning task. Moreover, the error classification sum function can correspond in the case of the console application. There is a problem, though, in the method of acquiring the error for the GUI application.

10.7   Function to Understand Programming Behavior for TAs

10.7.1   Analysis for TA

Data mining, as a smart educational technology, can support the TA and help in the realization of a smart learning environment.

During programming exercises, the main role of TAs is to assist students in correcting their errors. The programming behavior of a particular student, which includes his or her programming style, is different from other students' behaviors. For this reason, it is hard for TAs to provide appropriate guidance other than error handling for each student [25]. The lecturers understand students' learning contexts due to their prior teaching experiences. TAs have greater difficulty understanding students' learning contexts because they do not have enough teaching experience to do so. Therefore, the present study proposes a function to help TAs understand student programming behavior and common student difficulties. The factors of problem solving in programming include how much a particular student follows the programming codes, understands the grammar of the programming language, and uses the compiler. This section reports on the results of our data mining, which focused on the programming mode. The programming mode is the basic attitude that represents how much a particular student follows the programming codes that they are encouraged to follow at our institute. Furthermore, by focusing on the programming mode, we can collect students' behavioral data in real-time. This is done through the programming exercise support system that we have developed [19].

The present study classifies the features of the students' programming behaviors in order to infer the characteristics of each student. Through the classification, we have examined the relationship between the results of the questionnaire toward the programming mode and the programming behaviors. The behaviors include the number of compilations, the number of trial executions, the number of errors, the number of repetitions of the same errors, the average interval of the compilations, and the average intervals of the executions. The programming codes are shown in Table 10.7. We found that each student had a particular programming trait. We also found that we can measure that trait by observing how much a particular student follows the programming codes.

Table 10.7 The programming codes

Code #	Programming code details
1	When the grammar is ambiguous, examine it in the texts or manuals
2	Add one line of sentence and compile
3	When compile errors appear, deal with the first error
4	Construct programs from the skeleton
5	When the execution result is not correct, trace the execution process
6	Insert spaces after keywords so that they are highlighted
7	Write output sentences first so the behaviors of the program can be observed
8	Make and try several solutions to solve the errors
9	Insert spaces after commas, so that they are easily seen
10	When inserting an opening parenthesis, insert the corresponding closing parenthesis immediately
11	Write meaningful comments so that the semantics of the program can be understood
12	Choose meaningful variable names
13	When the usage of the instruction is ambiguous, refer to the samples
14	Insert space lines so that blocks in the program are clearly seen
15	Indent the codes so that the structure of the program can be clearly seen
16	Insert spaces before and after operators so that they are highlighted
17	When modifying the program, leave the old source codes as comments
18	When the program behaves strangely, print out intermediate variables
19	Write many more programs until you can write them comfortably
20	Use the patterns of program codes

10.7.2   Classification of Programming Behaviors

In this section, we classify the features of programming behaviors. In order to do so, we have performed a cluster analysis of the records of the programming behaviors. Clustering is a process of grouping objects into classes of similar objects [26]. It is an unsupervised classification or partitioning of patterns (observations, data items, or feature vectors) into groups or subsets/clusters based on their locality and connectivity within an n-dimensional space. In the present study, we have performed a cluster analysis over the submitted programs that solve the assignments. The total number of the subjects was 80 and we employed seven feature variables. Figures 10.14 and 10.15 show the results of the cluster analysis.

We employed Ward's method of hierarchical clustering in the cluster analysis [27]. We also employed k-means for the non-hierarchical clustering. Ward's method is a criterion applied in hierarchical cluster analysis [28]. K-means is an algorithm that clusters objects based on attributes in k partitions [29]. The result of the k-means analysis depends on the initial values. Therefore, we have chosen initial values as the best values produced by Pseudo-F, where the number of clusters is 4, in our preliminary experiments. Pseudo-F is an evaluation criterion in cluster

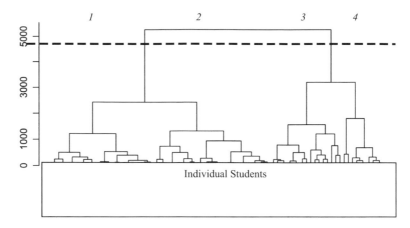

Fig. 10.14 Cluster analysis using Ward's method

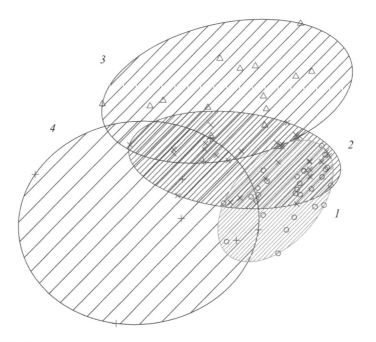

Fig. 10.15 Cluster analysis using k-means

analysis [30]. A good Pseudo-F value means the resultant clusters have little overlapping. In addition, the density of each cluster is high.

The present study examined the relationship between the students' programming behaviors and their programming modes. Tables 10.8 and 10.9 show the results. The numerical values in Tables 10.8 and 10.9 are mean values of the number of

Table 10.8 Breakdown of dotted line upper floor layer in Fig. 10.1

Cluster	Solvingtime	Compilation interval	Execution interval	Compilation frequency	Execution frequency	Number of errors	Number of same errors[a]	Codes of programming	Number of students
1	528	209	267	7	5	1	1	10	25
2	812	379	606	6	3	3	3	6	28
3	1405	590	1160	7	3	4	3	5	17
4	1506	228	337	18	13	5	4	4	10

Table 10.9 Cluster breakdown of Fig. 10.2

Cluster	Solving time	Compilation interval	Execution interval	Compilation frequency	Execution frequency	Number of errors	Number of same errors[a]	Codes of programming	Number of students
1	514	239	289	6	5	1	1	9	27
2	884	364	589	7	4	3	3	8	30
3	1398	599	1216	7	3	4	3	7	15
4	1694	225	378	20	14	6	5	5	8

where [a] the same person making the same error multiple times individuals in each cluster. The numerical values of the programming modes are the answers of four-stage evaluation (+1 done and −1 not done). Moreover, we have performed the correlation analysis with the duration time for problem-solving and programming mode. We have observed positive correlations (0.24) in 18 of the programming codes.

The dotted line of Fig. 10.14 indicates the middle of the dendrogram, that is, 2500. We can observe that Tables 10.8 and 10.9 are similar. We can conclude that there were four clusters, as follows:

- **Cluster 1**: Duration time of problem-solving is short. The score of the programming mode is high. The intervals of the compilation and the intervals of execution are short. The compilation frequency is few. The students understand the contents of the errors and what they are doing in the program.
- **Cluster 2**: Duration time of problem-solving is shorter than cluster 3. The score of the programming mode is low. A lot of errors exist and many are similar errors. The intervals of the compilation and the intervals of execution are shorter than those of the cluster 3. The students are doing the programming without understanding the contents of the errors.
- **Cluster 3**: Duration time of problem-solving is long. The score of the programming mode is low. The students are repeating the same error. The students are compiling without understanding the contents of the errors.
- **Cluster 4**: Duration time of problem-solving is long. The score of the programming mode is low. In Table 10.8, there are many compilation frequencies and execution frequencies. The students compile frequently and are committing many errors. In Table 10.9, the compilation frequency and the execution frequency are low. Surprisingly, these students are submitting the correct solutions to the problems. They have likely copied the correct answers from cluster 1 students.

10.8   Evaluation of the Smart Classroom by TAs

We set up a hypothesis based on the considerations of the previous section. The hypothesis is that the programming behavior appears in the programming mode.

Experiments. In this section, we verify the hypothesis about the programming mode and programming behavior. This was to what degree the understanding level in the programming codes corresponds to the programming behavior. Moreover, the present study verifies the differences in the results of the questionnaires about the subjective understanding level of the students before and after the experiment. The TA instructs the students based on the results of the questionnaires completed before the experiment. The TA guides the basic attitudes within the low numerical value for understanding level based on the questionnaires.

The verification method is to evaluate the duration time of the problem solving of both the experimental group and the control group. The group consists of forty students who all submitted their solutions late. These students were split into two groups at random to create the control group and the experimental group. The procedures of the experiment were as follows:

1. Perform understanding level inquiry of the programming mode the first time. Acquire the duration time for problem solving before the experiments.
2. First experiment: guidance from TA.
3. Second experiment: guidance from TA.
4. Third experiment: guidance from TA.
5. Fourth experiment: no guidance from TA. Perform understanding level inquiry, the second time, of the programming mode. Acquire the duration time for problem solving after the experiments.

Figure 10.16 shows the change in the problem solving time. We can observe the improvement of the duration time of the problem solving in the experimental group; however, there was no significant difference between the two groups.

Table 10.10 shows the result of the understanding level inquiry. The understanding frequency value of the experimental group rose by 5% compared with the control group. The result of the questionnaire, according to the cluster, is shown in Table 10.11. In Table 10.11, the gray background shows where the improvements in student understanding levels are remarkable.

Discussion. The problem solving time may have been shortened due to the understanding level of the programming codes corresponding to the programming behavior. Data mining of students' behaviors enabled the TAs to provide effective guidance. Moreover, the results of the data mining provided the TAs with the features of the students' programming behaviors.

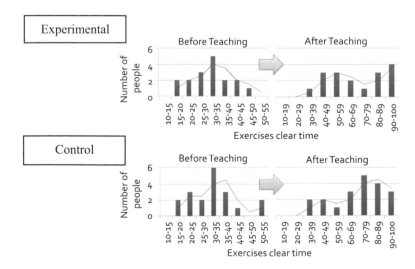

Fig. 10.16 Changes in problem-solving time

Table 10.10 Results of the questionnaire of understanding level by student subjectivity

Group	Understanding frequency value			
	Before	After	Effective	None
Experimental	6.66	9.7446% UP	13 students	4 students
Control	5.84	8.241% UP	13 students	6 students

Table 10.11 Results of the questionnaire in each cluster

Cluster	Matter	1	2	3	4	5	6	7	8	9	10	11
1	Before	0.1	0.8	0.7	0.0	0.3	0.1	0.6	0.4	−0.2	0.1	0.7
	After	−0.2	1.0	0.9	0.1	0.4	0.1	0.5	0.6	0.0	0.1	0.9
2	Before	−0.1	0.3	0.7	0.2	0.3	0.3	0.8	0.4	−0.4	0.6	0.7
	After	0.2	0.7	0.7	0.2	0.3	0.7	0.9	0.9	0.1	0.6	0.9
3	Before	0.1	0.2	0.8	−0.2	0.5	0.0	0.7	0.5	0.0	0.3	0.7
	After	−0.2	0.6	0.9	0.1	0.3	0.6	0.8	0.8	0.1	0.5	0.7
4	Before	−0.3	0.3	0.7	−0.1	−0.2	0.5	0.6	0.4	−0.3	0.6	0.9
	After	0.1	0.6	0.6	−0.3	0.1	0.6	0.8	0.3	−0.1	0.6	0.7

Cluster	Matter	12	13	14	15	16	17	18	19	20	Students
1	Before	0.8	0.8	0.8	0.7	0.8	−0.3	0.5	−0.4	1.0	5
	After	0.5	0.9	1.0	0.3	0.8	−0.1	0.5	0.4	0.8	
2	Before	0.1	1.0	0.9	0.3	0.6	−0.2	0.3	−0.3	0.8	13
	After	0.2	0.7	0.8	0.3	0.7	0.0	0.5	−0.2	0.9	
3	Before	0.3	0.3	0.8	−0.2	0.3	−0.4	0.1	−0.4	0.7	11
	After	0.3	0.6	0.8	0.2	0.6	0.0	0.4	−0.4	0.9	
4	Before	0.0	0.8	0.7	0.4	0.9	−0.5	−0.2	−0.6	1.0	7
	After	0.3	0.5	0.9	0.5	0.5	0.2	−0.2	−0.3	1.0	

We possibly did not observe significant difference between the two groups because of the close connections between friends. For example, students may have shared the advice of the TA with their friends.

The guidance effects on programming mode are shown in Table 10.12. The data mining of students' behaviors enables TAs to give advice beyond just simple error correction.

Table 10.12 Effects of programming mode

Cluster	Effective	Ineffective
1	Examine the grammar Write the comment Make a lot of programs	None
2	Adjust the appearance Write the comment	Make a lot of programs
3	Adjust the appearance	Compile for each line
4	Compile for each line Deal with the first error Copy the sentences that work	Examine the grammar Write the output sentence first

10.9   Scaling-up to Smart University

We have experimented on subjects in programming exercises. We would like to apply it fully to other subjects.

We are currently trying the proposed method in language classes [31]. We are studying how the students can reflect even in the face-to-face class. Typically, we have implemented Web exercises using Google Forms for continual self-reflection. We have performed the text mining to "Devised it" of the Web exercises. We have observed the transformation from the mentally passive word "Do" to the active words "Examine it." We found that Google Forms motivates students' self-regulatory learning. Based on these findings, we also present a prospect for a lesson that draws out the subjectivity of the students.

Furthermore, we are studying which tasks effectively motivate students. Currently, it is not very smart, but further study of deep learning in programming classes will make it possible to extract issues that increase motivation. The smartness level can be improved by developing these areas.

There is manual work in realizing smartness and it would be a restriction when considering large-scale deployment. By using more sophisticated machine learning, we can realize a smart learning environment. Moreover, we have analyzed the relation between programming behavior and programming mode. The results were related to the behavioral features and the programming mode. The authors have reached the hypothesis that the duration time of problem solving could be reduced as a result of the effective guidance of the TAs concerning programming mode. Then, the authors performed the assessment experiments of guidance by TAs. As a result, the duration time of the problem solving of students who were taught has been shortened. Therefore, the proposed technique enables TAs to provide effective support for students because they can better obtain a deep understanding of the learning situation of each student. Therefore, we can conclude that the proposed function enables TAs to effectively support students when learning and programming.

10.10   Conclusions

This chapter described the achievement of the learning situation assessment function in the programming exercises for entry-level programming classes. This function provides an assessment of the learning situation of individual students as well as the class as a whole. The function for students indicates who is behind in their work and who needs help. The function for the class displays the sum of the work progress from the beginning of the exercise until submission. Moreover, the function provides the error classification summary by identifying compile error lines based on the example answer program. We found that the function applied to the Java programming exercises actually produced accurate output. Moreover, the function identifies students that are behind in their work via outlier analysis of class work progress. Because the instructor appropriately sets the standard value of the outlier, this function can identify students who are behind in order to provide them with assistance before it is too late. In the assessment experiment in an actual lesson, we observed appropriate guidance being presented using information from the proposed function. We, thus, can conclude that the proposed function is effective for assessing student learning situations in programming exercises for beginners.

In summary, we have achieved a smart educational environment through the use of this function. The environment supports instructors and TAs so that they can provide smart pedagogy to students.

Acknowledgements This work was supported by Japan Society for Promotion of Science (JSPS), with the basic research program (C) (No. 15K01094 and 26240008), Grant-in-Aid for Scientific Research.

References

1. Macfadyen, L.P., Dawson, S.: Mining LMS data to develop an "early warning system" for educators: a proof of concept. Comput. Educ. **54**(2), 588–599 (2010)
2. Open University of Japan: H21-22 Survey on the Promotion of ICT Use Education. Center of ICT and Distance Education (2011)
3. Klosgen, W., Zytkow, J.: Handbook of data mining and knowledge discovery. Oxford University Press, New York (2002)
4. Romero, C., Ventura, S., Garcia, E.: Data mining in course management systems: moodle case study and tutorial. Comput. Educ. **51**, 368–384 (2007)
5. Horiguchi, S., Igaki, H., Inoue, A., et al.: Progress management metrics for programming education of HTML-based learning material. J. Inf. Process. Soc. Jpn. **53**(1), 61–71 (2012). (In Japanese)
6. McCartney, R., Eckerdal, A., Mostrom, J.E., Sanders, K., Zander, C.: Successful students' strategies for getting unstuck. SIGCSE Bull. **39**(3), 156–160 (2007)
7. Sagisaka, T., Watanabe, S.: Investigations of beginners in programming course based on learning strategies and gradual level test, and development of support-rules. J. Japan. Soc. Inf. Syst. Educ. **26**(1), 5–15 (2009). (In Japanese)
8. Igaki, H., Saito, S., Inoue, A., et al.: Programming process visualization for supporting students in programming exercise. J. Inf. Process. Soc. Jpn. **54**, 1 (2013). (In Japanese)
9. Kurasawa, K., Suzuki, K., Iijima, M., Yokoyama, S., Miyadera, K.: Development of learning situation understanding support system for class instruction in programming exercises. Inst. Electron. Inf. Commun. Eng. Technol. Rep. ET Educ. Eng. **104**(703), 19–24 (2005). (In Japanese)
10. Colthorpe, K., Zimbardi, K., Ainscough, L., Anderson, S.: Know thy student! Combining learning analytics and critical reflections to develop a targeted intervention for promoting self-regulated learning. J. Learn. Analytics **2**(1), 134–155 (2015)
11. Ryan, C., Michael, C.L.: Debugging: from novice to expert. ACM SIGCSE Bull. **36**(1), 17–21 (2004)
12. Alex, G., Johan, J., Bastiaan, H.: An interactive functional programming tutor. In: Proceedings of the 17th ITiCSE 2012, pp. 250–255. ACM (2012)

13. Serral, E., De Weerdt, J., Sedrakyan, G., Snoeck, M.: Automating immediate and personalized feedback taking conceptual modelling education to a next level, In: 2016 IEEE Tenth International Conference on Research Challenges in Information Science (RCIS), pp. 1–6. IEEE (2016)

14. Truong, N., Roe, P., Bancroft, P.: Automated feedback for fill in the gap programming exercises, In Proceedings of the 7th Australasian conference on Computing education, vol. 42, pp. 117–126. Australian Computer Society, Inc. (2005)

15. Uskov, V.L., Bakken, J.P., Pandey, A., Singh, U., Yalamanchili, M., Penumatsa, A.: Smart university taxonomy: features, components, systems. In: 2016 Smart Education and e-Learning, pp. 3–14. Springer, Cham (2016)

16. Hwang, G.J.: Definition, framework and research issues of smart learning environments–a context-aware ubiquitous learning perspective. Smart Learn. Environ. a Springer Open Journal, 1:4, Springer (2014)

17. Friend, M., Bursuck, W.: Including students with special needs: a practical guide for classroom instructors, Chap. 5, pp. 164–165. Prentice Hall, Saddle River (2006)

18. Watanabe, H., Arai, N., Takei, S.: Case-based evaluation support system of novice programs written in assembly language. J. Inf. Process. Soc. Jpn. 42(1), 99–109 (2001). (In Japanese)

19. Kato, T., Ishikawa, T.: Design and evaluation of support functions of course management systems for assessing learning conditions in programming practicums. Int. Conf. Adv. Learn. Technol. 2012, 205–207 (2012)

20. Tanaka, K.: Yokuwakaru Jyugyouron. Minervashobo (2007) (In Japanese)

21. Japan Society for Educational Technology: Kyouiku Kougaku Jiten. Jikkyo Shuppan (2000) (In Japanese)

22. Diffutls. http://www.gnu.org/software/diffutils/diffutils.html. Accessed 14 Sept 2016

23. Ueno, M.: Online outlier detection for e-learning time data. J. Inst. Electron. Inf. Commun. Eng. J90-D 1, 40–51 (2007). (In Japanese)

24. Bull, C.R., Bull, R.M., Rastin, B.C.: On the Sensitivity of the chi-square test and its con-sequences. Meas. Sci. Technol. 3, 789–795 (1992)

25. Yasuda, K., Inoue, A., Ichimura, S.: Programming education system that can share problem-solving processes between students and teaching assistants. J. Inf. Process. Soc. Japan 53(1), 81–89 (2012). (In Japanese)

26. Jain, A.K., Murty, M.N., Flynn, P.J.: Data clustering: a review. ACM Comput. Surv. 31(3), 264–323 (1999)

27. Kamishima, T.: A survey of recent clustering methods for data mining (Part 1): try clustering! J. Japan. Soc. Artif. Intell. 18(1), 59–65 (2003). (In Japanese)

28. Michael, R.A.: Cluster Analysis for Applications. Academic Press, New York (1973)

29. MacQueen, J.: Some methods for classification and analysis of multivariate observations. In: Proceedings of the Fifth Berkeley Symposium on Mathematical Statistics and Probability, California, USA, vol. 1, pp. 281–297 (1967)

30. Calinski, T., Harabasz, J.: A dendrite method for cluster analysis. Commun. Stat. 3, 1–27 (1974)

31. Kato, T., Kambayashi, Y., Kodama, Y.: Practice for self-regulatory learning using google forms: report and perspectives. Inf. Eng. Express Int. Inst. Appl. Inf. 2016 2(4), 11–20 (2016)

Part IV
Smart Universities: Smart Long Life Learning

Chapter 11
The Role of e-Portfolio for Smart Life Long Learning

Lam For Kwok and Yan Keung Hui

Abstract Life-long learning and whole person development are becoming more critical than just academic performance, particularly in career development perspective. Employers are looking for soft skills more than hard skills from employees. It is one of the responsibilities of universities to develop students' work readiness. A smart university, as part of a smart city, is no longer limited to provide technologies inside and outside classrooms. In this chapter, we discuss how a smart university may facilitate self-regulated learning of learners through the introduction of personal development e-Portfolio, which assists learners in planning their development path and reflecting upon their own learning. An implementation example in the City University of Hong Kong is reviewed. Also, the way of extending it to lifelong and professional development is discussed.

Keywords Smart university · e-Portfolio · Whole person development · Outcome based learning · Self-regulated learning

11.1 Introduction

"We now accept the fact that learning is a lifelong process of keeping abreast of change. And the most pressing task is to teach people how to learn." [1]. It is critical for learners to acquire new knowledge and skills in an ever-changing work environment. Lindeman, an American educator notable for his pioneering contributions in adult education, stated that [2] "not merely preparation for an unknown kind of future living... The whole of life is learning, therefore education can have no endings". A similar view was found in recent years [3]: "Lifelong Learning is a

L.F. Kwok (✉) · Y.K. Hui
Department of Computer Science, City University of Hong Kong,
Kowloon Tong, Hong Kong
e-mail: cslfkwok@cityu.edu.hk

Y.K. Hui
e-mail: john.ykhui@cityu.edu.hk

© Springer International Publishing AG 2018
V.L. Uskov et al. (eds.), *Smart Universities*, Smart Innovation,
Systems and Technologies 70, DOI 10.1007/978-3-319-59454-5_11

feature of modern life and will continue to be so. Change is everywhere and we need to learn to cope with it in different aspects of our lives. Jobs are changing with continually developing technology and pressures to keep up with foreign competitors. Daily life is changing with faster communications and more technology in our homes..." The practicing professionals, such as engineers, face similar challenges throughout their career life and, therefore, must remain competent throughout their working careers in order to carry out their duties properly.

Learning may happen anytime and anywhere, consciously or unconsciously. A person's mind may be triggered upon watching an occurring event, participating in discussions or activities, reading, or simply listening to others. These mental activities may result in learning, but it is always a difficult task to measure what a learner has learnt or achieved, especially referring to skills and knowledge that are beyond the curriculum; however, such skills in communication, collaboration with others, learning, etc., are attributes frequently sought when one is looking for a job. The National Association of Colleges and Employers (NACE) [4] points out that about 80% of employers are looking for evidence of leadership skills and 78.9% are looking for the ability to work in a team. Communication skills, problem-solving skills, work ethic, initiative, analytical/quantitative skills, and flexibility/adaptability are the other competences/attributes that more than 60% of employers are looking for. In addition, these figures also show that leadership skills and being involved in extracurricular activities are having higher influential power in job-hunting than just having a high grade point average (GPA) in academic results. Similar results can also be found in both Hong Kong [5] and Mainland China [6] that communication skills, teamwork spirit, and creative problem solving are the major attributes that organizations are looking for from the applicants; however, most graduates only have a deep knowledge in a specific discipline that may prevent them from having immediate success in the global market [7]. Gaps in the needs between employees and companies are increasing. Managing to learn in these aspects is thus an important process in whole person development.

Outcome-based learning is a learning paradigm that places focus on defining the learning outcomes at the beginning of a learning process, designing the curriculum with linkages of its associated learning activities to the defined outcomes, and measuring achievements of students according to the defined learning outcomes [8]. Although outcome-based learning is initially considered for those courses with a set of more definite goals, it does provide a possible structured approach for the whole person development of a student, which refers to the development of a variety of skills and knowledge beyond normal courses.

Portfolios have been used in some disciplines, including education, to organize and present works; to provide a context for discussion, review and feedback from teachers, mentors, colleagues, and friends; and to demonstrate progress and accomplishments over time [9]. Traditional paper-based portfolios are bulky and, hence, not readily portable and accessible when needed. On the other hand, electronic portfolio systems can be used to organize artifacts in different media in order to demonstrate what a student has achieved.

As described by Lombardi et al. "education becomes an important component and has been used as one of the important criteria in performance evaluation of smart cities" [10]. A smart university, as part of a smart city, provides learners with a smart learning environment, transforming them into the smart workforce. A city requires smart people to carry out their jobs smartly and innovatively in order to maintain its classification as a smart city.

A smart university is a learning environment where knowledge is shared among various stakeholders including teachers, learners, and employees in a seamless way [7]. Being smart is merely not only focusing on technological aspects in facilitating learning by setting up more sensors, building better networks, and providing more equipment in the classroom; it is also focused on management aspects of learning by allowing all stakeholders know what a learner has learnt, how competent a learner is, and what learning path a learner is to be suggested. Through self-reflection, learners are able to plan for their own learning paths, evaluate their learning progress, and be able to match their career needs so that they are work-ready at the time they get a new job opportunity, either after graduation or at different stages in their lifelong career development.

This chapter discusses why current e-portfolio systems cannot support the whole-person development, explains how the concepts of learning design may help, proposes a new approach to building a portfolio to record the whole person development, and illustrates an example of a practical implementation in the City University of Hong Kong (CityU). Finally, the similarity between lifelong learning/professional development and whole personal development is identified and, thus, the development of e-portfolio could possibly be extended to lifelong learning and professional development. Potential contributions are discussed followed by the conclusion.

11.2   Literature Review

11.2.1   Smart University

What is smart? There are many definitions for this word, but we tend to define "smart" as "very good at learning or thinking about things, showing intelligence or good judgment or quick in action in handling problems" [11]. When a system can provide what one needs according to the dynamic need of users, it can be considered as smart.

Smart University. A smart university is part of a smart city. A smart university is defined as "a platform that acquires and delivers foundational data to drive the analysis and improvement of the teaching & learning environment" [12]. It is suggested that "a smart university should have tools, similar to those suggested in the European Competence Framework (ECF) framework, to build educational profiles and consequently, curricula and courses that both adhere to the standards required by the scientific and professional communities (e.g., IEEE, ACM)" [7].

The rapid development of technologies, especially the Internet and mobile devices, introduces new challenges to teaching and learning. "Being smart" should not be confused with "being digital" [7]. A smart university provides a learning environment that does not only enable learners with access to digital resources and interaction with learning systems in any place and at any time, but also actively provides the necessary learning guidance, hints, supportive tools, or learning suggestions to them in the right place, at the right time, and in the right form [13]. It is also important to provide personalized learning support based on the learning status and personal factors of a learner such as learning progress, knowledge levels, learning styles, cognitive styles, and preferences. These enable learners to set their learning goals and to construct their learning plans in supporting self-regulated learning.

11.2.2 Stakeholders in Smart University

Education involves numerous closely related stakeholders including teachers, students, parents, institution administrators and managers, and the Government. Stakeholders in education will have different requirements on intelligence that, in turn, require various supporting technologies and, therefore, may have their own expectation on what smartness they need [11].

Students. Students could potentially want their studies to be more interesting and their learning methods to be efficient. If motivated, the goal would most likely be to be admitted to a better school or university or get a better job. They desire to be self-managed and self-controlled when it comes to their own learning.

Teachers. Teachers want to know the learning progress and status of students' knowledge in order to apply corresponding teaching methods to motivate and stimulate the students' learning interest. They also want to know the latest developments in the subject matter in order to give instructions more effectively and efficiently. The effectiveness can be measured by the level of student engagement in the class as well as in deeper learning. They also need to understand the overall performance of students in order to better communicate with parents and improve collaboration between the institution and the students' parents. Teachers may also expect to reduce their administrative work or streamline the administrative duties in a smart university.

Parents. Parents may want to monitor their children's learning progress and performance. In addition, they could potentially expect to receive ample warning when their children's behavior deviates from normal.

Institutions. From the perspective of an institutional management team, they might want to make decisions based on data. In order to achieve that, it may be necessary to know the operational efficiency of the institution such as performance indicators of teachers and students and ranking data as compared to other institutions. It is desirable to promote student success, such as getting better further study opportunities, better personal development, and work readiness, through early prediction of students' performance. It may also be necessary to report to the local education authorities on various key performance indicators.

Local Educational Authority. The local educational authority keeps track of accurate and reliable performance indicators from institutions, which can support their decision making in defining/refining the education policies and plans. Data is collected for the purposes of monitoring and analyzing institutional performance.

11.2.3 Lifelong Learning

"The illiterate of the 21st century will not be those who cannot read and write, but those who cannot learn, unlearn, and relearn" [14]. Lifelong learning is well believed to be an important way of continuous success or survival in career development. The need for lifelong/professional learning is increasing as a result of globalization, rapid technology development, and moving to a knowledge-based economy. This requires employees to acquire new knowledge and skills [15]. It is particularly true for professionals such as engineers who have a need to acquire new knowledge and skills when performing their duties. Continuing professional development (CPD) is, therefore, a commitment by all professionals to continually update their knowledge and skills in order to remain professionally competent [16].

Competency, consisting of knowledge, skills, and behavior, is a standardized requirement for performing a specific area of jobs. Competency is defined as "a specific, identifiable, definable, and measurable knowledge, skill, ability and/or other deployment related characteristic (e.g. attitude, behavior, physical ability) which a human resource may possess and which is necessary for, or material to, the performance of an activity within a specific business context" [17].

Competency requirements become more common in lifelong learning, specifically in the area of training and performance improvement. Learning outcomes, with respect to the required competence, are clearly defined. Therefore, a professional engineer's competency level can be measured with the corresponding discipline of engineering. Professionals can begin their learning processes by defining the learning goal with intended learning outcomes as well as designing a customized learning path according to professional needs.

The major challenge for standardizing competency requirements of a specific range of job functions across an industry is that we need to clearly define learning outcomes for each unit of competency in order to facilitate the establishment of the progressive learning paths.

In traditional lifelong learning, courses or training materials are organized in a structured, but rigid, way. Learners may only follow predefined paths in achieving the learning goals. When there is no explicit mapping between the courses with intended learning outcomes and no properly maintained e-portfolio, it is hard to recommend the most suitable learning paths to learners and, thus, adaptive lifelong learning is difficult to achieve [18]. It is particularly true in the information age that learning courses, learning materials, and learning activities come from various sources without any explicit or predefined structure. Identifying meaningful learning opportunities through these unstructured resources becomes unclear and, as a result, is one of the major challenges in smart lifelong learning [19]. It is expected that smart lifelong learning should be able to support learners in exploring, identifying, and seizing both structural and nonstructural learning activities, as well as proposing learning paths based on learners' learning history, learning goals, and intended learning outcomes of the corresponding learning activities available to the learners [20]. Self-directed lifelong learning relies on the matching of available learning activities with learners' personal needs, learning history, prior competent levels, and preferences in a quick and effective way [21].

11.2.4 Whole Person Development

Setting learning goals, planning learning activities, executing the learning plan, and reflecting on the learning outcomes are essential parts in personal development. A learner establishes learning goals and will need to plan a series of learning activities in order to achieve these goals. After the plan is executed, that is, participating in the chosen learning activities, a learner adjusts the goals and refines the plan based upon reviewing and reflecting on one's own learning achievements.

Personal development can be regarded as the learning process of a person. According to Kolb's experiential learning theory explanation [22], learning is not conceived in terms of outcomes. Rather, learning is an experience-based continuous process that requires interaction between a learner and the environment. Knowledge is created through the interaction that resolves the conflicts between abstract concepts and concrete experience under certain situations. For example, two people may have a different experience regarding the same issue when the issue happens in different situations. In other words, every person who participates in the same learning activity may have a different learning experience. This, however, does not necessarily mean that the learning experience of a person is immutable. Learning experiences may be mutated when certain situations apply to it. That is to say, learning is not a single execution process. The learning process, according to Kolb [22], "can be described as a four-stage cycle involving four adaptive learning modes —concrete experience, reflective observation, abstract conceptualization, and active experimentation".

The learning process cycle from the experiential learning theory perspective is visualized in Fig. 11.1, where abstract concepts are transformed into the concrete experience through experimentation under different situations. The concrete learning experience is gathered and transformed into new abstract concepts through reflection and observation based on the experience or situation. To apply this theory

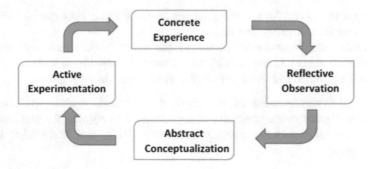

Fig. 11.1 Experience learning cycle

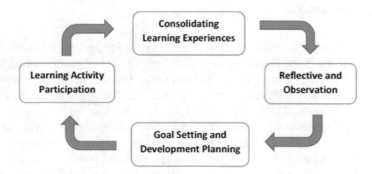

Fig. 11.2 Personal development cycle

to personal development, an example from the Quality Assurance Agency for Higher Education (QAA), UK is shown in Fig. 11.2. From QAA guidance [23], abstract conceptualization refers to the process of identifying learning goals and creating development plans. Through executing the plan under different learning situations, concrete experiences and achievements can be obtained. This is based on which reflection and observation is carried to identify new learning needs.

11.2.5 Outcome-Based Education

In the 1950s, Ralph W. Tyler began the discussion of outcome-based learning. Throughout the following decades, different definitions have been proposed [24–29]. In the year 2002, Harden [24] characterized outcome-based education as:

(1) Developing and publishing clearly defined learning outcomes that must be achieved before the end of the course;

(2) Designing a curriculum, learning strategies and learning opportunities to ensure the achievement of the learning outcome;
(3) Matching the assessment process to the learning outcomes and assessing individual learner to ensure that they achieve the outcomes; and
(4) Providing remediation and enrichment for learners as appropriate.

Despite different meanings of wordings and its original target for well-structured curriculum, it matches the personal development cycle very well for extracurricular activities, achievements for developing various skills, and knowledge beyond normal courses.

11.2.6   e-Portfolio

Portfolio intends to "contain material collected by the learner over a period of time; that the portfolio is the learner's practical and intellectual property relating to their professional learning and personal development… the learner takes responsibility for the creation and maintenance of the portfolio and if appropriate, for the presentation of the portfolio for assessment" [30]. The materials collected are related to an individual's personal development, such as transcripts, recognitions, and certificates. There are different definitions of e-Portfolio varying in the context of use. Electronic portfolio refers to a digital form of a physical portfolio that documents individual learning and personal development. A highly profiled definition of e-Portfolio was suggested by Educause NLII in 2003 [31], saying that e-Portfolio is "a collection of authentic and diverse evidence, drawn from a larger archive, that represents what a person or organization has learned over time, on which the person or organization has reflected, designed for presentation to one or more audiences for a particular rhetorical purpose." Strictly speaking, an e-portfolio is a particular perspective of the personal development archive for the specific context of which the portfolio in use. Some example contexts include: assessing the development; presenting the works during the development; guiding, tracking, and reviewing; and planning the development.

Assessment e-Portfolio. It is used to demonstrate an individual's learning outcome by relating evidence to certain defined learning objectives. The assessment would be made based on defined rubrics [32]. It is the perspective used for assessing the degree of learning in a learning process.

Presentation e-Portfolio. It is used to show an individual's learning, achievements, and recognition to a general audience [32]. It can be in any form, but the most common is a resume. Profiles can also appear on social networking sites.

Learning and development e-Portfolio. It is a life-long portfolio, which helps to document, guide, and advance an individual's learning over time. There are planning and reflection modules [32]. A personal development portfolio is an example of this. Personal development planning (PDP) is defined as "a structured and supported process undertaken by an individual to reflect upon owns learning, performance and/or achievement and to plan for their personal, educational and career development" [23].

Learners may assume responsibility for improving themselves through iterations of goal setting, action planning, activities participation, affirmation of achievements, self-reflection, and refining the intended learning outcomes. This process may help enhancing their critical thinking and meta-cognitive capacity, meaning they know what they know. Most importantly, their self-esteem could be enhanced through seeing their own growth. The dynamic portfolio process consists of four functions [33]:

(1) Setting the learning goals. There are different learning goals for different courses or learning activities. There are also different learning goals at different stages of the personal development. It is important to set up clear goals at each stage.
(2) Planning the learning activities. Once the learning objectives are defined, there should be an action plan set out to achieve the learning outcomes. The action plan involves defining a series of activities relevant to achieving the learning outcomes.
(3) Recording the activities and linking achievements to learning outcomes. Details of participating in activities are recorded. The time dimension of the portfolio is maintained to demonstrate the personal development history; however, information about the impact of incidences on learners, the experience gained from an activity, the feedback from peers and teachers, and the reflection details during the whole process could be lost as a result of frequent updates. The collection of this formative evaluation can help learners to know how they learn, which is useful in planning their further development. An additional aspect is to give learners the ability to map their achievements to activities as well as learning outcomes.
(4) Reflection. Besides the organization of portfolio content and the audit on the learning process, the proposed framework should allow external comments to be captured. Portfolio sharing and peer commenting are desirable features in the reflection process. Affirmations and comments from instructors and peers would boost learners' momentum for advancement.

Learning profile provides information on identity, preference, hobbies, etc., of a learner. Learning goals represent learning objectives of a learner in the form of a list of focused and achievable targets or attributes. These learning objectives can only be achieved by a series of learning activities and, thus, a substantial amount of information depicted in the portfolio focuses on the participation of activities. Obviously, an activity is a visible act that is indispensable in a portfolio. The actual evidence relevant to accomplishing a specific learning outcome is the achievements induced from participating in that activity. It is important to distinguish an achievement from an activity due to the fact that the latter solely provides an opportunity for development of valuable attributes and perceptible contributions. This helps to avoid learners spending much time on activities aimlessly without making any significant achievement relevant to the intended learning outcomes. Achievements could be a quantifiable performance evaluation of domain specific knowledge and skills in addition to generic skills. In terms of information structure, each of the intended learning outcomes should be connected to multiple instances of activities. Similarly, learners can benefit from an activity in more than one aspect. Consequently, an activity usually links to multiple achievements and, thus, forms a networked path echoing back to the individual learning outcome. Reflection is the key to learning. A list of activities and achievements alone can only give a superficial account of efforts made by learners. In a reflective process, learners integrate their experiences in various activities and achievements so as to match with the designated learning objectives and search for the direction for improvements [8].

11.2.7   Self-regulated Learning

The Self-regulated Learning. Self-regulated learning (SRL) is related to goal-directed, self-controlled learning behavior in learning processes [34] and is defined as [35]: "a process in which individuals take the initiative, with or without the help of others, in diagnosing their learning needs, formulating learning goals, identifying human and material resources for learning, choosing and implementing appropriate learning strategies, and evaluating learning outcomes". Zimmerman [36] described that "self-regulated learners plan, set goals, organize, self-monitor, and self-evaluate at various points during the process of acquisition" and become self-masters of their learning and directly enhance their learning outcomes. It has a learning cycle similar to the personal development cycle and is a closed iterative loop to distinguish self-evaluation, planning and goal setting, and learning.

Self-regulated Learning and Smart University. A smart university has the ability to support self-regulated learning by automatically obtaining, acquiring, or formulating new or modifying existing knowledge, experience, or behavior in order to improve its operation, business functions, performance, effectiveness, etc. [37]. It also has the ability to suggest various learning paths for learners through various prediction models. Smartness relies on the availability, accuracy, and comprehensiveness of data. Intelligence is based on known procedures and through integration of existing information systems merging with human intelligence.

An example of evidence of outcome-based education is the achievements of students in different aspects. These are not limited to course-embedded assessments, but also include participation in extra-curricular activities as well as self-reflections with respect to the learning goals. Data can also be obtained in many other situations such as teaching and learning, extracurricular activities, and performance assessments of teachers and students. Given the scale of "big data", it is crucial for the educational authority, at a higher hierarchy, to define the standard for data collection from various sources based on the need of planning and monitoring.

A unified standard enables data to be collected and compared in the same format [11].

Self-regulated Learning and e-Portfolio. It is found that e-portfolios are increasingly being used to support SRL [38] and is also found to motivate students more so than paper portfolios [34]. E-portfolios are used to facilitate acquiring skills relating to SRL including self-assessment of performance, formulation of learning goals, and selection of future tasks. E-portfolios normally include learning plans with goal settings and action plan to meet the goals [39]. Learners may perform self-evaluation to verify their progress against the goals, and to reflect their learning experiences. SRL is achieved through these self-managing and self-reflective tasks [40].

11.2.8   The Challenges

There are many challenges in building e-portfolio systems for OBE, both technical and administrative procedural challenges.

Technical Challenges. Specifying outcomes clearly is a difficult task, particularly for unstructured learning activities. It is always a difficult task to manage unstructured data electronically. In addition, a learning activity may have similar or even a small set of common attributes contributing to the learning outcomes. A learner may want to achieve different sets of learning outcomes at different stages.

There may be a lot of achievements of a learner at different stages, especially in terms of whole person development and lifelong learning. A learner may not realize what specific attributes are contributing to the set of achievements or learning outcomes. The ignorance in understanding learning attributes may lead to a lack of reflection during the learning process and lead to difficulties in linking evidence of achievements to the learning goals. As personal development involves many learning goals, development plans, and learning situations, recording the details of a learning process could be helpful for reflection and observation.

Although the above-mentioned e-portfolio systems may provide entry points for recording some of the essentials, they mainly assist learners in executing e-learning activities and reflecting upon their learning. They do not have a process model or well-defined approach to assist the complete personal development cycle as described earlier. In particular, supports for goals setting and planning are missing. Existing systems may not have an approach to assist planning where the plan of the next learning process directly influences the quality and outcomes of learning at next stage. Facilities to assist the entire cycle of personal development are, thus, desired to be integrated into existing e-portfolio module of an e-learning system, including goal setting and learning plan design [41].

A university may face challenges to cope with novel needs of learners, and to provide a seamless integration between the education production system (i.e., the education of the future workforce) and jobs, firms, industries, and organizations, which are requesting a multi-disciplinary education with complementary competencies and skills ranging from humanities to technologies [7].

Administrative Procedural Challenges. With a historical failure rate of more than 65% of enterprise systems implementations worldwide [42], information technology services/projects that are built around discontinuous, unique, and complex deliveries of projects require careful management and maintaining good client satisfaction and relationship [43]. The successfulness of the system depends on the affective attitude towards a specific computer application by someone who interacts with the application directly [44, 45]. Gaining user satisfaction and acceptance is the key successful factor of the system. In relationship marketing, trust and commitment are necessary for constructing and maintaining the relationship between the project team and other stakeholders [46–48]. Adopting e-portfolio and outcome-based learning in personal development, particularly for extracurricular activities, is clearly requiring respectively new administrative procedures to gain buy-in, acceptance, and satisfaction from various stakeholders.

11.3 The Design

11.3.1 The Design: Challenges

While personal development highlights the importance of learning in both structured curriculum and unstructured learning histories, traditional learning management systems fail to capture information about extracurricular activities in association with learning outcomes and attributes/competence expected from employers. Major challenges are identified below:

(1) System architecture: Too many standalone systems induce non-standard information definition and difficulties in the information exchange. There are many information systems available in higher education, including student information system, learning management systems, administrative management systems, space and facility management systems, etc.

(2) Information model, goal setting, and development planning: There is no standard set and meaning of a list of learning outcomes and attributes, no linkage between extracurricular activities and achievements, and no linkage between extracurricular activities/achievements and learning outcomes/attributes. Therefore, a well-defined information model is missing and thus goal setting and development planning are infeasible.

(3) Learning activity participation: Not all extracurricular activities are captured electronically in a single repository and no non-academic achievement record is recorded or vetted by the institution.

(4) Consolidating learning experiences: Attendance and feedbacks of participating in learning activities are not captured and thus learning experiences cannot be traced [49].

(5) Reflection and observation: It is difficult in drawing statistical reports on extracurricular activities because of the missing a well-defined information

model with corresponding information in non-academic activities, and thus reflection and observation are hard to be performed [33].

11.3.2   The Design: System Architecture

The first challenge for building a central repository is the selection of system architecture and platform; however, there is no quick solution to transform a university from being digital to being smart. Smart systems will not be built from scratch, but evolved on existing systems and infrastructures.

The inevitable trend in building smart systems in smart universities is to enhance and improve existing systems features and increase the level of automation gradually. It is a long evolving process of developing a smart university by making a digital campus work cleverer as well as smarter [11]. Fortunately, most institutions have a centralized database for storing the academic learning histories covering admissions, enrollment, registration, academic advising, degree audit, program and course registration, academic results and feedbacks, academic achievements, and academic transcripts. They also keep major student life activity records and processes, such as exchange, residence hall history, internship/campus work schemes, and student development courses. These platforms normally allow the building of self-service functions on top of the existing student information systems by sharing both the database and application layers. Taking CityU as an example, the student information system (SIS) is based on an off-the-shelf product (BANNER) with capabilities in building self-service functions, which form an integrated platform. The platform is called "administrative information management system" (AIMS) [50].

Figure 11.3 shows the high-level system architecture of an e-portfolio system, where the database layer and the application layer of the existing system are being used as the basis. The functional layer is re-defined in three groups:

(1) Administrative module for maintaining common code lists, which is a list of attributes with codified meaning as shown in Table 11.2.

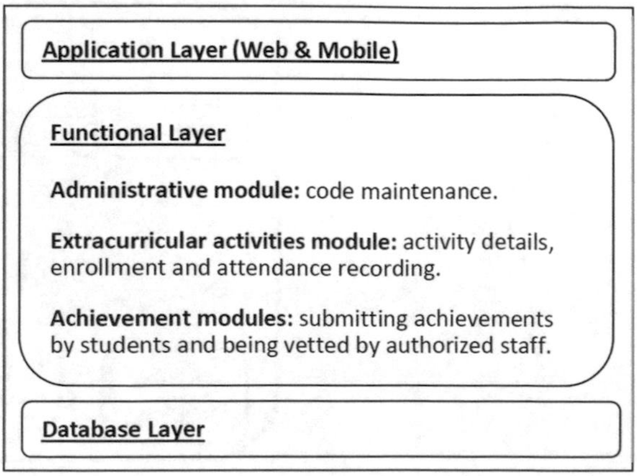

Fig. 11.3 System architecture of the proposed system

Application Layer (Web & Mobile)

Functional Layer

Administrative module: code maintenance.

Extracurricular activities module: activity details, enrollment and attendance recording.

Achievement modules: submitting achievements by students and being vetted by authorized staff.

Database Layer

(2) Extracurricular activities module for managing activity details, enrolments, and attendances.
(3) Achievement module for submitting achievements by students and being vetted by authorized staff.

A seamlessly integrated system provides a standardized look and feel of system functionalities to all users and thus it may gain better user satisfaction because of technology acceptance and perceived ease of use. Moreover, it may shorten the implementation period and implementation cost by reducing system complexity and increasing the system stability because all these hardware and network architecture, as well as application architecture, can be re-used.

Mobile technology cannot be ignored when engaging learners in universities. Tecmark's recent research indicated that a person looks at the smartphone more than 200 times a day [51]. Most learners in universities are around the age of 20, when mobile devices are "something that they couldn't live without" [52]. Therefore, the mobile-friendly design is a major consideration when choosing the application platform.

11.3.3   The Design: Information Model

Figure 11.4 shows the proposed conceptual information model of a new type of portfolio system in which learning outcomes and attributes are important elements. Activity organizers and learners, however, do not have a standardized or consistent way of defining and using the attributes, learning outcomes, and their mappings.

In fact, code lists and mappings of learning outcomes and attributes can be defined at the institutional level by involving representatives from all departments. These representatives will organize extracurricular activities and define a standard

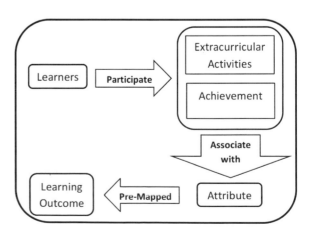

Fig. 11.4 Information model of the new portfolio system

Fig. 11.5 Graduate
outcomes of City University
of Hong Kong

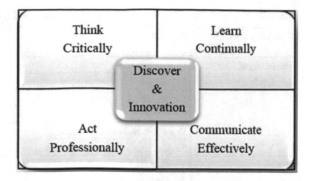

set of attributes as a result of reviewing the local employer survey and the institution's strategic plan.

In 2010, the Hong Kong Government defined attributes based on the employer
survey conducted [5], the strategic plan of the institution [53], and literature review
relevant to personal development. Attributes can be grouped in different categories
including spiritual, intellectual, physical, social, psychological/emotional,
professional/career, and international. Professional/career related attributes may
include skills such as career planning, leadership, organization of work, language
proficiency, professional/business ethics, and application of professional knowledge, while social related attributes may include skills such as team work, inter-
personal skills, negotiation, conflict resolution and communication skills, and civil
responsibility.

Learning outcomes and corresponding mapping with attributes can be identified
from graduate outcomes of the institution, as shown in Fig. 11.5. Tables 11.1 and
11.2, list out learning outcomes and the mapping with corresponding attributes
based on the graduate outcomes of CityU [54].

Table 11.1 List of learning outcomes

Learning outcome	Description
GO1	Apply effective communication, language, numerical and IT skills to a variety of professional settings
GO2	Apply multi-disciplinary critical thinking skills to solve problems and create new ideas
GO3	Relate cultural awareness to collaborate effectively in a broad range of teamwork situations
GO4	Generate a positive and flexible approach to lifelong learning and employability
GO5	Reflect on the ethical and social responsibility required for professional citizens in global society

Table 11.2 Sample mapping between learning outcomes and attributes

Category	Attribute	Learning outcome
Professional/Career	Career planning	GO4
	Leadership	GO3
	Organization of work	GO3
	language proficiency	GO1
	Professional/business ethics	GO5
	Application of professional knowledge	GO2
Social	Team work	GO3
	Interpersonal skills	GO1
	Negotiation, conflict resolution and communication skills	GO1
	Civil responsibility	GO5

11.3.4 The Design: Personal Development Cycle

Learning Activity Participation. The major challenge of recording learners' participation in extracurricular activities is that the expected learning outcomes of an activity are sometimes difficult for learners to know, but can be purposely designed by activity organizers. In addition, various organizing departments have different ways of announcing activities, receiving applications and recording feedbacks and attendance records. They may use standalone systems, Excel spreadsheets, or even hardcopy forms.

In order to solve this design issue, not only the attributes, learning outcomes, and the mapping between them should be defined clearly, but a centralized application module in the same student information system for organizers to create and maintain extracurricular activities, should also be developed. Activity information can then be synchronized to the institutional announcement system so that learners can receive personalized announcements from a single channel and perform online registration effectively.

Trainers/supervisors can input feedbacks of a specific learner on a specific activity and organizers can record the attendance as well. Learners' participation in extracurricular activities and achievements can be recorded centrally and structurally.

Consolidating Learning Experience. The challenge of consolidating learning experience is due to the missing link between extracurricular activities and achievements as well as between attributes and learning outcomes. Low incentive from learners on reporting achievements is another major challenge to keep information complete in a centralized repository.

Each organizing department is being requested to define the related attributes of an extracurricular activity when it is created. All organizers receive training to ensure a common understanding to interpret attributes that are mapped to an extracurricular activity.

Furthermore, having the system architecture, it becomes feasible to consolidate learning experience because information models and the learning activity participation records can be stored in a central repository. A set of static reports, as well as a generic reporting tool that allows flexibility in defining the search criteria, output fields, and sorting/grouping orders, are implemented. These allow learners to analyze their learning history from the point of view on either learning outcomes or attributes and allow organizers to analyze data for facilitating decision-making.

By building capabilities of reporting achievements by students and having authorized staff to perform vetting on the reported information is one of the major ways of motivating students to self-report their achievements to the system. Usability, achieved in the same system architecture as mobile friendly devices, is another key factor to improve student motivation.

Reflection and Observation. The challenge of performing reflection and observation in the personal development cycle is mainly due to a lack of complete information being captured.

With the central repository being built, advisors and learners can conduct a review of the learning history across different attributes and learning outcomes. In addition, they can holistically review both academic and non-academic activities being mapped to the same set of learning outcomes and attributes. Strength and weaknesses can be easily identified and thus proper suggestions to students on further improvement can be provided.

Goal Setting and Development Planning. The challenge of goal setting and development planning in terms of extracurricular activities is its non-structured nature. Any departments at any time for varying reasons may organize activities. Achievements may appear in many forms. In contrast, an academic curriculum has much clearer learning outcomes together with a series of pre-defined courses and activities for learners.

An academic advisor can suggest a study plan and discuss with each learner at the very beginning and monitor the progress from time to time on academic aspects. Alerts can be issued to the learner and the academic advisor when there is an indicator of potential deviation from the plan. Study plans can be revised to rectify issues promptly, but this does not apply to extracurricular activities and achievements.

With the proposed information model, goal setting and development planning become feasible. Although it cannot be performed from the very beginning of a learning lifecycle, it can be reviewed and defined from time to time based on the accumulated learning history, the gained soft skills and abilities, and available extracurricular activities instead of purely based on a learner's personal interest and time availability. At the same time, activity organizers in departments of a smart university may then plan for the most desirable activities in the future based on historical statistic instead of simply repeating the same activities regularly.

11.4   Implementation Review

CityU has recently launched a system, CRESDA, which is built into the student information system to record the extra-curricular activities and self-achievements in a centralized repository according to the design described in the previous section. It was implemented based on the concepts of personal development, outcome-based learning, and e-portfolio to centrally record learners' extracurricular activities with linkage to the desired attributes/competences according to the graduate learning outcomes as defined. It aims to assist learners in becoming smarter through self-regulated learning in a smart university. An implementation review is, therefore, important before lifelong learning/professional development is further explored.

11.4.1   Acceptance and Satisfaction

As a result of gathering comprehensive user requirements from academic departments and academic support units, a set of mappings between the learning outcomes and attributes have been developed. These are readily available for activity organizers and learners to link the learning attributes to extracurricular activities and achievements correspondingly. The following key approaches were adopted during the implementation with consideration of the implementation challenges as discussed:

(1) Representatives of all user groups are involved from the very beginning throughout the project [55]. Building trust and commitment to such a project is a long process. It is essential to pull core stakeholders into the project from the very beginning and empower them throughout the project lifecycle from user requirement gathering, design, implementation, and evaluation of the system.

(2) Representatives of all user groups are involved during the user acceptance test (UAT) and user training [56–58]. The involvement of user representatives in the project early may also help the system to be accepted widely by all user groups.

(3) The prototype methodology is adapted as "seeing is believing" [59]. With a prototype that is expected to be identical to the future system, in both look and feel of all functional design and screen navigation, stakeholders can make comments and provide feedback more easily and actively during the design stage. Problems identified during the design stage can be solved much easier than problems identified at a later stage of the project because "prevention is better than cure" [60].

(4) All types of potential users carried out the user acceptance test in order to increase the success of the system [55] because involving representatives from all potential user groups in testing may be treated as a part of user training and may make it widely accepted by all user groups.

(5) Psychological and emotional barriers or "Cognitive lock-in" [61] is one of the major reasons of being resisted to a new information system. Phase approach by delivering system functionalities in batches and gaining quick and small wins can reduce the cognitive lock-in of an end user as well as an increase in the possibility of gaining satisfaction through perceived ease of use and technology acceptance [56]. As such, CityU has finished the implementation of Phase 1 and is reviewing the continuous improvement of next phases.

(6) Perceived ease of use and technology acceptance has a positive influence on satisfaction, trust, and commitment of an information system [56–58]. CityU has designed a series of promotion activities tailored for different user groups. Hands-on classroom training is designed for administrative staff while seminars and demonstrations are designed for students for gaining perceived ease of usage and technology acceptance, thus gaining the acceptance, trust, and commitment. Other events, such as web-story and opening ceremony, are being used to enhance awareness and buy-in.

11.4.2 Strengths and Areas for Improvement

The Phase 1 implementation was completed in early 2016 with a total of 27 user representatives from all user groups participated in and endorsed the user acceptance test. These user representatives developed test cases while technical staff set up the testing environment with pre-requisite data. A pre-UAT briefing was conducted in order to ensure user representatives familiar with the testing procedures and methodologies. Each user representative can test the system using different simulated user privileges. Major positive feedback includes:

(1) The available learning histories, together with the associated learning attributes associated with extracurricular activities, make learners think in terms of achieving goals and learning attributes from activities.
(2) The predefined list of attributes and learning outcomes make the definition be transparent and standardized.
(3) A central repository can facilitate decision making by obtaining an overall picture with using the generic reporting tools.
(4) The available data in a central repository can help learners review their learning history, plan for future learning activities, and facilitate their resume building for job seeking so that learners may have more control over their learning.
(5) Mobile friendly design can encourage user acceptance.

The following enhancement opportunities were identified:

(1) For a better self-reflection by each fresh graduate, it is required to be facilitated in generating customized resumes from their learning histories by matching the attributes that employers are looking for.

(2) For assisting learners in learning plan development, it is required to provide an automatic enrolment feature or alert feature based on the predefined criteria to better facilitate self-regulated learning.
(3) Different learners participating in the same activity may gain a different level of knowledge and skills [5] and therefore it is required to build a linkage between extracurricular activities and achievements in order to make them traceable.

The implementation provides an e-portfolio tool to supplement existing systems of academic learning activities covering also the extracurricular learning activities and facilitating learners to plan their learning based on the required graduate outcomes. Phase 1 implementation was completed and is being planned for university-wide launching followed by a quantitative review [37, 38].

11.5 Extending to Lifelong Learning and Professional Development

The need for CPD. Professionals, such as engineers, need to take opportunities to update their depth and breadth of knowledge and expertise by means of continuous professional development (CPD) and to develop those personal qualities required to fulfill their roles in industry and in society. Therefore, the practicing professionals should aim to remain competent throughout their working careers so that they can properly carry out their various duties [16].

CPD covers not only technical matters, but broader studies of importance to professional engineers in advancing their careers such as communication, environmental matters, financial management, leadership skills, legal aspects, marketing, occupational safety, and health and professional ethics. Similar to learning in a university, the format of CPD activities are flexible and may include participating and organizing courses, lectures, seminars, conferences, presentations, workshops, industrial attachment and visits, e-learning, and other professional activities.

Challenges in CPD. While a lot of professional bodies in engineering are adopting the approach of competence-based assessment of experience, the major difficulty for standardizing professional competency requirements for all disciplines is the need to clearly define learning outcomes in each knowledge domains and define benchmarks for each unit of competency. This is very similar to the challenge of how extracurricular activities are mapped with outcome-based learning in the context of a smart university. In a continuous professional development, soft skills building through unstructured curriculum occurs even more than in a university. A major challenge is to facilitate self-regulated learning for professionals and allow more flexibility in matching the CPD requirements. With the recent advancement of technology, learners may learn from various open sources such as massive open online course (MOOC). Therefore, MOOC may be a potential source of information for building the e-portfolio in facilitating learners' goal setting, planning, and self-evaluation through enhancing their self-regulated learning.

Can MOOC help? There are no formal definitions for a MOOC. The Oxford Dictionary defines it as a course of study made available over the Internet without charge to a very large number of people; Wiki defines it as an online course aimed at large-scale interactive participation and open access via the Internet. In addition to traditional course materials such as videos, readings, and problem sets, MOOCs provide interactive user forums that help build a community for the students, professors, and teaching assistants (TAs). Of course, there is still room for improvement on the level of interactive participation for the time being.

Such a platform may accommodate a large amount of course materials and support a large amount of users, which may have a good potential to support CPD for the professional community. The content may include courses at various levels, seminars held at various times and places, live demonstration of specific skills, etc. The user group may include professionals such as engineers; students who may want to become a member of a professional body; and even the general public who may want to acquire skills and knowledge in the certain discipline of professionals. In this sense, the board concept of MOOC can certainly help.

The content for supporting the professional development can be at various levels. At the university level, the academic programs in engineering disciplines, for example, that have gone through the accreditation process can be put on the MOOC platform. These may help engineering students enrich their choice of electives which may not be offered in one university but can be found in others. If this concept can be extended to other engineering programs under various recognition schemes like Washington Accord, the choice of engineering courses may be widely open and this can enhance the engineering education in general. This concept may be applied to other professional disciplines.

At the CPD level, this can be even more flexible. Currently, there are numerous seminars organized by various organizations. Using CPD for professional engineers, as an example, the contents of these seminars and talks can be put online. This means professional engineers may now have more options to fulfill their needs and requirements in professional development. For example, when having difficulties in finding CPD training in a particular type of soft skill such as safety, a generic essential training element for all engineering disciplines, in their areas may find it easily in other engineering disciplines. Thus, putting seminars and talks of CPD nature on MOOC may help professional engineers to further develop themselves according to their needs and fulfill the CPD requirements as defined by the corresponding professional association they belong to.

Issues when adopting MOOC for CPD. When adopting the concept of MOOC in supporting the training and CPD requirements of professional engineers, though, there are certain areas that need to be addressed before it can really help [16].

(1) In order to make the contents of MOOC searchable by learners, these content materials can be treated as learning objects. Each of these learning objects will need to be accompanied by a set of learning object metadata (LOM). The IEEE Learning Object Metadata (LOM) standard [62] defines a learning object as any entity, digital or non-digital, that may be used for learning, education or

training. It aims to facilitate the searching, evaluation, acquisition, and using of learning objects by learners, instructors, or automated software processes. Metadata is the data about a learning object, including attributes about the metadata itself, the measures and controls of the contents, its physical properties, description of the contents, and its contextual use. This information may help learners obtain relevant learning resources easily [63, 64]. The sharing of learning resources on MOOC is more convenient and practical with the use of metadata.

(2) Although seminars and talks are relatively smaller units than courses in terms of length and materials to be covered, it is worth specifying its intended learning outcomes with descriptions of its content before putting it on the MOOC platform. Such a mental exercise from the authors or organizers make it a natural screening process for putting only valuable contents with clear learning objectives on the MOOC platform.

(3) A competence model is required to clearly define the requirements of a professional discipline. The generic model of competence for all disciplines defines the general educational requirements and the number of years of relevant experience. Each discipline may need to define the domain knowledge contained and its level of attainment. The representation of professional competence requirements in ontology helps professionals clearly understand the domain knowledge and skills that they must acquire. Discipline panels may maintain and update these competence components and knowledge domains. The competence model provides a basis for accreditation of engineering degrees and for a professional assessment.

(4) A training path can be considered as a learning sequence in which a learner may study the courses, seminars, talks, etc., according to a set of pre-defined rules. A training path is established from the training goals and training history of a learner [15]. The setting of the training goal by professionals provides a better control of training by themselves so as to achieve self-regulated learning.

A mechanism is required to categorize the competency requirements of professionals, like engineers, using ontology for different disciplines, to automatically record their learning progress and qualifications, knowledge, and skills they gained, and to propose personalized learning paths based on these information. Professionals may first set up their respective training goals. With reference to their training history and the competence model, they may identify their training needs from the course and seminar information available on the MOOC platform. The stated intended learning outcomes on the MOOC for corresponding courses and seminars can be used for matching against the competency level of respective engineering disciplines. Possible technologies to be adopted include Semantic Web ontology technologies and IEEE LOM standards to define the category and details of each competence components, course information, and training record ontology for learners. Supervisors or human resource managers may define the training targets with learners based on learners' training history and propose personalized training paths to learners.

A professional engineer may need to keep records of qualifications, knowledge, or skills one possesses to identify a clear training goal. With reference to a clear goal and the corresponding training logs, learning advice would be more concise and relevant by matching course information against ones' current competency level. The courses to be suggested to a learner and the sequence of these courses form the training path of the learner. The sequence of the suggested courses are set based on a set of rules in order to ensure the suggested learning path fits the learning goal the most.

(5) After the competence components for engineering disciplines and the intended learning outcomes of training materials have been clearly specified, it is possible to build an individualized CPD portfolio for engineers. The building of such a portfolio can be part of a larger personal development portfolio that can be used to demonstrate the fulfillment of CPD requirements. There are many challenges in building such a portfolio system. Specifying outcomes clearly is a difficult task. A program of study, a specific course, a seminar, or even a visit, may have a series of goals or outcomes for a learner to achieve; however, the outcomes have to be specific before they can be measured. The information to be collected in such a portfolio includes several major entities: learning outcomes, activities and achievements, reflections, and abstracts. This information may help learners organize their learning objectives and track evidence of their achievements and competence [33]. The portfolio should reflect the learning process according to the selected learning path dynamically. Through iterations of action planning, activities participation, affirmation of achievements, self-reflection, and refining the intended learning outcomes, professional engineers should assume responsibility for improving themselves. The process enhances their critical thinking and meta-cognitive capacity, meaning they know what they know. The dynamic portfolio process consists of four functions: planning the goals, recording the activities, linking the achievements to learning outcomes, and reflection.

There are a lot of accomplishments of a learner at different stages, especially in intensive training as a trainee. The ignorance in understanding learning attributes may lead to a lack of reflection during the learning process and lead to difficulties in linking evidence of achievements to the learning goals. The outcome-based portfolio construction is focused on defining the intended learning outcomes, recording the learning activities, collecting learning achievements, and linking them to intended learning outcomes.

(6) In order to maintain training contents on MOOC, a vetting mechanism is desirable. Such a vetting mechanism will rely on the setting up of a panel, which consists of domain experts in the professional disciplines and in areas relevant to the CPD requirements. The main focus of such a mechanism is to check whether the specification of the intended learning outcomes of a training unit has been properly done and check whether such intended learning outcomes can be mapped to the competence components of respective disciplines. The requirement on such a clear specification of intended learning outcomes may help the content providers think carefully as to how the courses provided

can map to the competence component at varying level. This may, in turn, help professionals find the right courses fulfilling their training requirements.

Continuing professional development (CPD) is a commitment by all professionals to continually update their skills and knowledge in order to remain professionally competent. The challenge of defining a competence-based approach in assessment is similar to that of defining an outcome-based approach in extracurricular activities in a smart university context. The issues in adopting MOOC in helping professionals meet the training and CPD requirements are similar to the implementation of CRESDA in CityU. The common issue is to solve difficulties in providing learners the abilities in setting their learning goals, planning their learning activities, evaluating their learning histories, and providing assistance in advising proper learning materials and learning paths. These again rely on their self-regulated learning ability, which can be enhanced through the development e-portfolio. Keeping these records in a portfolio of professional development may help them further develop themselves professionally in the long run.

11.6   Discussion

Although the implementation of CRESDA is in a university setting, it can also be extended to handle lifelong learning because it shares the same personal development cycle and challenges as what learners in universities are facing for personal development. This session discusses some areas requiring further studies and some potential contributions in different aspects:

Further studies. The proposed design can contribute to research domains, including smart university, lifelong learning/professional development, e-portfolio, outcome-based learning, whole person development, and learning analytics.

(1) Whole person development involves not only the academic knowledge, but also the skills and knowledge to be gained through extracurricular activities and self-achievements. Extending the data landscape of learners from academic activities and achievements only to also covering non-academic activities and achievements can fill the gap between learners and companies' needs. This data landscape needs to be further refined.

(2) The learning outcomes can be applied to link up the extracurricular activities. This linkage may provide an overall picture of a learner on both academic and extracurricular achievements, joining a series of learning activities linked with the same set of learning outcomes under learners' self-control. Defining a set of clear learning attributes for various learning activities in achieving the intended learning outcomes becomes an important issue to address.

(3) Learning analytics is an emerging research area, where the majority of the researches were focused on the prediction of student success from students' academic performance, demographic, and psychological information. Only

30% of the variance in student success can be explained [65]. It was validated that including non-academic skills and achievements can enhance the accuracy of classification models in learning analytics for predicting student success [66, 67].

(4) Including the non-academic skills and achievements into e-portfolio can enhance the portfolio and thus improve the prediction power of learning analytics [68]. Measuring the level of competence of soft skills, however, requires further studies.

(5) One of the indicators in a smart learning environment is the ability to provide capability and flexibility in learning activities, planning by both the learners and designers, and in goal setting, for lifelong learning and professional development.

Potential Contributions. In a practical point of view, this can contribute to various stakeholders in the teaching and learning processes:

(1) Institution: can make use of the information for better decision-making. It also facilitates the institution in providing a better learning environment for the teaching and learning. While the institution and ranking agencies put more focus on internationalization and graduate employability [53, 69], having a centralized repository for related activities can facilitate record keeping, activity planning, and analysis in order to create a better policy to improve the level of internationalization and students' skills for better employability.

(2) Curricular designers: can review and revise their learning activity designs for enhancing student success with the extension of data landscape of learners since student development outcomes through non-academic activities and achievements, are critical for research on the course and curricular design [65].

(3) Activity organizers: can make use of the information to enhance activity contents, frequency, timing, and targeted students for the purpose of enhancing students' whole person development as well as employability. Instead of organizing activities in a routine way following previous years' practices, activities can be planned in a much better way based on various factors such as what activities can improve students' learning outcomes and employability, when will be the best time for an activity, which activities are most popular, etc.

(4) Teachers or advisors: can monitor and evaluate student's learning progress in both academic and non-academic activities and achievements in order to provide more accurate suggestions to student's study paths and provide intervention to at-risk-students timely and effectively. Proper alerts and intervention are useful for teachers and advisors to provide personalized learning activities and materials to students in academic side. Compared to assessing and giving advice purely based on students' academic results, teachers/advisors can now provide suggestions with overall considerations.

(5) Learners: can better know their strengths and weaknesses during self-reflection and self-awareness with consolidated information on both academic and non-academic achievements. It helps students to review and revise their study

plan and also enhance their self-regulation level. This further improves their engagement and, thus, their learning outcomes. Instead of applying activities simply based on interest and time availability, students can plan their participation based on accumulated learning outcomes that they have already gained and compare this with the actual learning outcome to be achieved.

(6) Employers: can better match potential candidates against recruitment criteria based not only on academic performance, but also on evidence of what soft skills candidates have developed and achieved.

(7) Professional assessment body: can assess one's ability in a more structured and systematic way.

11.7 Conclusion

In this fast changing environment caused by digital enablement, the amount of time for face-to-face lectures is decreasing; hence, the amount of time for self-regulated learning will need to be increasing with more and more self-learning activities followed by self-evaluation. Universities must maintain their leading role in providing a more flexible learning environment in order to facilitate better self-regulated learning and transform from a smart university into an even smarter one. Due to this, both teachers and students may enjoy better learning through higher participation in learning activities more efficiently and effectively, with collaboration to achieve a common objective of better learning [7]. A smart university can enhance learners' self-regulated learning and allow them to plan and evaluate their learning in a more effective way and extend it to their lifelong learning and professional development throughout their career life.

The proposed CPD model and the CRESDA system provide a mechanism in associating intended learning outcomes to learning courses and learning activities, which facilitates learners in selecting the most desirable learning courses and learning activities dynamically according to their personal needs in achieving the various goals at different stages in the lifelong learning process. Thus, they have more control in managing their own lifelong learning process. A smart university may facilitate learners in achieving smart lifelong learning in terms of adaptation to the dynamic needs of learners to allow a greater degree of self-directedness and self-organization with the help of suitable learning technologies. When learners establish personal learning goals by reviewing their learning history in the e-portfolio and reflecting on their current learning activities, they may select learning courses or learning activities recommended by the proposed CPD model and the CRESDA system by means of mapping the learning outcomes of learning activities to desirable attributes to be achieved. Learning path recommendation becomes feasible and, thus, learners may manage their learning in a self-directed and self-organized manner. Since learners may have different requirements on selecting learning activities in terms of not only the learning outcomes, but also learning pace and learning style, the proposed CPD model and the CRESDA system may be adaptive to the needs of learners dynamically and thus may help the university to move one step forward towards a smarter university.

In this chapter, we reviewed smart universities and discussed the role of e-portfolio in both lifelong learning and personal development with the outcome-based education framework. Common challenges were identified, a system implementation was proposed, and an actual implementation in CityU was reviewed. A model combining training contents on MOOC with intended learning outcomes specified clearly, together with competence based definitions and personalized training portfolio, may help professionals fulfill their CPD and training requirements.

It is expected that the portfolio design proposed in this chapter may provide a useful reference for researchers and institutions in keeping learning history and promoting student success, including employability and continuous professional development.

References

1. Drucker, P.: BrainyQuote.com. https://www.brainyquote.com/quotes/quotes/p/peterdruck165702.html
2. Lindeman, E.C.: The Meaning of Adult Education, 1st edn. New Republic, New York (1926)
3. The Scottish Office: Opportunity Scotland: A Paper on Lifelong Learning. Stationery Office, Edinburgh, Scotland (1998)
4. National Association of Colleges and Employers: Job Outlook 2016: Attributes Employers Want to See on New College Graduates' Resumes (2015). http://www.naceweb.org/s11182015/employers-look-for-in-new-hires.aspx
5. Education Bureau—The Government of the Hong Kong Special Administrative Region: Survey on Opinions of Employers on Major Aspects of Performance of Publicly-funded First Degree Graduates in Year 2006 (2010). http://www.edb.gov.hk/attachment/en/about-edb/publications-stat/major-reports/executive_summary%20_fd_eng__april_2010.pdf
6. Guan, J.Q., Su, X.B., Guo, Y., Zhu, Z.T.: Design of primary school mathematics review teaching mode under the environment of e-schoolbag. China Acad. J. Electron. Publ. House 338, 103–109 (2015)
7. Coccoli, M., Guercio, A., Maresca, P., Stanganelli, L.: Smarter universities: a vision for the fast changing digital era. J. Vis. Lang. Comput. 25, 1003–1011. Elsevier (2014)
8. Kwok, L.F., Chan, K.S.: Building portfolio from learning plan. In: Proceedings of the 17th International Conference on Computers in Education (2009)
9. Barker, K.C.: ePortfolio: a tool for quality assurance (2006). http://www.futured.com/documents/ePortfolioforQualityAssuranceinEducationSystems.pdf
10. Lombardi, P., Giordano, S., Farouh, H., Yousef, W.: Modelling the smart city performance. Innov. Eur. J. Soc. Sci. Res. 25(2), 137–149 (2012)
11. Kwok, L.F.: A vision for the development of i-campus. Smart Learning Environment 2:2, a SpringerOpen Journal (2015)
12. Roth-Berghofer, T.: Smart University, the University as a Platform (2013). https://smartuniversity.uwl.ac.uk/blog/?p=100
13. Hwang, G.J.: Definition, framework and research issues of smart learning environments—a context-aware ubiquitous learning perspective. Smart Learn. Environ. 1(4). Springer (2014)
14. Toffler, A.: Future Shock, pp. 414. Random House (1970)

15. Kwok, L.F., Yeung, S.Y., Cheung, C.H.: A training path advisor for lifelong learning. In: Proceedings of the IASTED International Conference Web-based Education (2009)

16. Kwok, L.F.: Meeting CPD requirements—can MOOC help? Beijing Forum 2013: Global Engagement and Knowledge Sharing in Higher Education, pp. 78–90, 1–3 November 2013. Peking University, Beijing, PRC

17. HR-XML: Competencies   Schema   (2001).   http://xml.coverpages.org/HR-XML-Competencies-1_0.pdf

18. Ahmad, A., Omid, R.B., Somayyeh, M.: A novel approach for enhancing lifelong learning systems by using hybrid recommender system. US China Educ. Rev. **8**(4), 482–491 (2011)

19. Dascalu, M.L., Bodea, C.N., Mihailescu, M.N., Tanase, E.A., Pablos, P.O.: Educational recommender systems and their application in lifelong learning. Behav. Inf. Technol. **35**(4), 290–297 (2016)

20. Karoudis, K., Magoulas, G.D.: An architecture for smart lifelong learning design. In: Popescu, E., et al. (eds.) Innovations in Smart Learning. Lecture Notes in Educational Technology. Springer, Singapore (2017)

21. Drachsler, H., Hummel, G.K., Koper, R.: Recommendations for learners are different: applying memory-based recommender system techniques to lifelong learning. In: Proceedings of the Workshop on Social Information Retrieval in Technology Advanced Learning, pp. 18–26 (2007)

22. Kolb, D.A.: Experiential Learning: Experience as The Source of Learning and Development. Prentice Hall, New Jersey (1984)

23. The Quality Assurance Agency for Higher Education, UK: Personal Development Planning: Guidance for Institutional Policy and Practice in Higher Education (2009). http://www.qaa.ac.uk/academicinfrastructure/progressFiles/guidelines/PDP/PDPguide.pdf

24. Harden, R.M.: Developments in outcome-based education. Med. Teach. **24**, 117–120 (2002)

25. Spady, W.G.: Outcome-based instructional management: a sociological perspective. Aust. J. Educ. **26**, 123–143 (1982)

26. Brady, L.: Outcome based education: resurrecting the objectives debate. New Educ. **16**, 69–75 (1994)

27. Schwarz, G., Cavener, L.A.: Outcome-based education and curriculum change: advocacy. Pract. Crit. J. Curric. Superv. **9**, 326–338 (1994)

28. Guskey, T.: The importance of focusing on student outcomes. N. Cent. Assoc. Q. **66**, 507–512 (1992)

29. Spady, W.G.: Organizing for results: the basis of authentic restructuring and reform. Educ. Leadersh. **46**, 4–9 (1988)

30. Stefani, L., Mason, R., Pegler, C.: The Educational Potential of e-Portfolios: Supporting Personal Development and Reflective Learning. Routledge, New York (2007)

31. Educause, NLII Annual Review 2003. The New Academy (2003). https://net.educause.edu/ir/library/pdf/NLI0364.pdf

32. IMS Global Learning Consortium, Inc.: IMS ePortfolio Best Practice and Implementation Guide (2005). http://www.imsglobal.org/ep/epv1p0/imsep_bestv1p0.html

33. Kwok, L.F., Chan, K.S.: An outcome-based approach for building portfolio. In: Proceeding of 2nd International Conference on Hybrid Learning, pp. 169–176 (2009)

34. Beckers, J., Dolmans, D., Merrienboer, J.V.: e-Portfolios enhancing students' self-directed learning: a systematic review of influencing factors. Aust. J. Educ. Technol. **32**(2) (2016)

35. Knowles, M.S.: Self-directed Learning: A Guide for Learners and Teachers, Englewood Cliffs. Cambridge Adult Education, NJ (1975)

36. Zimmerman, B.J.: Self-regulated learning and academic achievement: an overview. Educ. Psychol. **25**(1), 3–17 (1990)

37. Uskov, V.L., Bakken, J.P., Pandey, A., Singh, U., Yalamanchili, M., Penumatsa, A.: Smart university taxonomy: features, components, systems. In: Smart Innovation, Systems and Technology, vol. 59, pp. 3–14. Springer (2016)

38. Kicken, W., Brand-Gruwel, S., van Merriënboer, J.J.G., Slot, W.: Design and evaluation of a development portfolio: how to improve students' self-directed learning skills. Instr. Sci. **37**(5), 453–473 (2009)
39. Crisp, G.T.: Integrative assessment: reframing assessment practice for current and future learning. Assess. Eval. High. Educ. **37**(1), 33–43 (2012)
40. Yang, M., Tai, M., Lim, C.P.: The role of e-portfolios in supporting productive learning. Br. J. Edu. Technol. **47**(6), 1276–1286 (2015)
41. Kwok, L.F., Chan, K.S.: Building an e-Portfolio with a learning path centric approach. J. Zhejiang Univ. SCIENCE C (Comput. Electron.) (2010)
42. Roses, L.K.: Antecedents of end-user satisfaction with an erp system in a transnational bank. J. Inf. Syst. Technol. Manag. **8**(2), 389–406 (2011)
43. Mainela, T., Ulkuniemi, P.: Personal interaction and customer relationship management in project business. J. Bus. Ind. Mark. **28**(2), 103–110 (2013)
44. Lee, Y.L., Hwang, S.L., Wang, M.Y.: An integrated framework for continuous improvement on user satisfaction of information systems. Ind. Manag. Data Syst. **106**(4), 581–595 (2006)
45. Somers, T.M., Nelson, K., Karimi, J.: Confirmatory factor analysis of the end-user computing satisfaction instrument: replication within an ERP domain. Decis. Sci. **34**, 595–621 (2003)
46. Bansal, H.S., Irving, P.G., Taylor, A.F.: A three-component model of customer commitment to service providers. J. Acad. Mark. Sci. **32**(3), 234–250 (2004)
47. Hennig-Thurau, T., Gwinner, K.P., Gremler, D.G.: Understanding relational marketing outcomes: an integration of relationship benefits and relationship quality. J. Serv. Res. **4**(3), 230–247 (2002)
48. Morgan, R.M., Hunt, S.D.: The commitment-trust theory of relationship marketing. J. Mark. **58**(3), 20–38 (1994)
49. Yang, H.H., Yu, J.C., Kuo, L.H., Chen, L.M., Yang, H.J.: A study of mobile e-learning-portfolios. WSEAS Trans. Comput. **8**, 1083–1092 (2009)
50. City University of Hong Kong: What are AIMS and Banner? http://www.cityu.edu.hk/esu/aims.htm
51. Tecmark: Tecmark Survey Finds Average User Pick Up Their Smartphone 221 Times a Day (2014). http://www.tecmark.co.uk/smartphone-usage-data-uk-2014/
52. Pew Research Center: The Smartphone Difference (2015). http://www.pewinternet.org/2015/04/01/us-smartphone-use-in-2015/
53. City University of Hong Kong: Strategic Plan 2015–2020 (2015). http://www.cityu.edu.hk/provost/strategic_plan/
54. City University of Hong Kong: City University Graduate Outcome (2016). http://www.cityu.edu.hk/qac/city_university_graduate_outcomes.htm
55. Sorum, H., Medagila, R., Kim, N.A., Murray, S., Delone, W.: Perceptions of information system success in the public sector. Transform. Gov. People Process Policy **6**(3), 239–257 (2012)
56. Scheepers, R., Scheepers, H., Ngwenyama, O.K.: Contextual influences on user satisfaction with mobile computing: findings from two healthcare organizations. Eur. J. Inf. Syst. **15**, 261–268 (2006)
57. Tsai, S.P.: Fostering international brand loyalty through committed and attached relationships. Int. Bus. Rev. **20**, 521–534 (2011)
58. Garbarino, E., Johnson, M.S.: The different roles of satisfaction, Trust, and Commitment in Customer Relationships. J. Mark. **63**, 70–87 (1999)
59. Practical Computer Application, Inc.: PCA Business Notes Series—Prototype Methodology: Seeing is Believing (2010). http://cdn.practicaldb.com/wp-content/uploads/2010/12/Prototyping-Methodology-Seeing-Is-Believing.pdf
60. Dromey, R.: Software quality—prevention versus cure. Softw. Qual. J. **11**, 197–210 (2003)
61. Johnson, E.J., Bellman, S., Lohse, G.L.: Cognitive lock-in and the power law of practice. J. Mark. **67**(2), 62–75 (2003)
62. IEEE Learning Technology Standards Committee WG12: Learning Object Metadata. http://ltsc.ieee.org/wg12/

63. Kwok, L.F., Cheung, C.H.: Managing learning resources in university undergraduate project study. In: Proceedings of IEEE 2007 International Workshop on Web 2.0 and Multimedia-enabled Education, 10–12 Dec 2007, Taichung, pp. 528–534 (2007)
64. Kwok, L.F., Cheung C.H., Tsoi, W.S.: Applying ontology in a project-based learning environment. In: Proceedings of 16th International Conference on Computers in Education (ICCE 2008), 27–31 Oct 2008, Taipei, Taiwan, pp. 551–555 (2008)
65. Dunbar, R.L., Dingel, M.J., Prat-Resina, X.: Connecting analytics and curriculum design: process and outcomes of building a tool to browse data relevant to course designers. J. Learn. Anal. 1(3), 223–243 (2014)
66. Burtner, J.: The use of discriminant analysis to investigate the influence of non-cognitive factors on engineering school persistence. J. Eng. Educ. 94(3), 335–338 (2005)
67. Lin, J.J., Imbrie, P.K., Reid, K.J.: Student retention modelling: an evaluation of different methods and their impact on prediction results. In: Research in Engineering Education Symposium, pp. 1–6 (2009)
68. Aguiar, E., Ambrose, G.A., Chawla, N.V., Goodrich, V., Brockman, J.: Engagement vs performance: using electronic portfolios to predict first semester engineering student persistence. J. Learn. Anal. 1(3), 7–33 (2014)
69. QS World University Ranking: Methodology. http://www.topuniversities.com/qs-world-university-rankings/methodology

Chapter 12
Towards Smart Education and Lifelong Learning in Russia

Lyudmila Krivova, Olga Imas, Evgeniia Moldovanova,
Peter J. Mitchell, Venera Sulaymanova and Konstantin Zolnikov

Abstract This paper describes the experience of introducing smart technology into the educational process at a Russian university between 2010–2016. Particular attention is paid to such innovative and smart techniques as training sessions, group teaching methods, role-playing games, the use of smart components, etc. The authors propose a method to enhance students' motivation for independent and life-long learning. This approach was trialed with students majoring in power engineering and the results are discussed. The main challenges in introducing smart technology are described, and areas for future development and improvement are identified.

Keywords Smart technology · e-learning · b-learning · Independent learning · Lifelong learning

12.1 Introduction

In the 21st century, engineers and university graduates need to be able to find information, making use of scientific and technological innovations, and interact with colleagues abroad. Moreover, innovative solutions to professional problems require a new approach. Universities should therefore focus on the individualization of professional training using smart technology. In this context, a new approach in the educational process (smart technologies) provides universities the opportunity to meet the requirements of consumers and employers. Smart education is a new educational paradigm that involves the implementation of an adaptive educational

L. Krivova · O. Imas · E. Moldovanova (✉) · V. Sulaymanova
Tomsk Polytechnic University, Tomsk, Russia
e-mail: eam@tpu.ru

P.J. Mitchell
Tomsk State University, Tomsk, Russia

K. Zolnikov
Institute of Strength Physics and Materials Science SB RAS, Tomsk, Russia

© Springer International Publishing AG 2018   357
V.L. Uskov et al. (eds.), *Smart Universities*, Smart Innovation,
Systems and Technologies 70, DOI 10.1007/978-3-319-59454-5_12

process using a range of smart information technologies. Smart education should provide the opportunity to benefit from the global information society, and to meet educational needs and interests [1].

The educational system in Russia is not immune to the influence of economic and geopolitical factors. One of these factors is the network economy that underpins international cooperation. In the context of globalization, academic mobility is one of the most crucial indicators of assessment for university performance, as well as the quality of education overall. Thus, globalization processes create a need for new forms of education, such as intercultural and transnational education. In Russia, the reforms envisaged by the Bologna process are now being felt by universities [2]. Internationalization of higher education requires the development and accreditation of modern curricula as well as new, more effective teaching techniques. Innovative technologies such as smart technologies determine the attractiveness of a university in the educational market. Their use meets the requirements of the educational environment, namely, to cater for a wide variety of all possible participants of the educational process.

The advantages of using smart education are in its unique ability to apply modern Information and Communication Technology (ICT) in: (1) the educational process; (2) new methodological and pedagogical educational approaches; (3) classrooms, (4) independent learning, etc. [3]. Smart technologies such as:

1. open educational resources for information (in addition to lectures, seminars, books, etc.);
2. new student-teacher and teacher-student relations;
3. a new testing and assessment approach;
4. a new approach to organization of communities (Internet community, professional community, etc.)

enabling restructuring of the educational process.

Rapidly evolving technologies lead to the rapid "aging" of knowledge (about 15–20% per year) [4]. Therefore, most obtained knowledge may become irrelevant within 3–5 years. In this case, the crucial goal of modern education is to develop self-learning skills and provide motivation for lifelong learning. Moreover, a smart approach to the educational process provides a wide range of possibilities for e-learning, blended learning (b-learning), etc.

The introduction of e-learning began in South Korea in 1997. Currently, there are 20 cyber universities with electronic educational services. Smart Education in South Korea is based on the concept of the country's competitiveness growth with limited natural resources. This is supported by the World Bank defining the category of "national wealth" as equal to 5% of natural resources and 77% of knowledge and skills.

In the modern world, traditional approaches may not be suitable for solving certain engineering challenges. Consequently, we need specialists who can quickly respond to any changes in the environment, to adapt to them, and to be engaged in continuous self-education. A modern specialist can be educated only through direct interaction with the professional community. A combination of group teaching methods, game teaching methods, interactive tutorials, and student collaboration connects students, teachers, and instruments, thus enriching the learning process. In this case, a smart environment and smart instruments support smart learning but are not sufficient in themselves. When introducing smart technologies to the educational process, the following aspects are especially significant:

- the readiness of a teacher to use smart technologies;
- the psychological and social features of students;
- a combination of student-centered, communicative, and socio-cultural approaches; and
- use of modern pedagogical principles.

A teacher should have creative skills, which allow him/her to leave behind professional stereotypes and find new solutions. In addition, in using smart technologies a teacher is able to minimize students' feelings of anxiety and uncertainty, which are typical for freshmen.

In our opinion, organizing a learning process based on teacher-student cooperation seems to be the most effective approach. Cooperative learning promotes student adaptation to the new environment, their personal growth, and professional development.

12.2   Literature Review

Issues of professional education have been widely discussed for an extended period of time. Since the middle of the 20th century, associations of engineers have been formed, followed by the establishment of engineering educational associations around the world. The latter have formulated requirements regarding graduates of engineering programs; however, the best ways to fulfill their sets of criteria are still discussed.

12.2.1   Student-Centered Approaches

It has been observed that the inclusion of scientific and engineering tasks in theoretical disciplines improves the learning outcomes of freshmen students [5]. Several universities have implemented Project-Based Learning in Engineering [6]. Universities are not always ready to undertake such drastic reforms, but they are forced to find and implement new approaches. For instance, in [7], the author emphasizes a new approach based on supervising students rather than traditional teaching. This method enhances student motivation to find correct answers and new ideas while broadening students' scope of interests. The author of [7] develops new social learning technology using existing ones: ontologies versus social tagging, exploratory search, trust, reputation mechanisms, etc.

Modern Digital Natives (students) make university faculty develop various distance learning platforms and online content. The new generation of students is accustomed to rapid satisfaction of their interests, so they strive to be involved in creative processes, gamification, and collaboration with fellow students and other participants of the educational process [8]. The authors believe that the socio-psychological characteristics of today's students require the creation of a new educational environment that meets their needs. In [9], principles were formulated that should be incorporated into the educational process to encourage students toward a lifetime of learning. It is assumed that lifelong learning skills are based on the skills developed while studying at school and university. It should be assumed that open, context-aware, ubiquitous learning environments are intended to be a foundation for supporting lifelong learning [10]. The authors develop criteria for a smart learning environment and note that the presence and promotion of new computer, communications, and sensor technologies are not sufficient to create such an environment. The creation of a smart learning environment should involve experts in different fields of education: educational theory, psychology, computer science, information technology, and, of course, college professors, academics, and school teachers. A properly designed educational environment is required for both online and face to face activities.

Issues of motivation have been widely discussed, both in Russia and abroad. For example, Kegan [11] developed a 6-level model of student motivation based on Maslow's 5-level model [12]. This model enables the determining of the initial level of student motivation and developing an individual learning path in order to enhance intrinsic motivation, which is an important factor in the success of educational outcomes.

12.2.2 Learning Systems

Nowadays, educational systems aim to make use of innovative approaches. One of the most popular forms of innovative education is smart education. Components of smart education (a smart environment, smart campus, smart learning, and smart classroom) are described in [3, 13, 14] in detail. Based on an analysis of traditional and distance learning, the authors conclude that properly organized e-learning and innovative tools (online counseling, webinar, digital white boards, online assignments, etc.) do not lead to a deterioration in the quality of learning outcomes. This is supported by a comparative analysis of ranking scores.

According to Tikhomirov [15], e-learning is becoming routine; however, the question of students' readiness for e-learning is still under discussion because e-learning is actually independent study. Education is a complex process comprised of the accumulation and transformation of knowledge, mindset formation, and formation of a personality that is ready for self-development. It is believed that innovations in the educational process improve the efficiency of cognitive processes and positively influence the quality of outcomes. A good example of innovative technology being implemented intensively throughout the world is blended learning. Concurrently, methods for the analysis and monitoring of learning outcomes are being developed [16].

Learning management systems (LMS) such as WebCT, Moodle, dotLRN, etc., are widely recognized as reliable control and management systems in education. In a study of blended learning and student satisfaction, it was shown that online teacher-student communication is not effective [17]. A survey reported in [18] shows that 84% of respondents noted that online feedback was unsuccessful and only 1% received too much feedback online from their teacher. Forty-six percent did not receive sufficient helpful online feedback from their teacher as opposed to 15% who actually received sufficient help. Students were satisfied, however, by the organization of the e-learning systems: 69% of students responded saying that they understood how the website for their unit related to the whole unit of study. In [19], over the course of 4 years, a blended model was examined in order to analyze its effects on the quality of education as well as the opinion of teachers and students regarding the model. The authors showed that both students and teachers had a positive attitude toward e-learning. The students involved had a unique opportunity to study subjects according to their own day plan, to use IT tools for acquiring skills, and to have free access to teaching aids. Moreover, they were emphatic that the knowledge assessment system became more understandable. While there were positive responses, the surveys also showed that the quality of communication between students and teachers was deteriorating. Forums and chats cannot substitute face-to-face communication. Some universities, though, such as Illinois State University and Idaho State are developing a variety of pedagogical concepts and technologies based on b-learning and implementing them at different levels of education (middle school, high school, university) [20].

12.2.3   Integrating e-Learning

In [21, 22], existing schemes of e-learning integration in the educational process are described in detail. It is noted that the development of e-learning courses for theoretical disciplines met some challenges. Although e-learning is an appropriate means of practicing lifelong learning, it does not provide future engineers with deep theoretical knowledge. This theoretical knowledge is developed only by personal communication with a teacher. The study [23] included a comparative analysis of a survey of the mobile learning preferences of faculty involved in teaching language and science courses. Aspects such as ease of use, continuity, relevance, adaptive content, multiple sources, timely guidance, student negotiation, and inquiry learning were examined. The authors found significant differences in the requirements for mobile language and science courses. Therefore, according to the language specialists' opinion, adaptive content is a key part of a mobile course. This enhances teaching effectiveness and improves students' learning outcomes, yet teachers of theoretical disciplines believe that a mobile learning environment and interaction or inquiry learning in the real world are more important. These conclusions are consistent with additional studies [17, 18]. Because of this, the authors of this paper suggest an individual approach and appropriate methodology for designing a mobile course. In addition, the structure and the content of a mobile course depend on the learning strategy (b-learning, e-learning, etc.). For instance, teachers are skeptical about studying higher mathematics using e-courses. This is explained by the fact that successful studying of mathematics is based on the development of skills and subjects, which can be achieved by adhering to the following components:

1. Acceptance: intention to learn mathematics;
2. Reaction: intention to participate in math activities;
3. Value: active acknowledgement of the practicality of mathematics and promotion of math activities;
4. Organization: integration of mathematics concepts into the student's personal value system; and
5. Confirmation of value: personal identification with math concepts and values [24].

Some e courses are used solely as an e library with teaching materials uploaded. Good examples of the successful introduction of an e-course are provided in [21, 25]. The authors utilized e-courses to teach a theoretical discipline to first-year students. The problem solving portion of the course was conducted online. In this case, the problem solving was checked by the teacher and automatically assessed. Only student counseling was arranged face-to-face. The students successfully completed all the online tasks before being permitted to proceed to the exam. Theoretical tasks and issues were discussed during lectures, understanding of which was assessed in the exam. The authors considered the experiment a success due to the fact that only those students who completed the e-course management to pass the exam with the highest grades.

Despite sustained discussions about the inclusion of industry representatives in the educational process as active participants, only a few papers are devoted to this issue. Most examples relate to medical social worker education and describe the experience of collaboration with organizations during senior undergraduate courses, as well as at graduate levels. For instance, [26] examines an interdisciplinary project in bioprocess production between graduate students and industry representatives. Such an approach not only motivated the students, stimulating their creativity, critical thinking, collaboration, and communication, but also allowed the students to gain expertise in allied professions, as well as establishing contacts with representatives of laboratories and companies producing biotech materials. The most appropriate scheme for engineering education is a gamification model. In [27], there is a comprehensive overview of existing models of gamification. The paper also describes in detail the author's model of using smart technology based on external motivational tools, which was implemented in the educational process. The authors conclude that the main goal in successful learning—motivation of students—was mostly achieved.

12.3 Goal and Objectives

The literature shows that the introduction and use of smart technology is still open to discussion. In higher education, there is no universal tool in the development of new approaches to smart education. Independent learning and motivation to lifelong learning, in our opinion, could be considered key elements of smart education. In addition, we offer a unique method for involving students in the work of professional communities from the earliest days of study.

Nowadays, educational systems are attempting to meet the needs of national economies throughout the entire world. Universities are forced to adapt to employers' requirements. In modern conditions, a stereotypical approach in solving professional problems may not be suitable in some cases. Consequently we need specialists who can quickly respond to any changes in the environment, and who are continuously engaged in their own professional development. Modern freshmen often have obscure ideas about their role in the educational process at their university, about the amount of new information they will obtain, and their future specialty. A properly organized educational environment would help students quickly adapt to their new lifestyle.

University years coincide with a period of intensive personal development in a physical, intellectual, social, and moral sense. While attempting to adapt to university life, the educational system and university relationships could prove challenging to new students. A successful start in the learning process requires students to have qualities such as self-discipline and self-organization. They became accustomed to parents' and school teachers' assistance to the point that finding themselves in a new environment leaves some freshmen diffident, burnt-out and passive, gradually losing grasp of the prospects for furthering their studies. However, not only students, but also faculty, are interested in effective student adaptation to university life [1]. In this case, one of the main challenges for academic staff is helping students during their social, psychological, and academic adaptation to university life; forming soft skills as well as necessary professional and general cultural competencies as part of student academic activities; and developing their personal academic pathway. Moreover, it is crucial to motivate independent learning and the development of special learning skills.

The problem of fostering motivation for independent and lifelong learning has been well studied. Some papers present a statistical analysis of low motivation problems and provide some recommendations for its improvement. Some methods for increasing motivation show a positive result, but are strictly limited to specific conditions. Modern education requires universal methods for developing student motivation for independent and lifelong learning. Let us note that, in Russia, priority is traditionally given to theoretical disciplines. Faculty attempt to resist replacing theory with professional disciplines. In the situation of decreasing face-to-face hours, faculty strive to preserve a deep theoretical education and often invite students to study theoretical and unresolved issues and problems independently. However, first-year students generally do not have the skills to gain the knowledge that they need. These skills should be taught, and the introduction of new educational technology should meet teaching objectives and be monitored at all levels of learning outcomes.

One of the key principles in education should be the organization of a smart environment based on the principles of "inverted" education, modified to develop motivation for independent learning ("role reversal education"). "Role reversal" education is an educational model where the students initially obtain information on a specific topic independently. The students then discuss this topic with their teacher and fellow students in the classroom. As a result of the discussion, this new information becomes a basis for further research. Moreover, some courses are taught in English as this model of education involves global educational content in order to acquaint students with the latest developments and achievements in their chosen profession. When introducing the principles of "role reversal" education, a teacher should "claim ignorance of the topic" and discover it together with the students, thereby creating an atmosphere of close collaboration and cooperation in cognitive activity. Using this approach, knowledge of theoretical disciplines will be a vital tool for the development of interdisciplinary problem-solving skills.

Thus, the goal of our study is to create an educational model that takes advantage of smart technology and the principles of lifelong learning in order to achieve the sustained involvement of students in education inextricably linked to professional skills. For the achievement of this goal, we have set the following tasks:

1. To organize an educational smart environment that requires free access to all educational resources, modern equipped classrooms, etc.;
2. To develop students' skills for independent learning;
3. To enhance students' motivation for lifelong learning; and
4. To create a curriculum that includes the interaction of students with professional communities.

12.4 Theoretical and Methodological Framework

Professional training of modern specialists can no longer be based on the traditional approach in higher education, when paradoxically the equipment and technology of the workplace are ahead of the educational program and its content. Companies are more interested in a specialist who knows not only the fundamentals of their specialty, but who are also able to break stereotypes, to innovate, and to create.

12.4.1   Organization of Smart Education

One of the major roles of a university is education. Nowadays, universities use the most progressive and sometimes even futuristic ideas to motivate students to acquire new knowledge and information, and to form the ability to find creative solutions to professional tasks. Thus, one of our objectives is to create a smart environment that will support a creative approach to professional activity. It is expected that this approach will lead to forming students' skills in generating innovative ideas, developments, and improvements. An integral part of such an educational environment is global technical equipment: unlimited access to the Internet at the university and campus, classrooms with projectors and smart boards, classrooms with student feedback, and modern equipped laboratories. For the organization of student independent work and cooperation, we use the LMS Moodle.

Cooperative learning. In this case, the role of the teacher becomes more complicated. Teachers are no longer "a translator of knowledge". Rather, they become effective managers of educational activity known as smart teachers. A smart environment and smart teacher are the main components of introducing a smart approach. In addition, a modern specialist should be properly educated in cooperation with the professional community.

Since 2010, at the Department of Electric Power Systems at Tomsk Polytechnic University (TPU), a group of 10 masters' students have been annually recruited to study on an individual educational pathway plan for a two year period. At the same time, power company specialists are involved in the educational process ensuring that classes are delivered not only by university staff, but also by the best specialists in this area. Laboratory work is organized using real equipment and the input data are approximated to reality. Practice is organized individually to involve students in solving professional tasks in lifelike conditions. As a result, we have a unique circle of cooperation that involves students, faculty, and the professional community in the smart environment (Fig. 12.1).

In addition to the above, the smart content is created, not only by faculty, but also with the participation of power industry specialists, as well as students (based on their questions and new ideas). Such an approach allows students to obtain

Fig. 12.1 Circle of cooperation

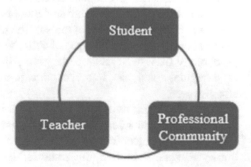

real-world knowledge that is applicable to their future professional activity. Thus, the university is becoming a place of professional independent learning.

Obviously, professional disciplines follow deep knowledge of theory. Therefore, interdisciplinary connection and a systematic approach are crucial for the successful development of professional skills. This should be achieved through specific tasks transformed from general theoretical knowledge to expertise. That is, students should always be sure that the knowledge obtained could potentially be applied in order to solve other more complicated and unresolved tasks. Group teaching methods and role-playing games are considered the most productive learning techniques. For example, the course "Power Stations (in English)" was delivered in classrooms with free Internet access and multimedia devices. The students were divided into three teams: one team of investors and two teams of power plant designers. The designers had to develop and present a new design of a power station without restrictions on the equipment and information used. The team of investors had to ask questions, discuss, and then choose one project for financial support.

Student motivation. To enhance student motivation to study professional disciplines in a non-native language, productive and reproductive types of tasks should be used. Unfortunately, not all students have sufficient language and professional skills, potentially leading to a number of difficulties with creative tasks. Simple tasks in combination with more sophisticated tasks will allow the potential of every student to be discovered and ensure a comfortable environment, developing professional language competence.

In addition to the basic educational program, students have the opportunity to participate in a double degree program. At TPU, an intensive language training program has been developed for students who would like to obtain a double degree diploma. This reinforces motivation for foreign language learning and for developing the professional competencies necessary for communication with colleagues from other countries. As noted above, independent learning is one of the main parts of the educational process. Nowadays, one of the most affordable methods for organizing independent learning at TPU is the LMS Moodle. To date, more than 1,000 courses have been developed and are being used for e-learning and b-learning. It should be taken into account that theoretical and professional disciplines require rather different approaches in terms of web-course development, as they are conducted at different levels of the educational process. Unfortunately, traditional and b-learning/e-learning models of education are not combined. This may be due to the unique set of tools of the discipline. A developer of a web-course has to not only be a creative specialist in the discipline, but also have knowledge of psychology from the content perception point of view. Therefore, the introduction of e-learning elements into the educational process requires analysis of their effectiveness.

To generate creativity and increase motivation for independent learning, the "role reversal" education method can be used. This method has been tested at the Department of Electric Power Systems for the discipline "Power Plants" for third-year students in the spring semester. Every student received an individual task for presentation on a new, not previously studied, topic. This new knowledge was then applied to power plant design. Each student had to find a reasonable way to present new equipment, switching devices, and current-carrying parts for his/her fellow students. At the same time, in some cases, students found a piece of information that was new for the teacher as well. Additionally, a number of new questions, to which the students were not used to, arose when they were searching for information for their presentation. Presentations were discussed in the classroom, corrected, and uploaded to the LMS Moodle. Thus, the educational content was created together by students and instructors. This approach is based on the principles of free education when teachers and students study the subject together. This provides psychological comfort while studying a new discipline and gives understanding that nothing is static or predefined in the professional area of knowledge.

The link between education and professional activity. It seems reasonable to study a professional discipline first in Russian prior to in English, expanding the students' boundaries of knowledge. The field of knowledge of professional disciplines (in Russian) and general English can be presented as two areas of knowledge. As a result of their overlapping, professional training in English is formed (Fig. 12.2).

The greater the bilingual professional skills—the greater the area of the circles' intersecting will be. Ideally, the training of a modern specialist means a full overlapping of these two sections. Here we face some difficulties, primarily, the low motivation of students to study professional disciplines in English. Student surveys show that only 5% of students have intrinsic motivation to study a foreign language, i.e. students study the subject only because it is included in the curriculum. At TPU, classes of "Professional English Training" amount to 32 h per week, giving 108 h of tuition each semester for first-year masters' students in Power and Electrical Engineering. Such a quantity of hours is sufficient for the special training to use English in their education and future career.

Participation, collaboration and co-authorship should become central in educational activities for faculty, following the principles of smart education. The suggested ways of increasing students' motivation for studying professional disciplines in English have been partially introduced and yielded positive results, but could be expanded and supplemented. Modern students have to know English to possess

Fig. 12.2 Field of knowledge

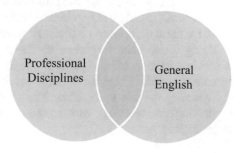

Table 12.1 Necessity of using english in professional activity

Questions	Yes, %	No, %
Is english required to obtain new information necessary for solving technical tasks?	85	15
Did you have any difficulties with modern equipment if the manual and technical documentation were in english only?	95	5
Do you have to communicate with specialists from foreign companies without an interpreter?	60	40
Were you satisfied by the interpreter (translator) service?	20	80

relevant information on the issue. It is obvious that translating into Russian and publishing scientific papers, new equipment design, instructions, etc. takes some time and such knowledge eventually becomes irrelevant. Table 12.1 shows the results of a survey for the management of Russian power companies (78 managers and engineers in total).

As the table shows, the vast majority of managers need effective and confident English for any aspect of their professional work. Even an interpreter does not solve the problem of free communication and normal comprehension with their partners from other countries.

Thus, modern students have to study, at minimum, one foreign language to a level that allows them to use the latest equipment, read and understand texts on a subject in the original language, communicate with foreign colleagues, participate in scientific conferences, etc.

Moreover, the majority of research and scientific papers are written in English. Smart education is implemented by using numerous components [3, 13]. Various smart devices, approaches, etc. support the education of future specialists in the power engineering industry. Table 12.2 shows smart components and their use in teaching theoretical and professional disciplines at the Department of Higher Mathematics and the Department of Power Plants at TPU.

12.4.2 Supporting Smart Education

Each educational institution has a unique system of activities to develop students' personalities [15]. At TPU, guidance counseling plays a key role in providing the basis for the development of students' general and professional competences [28].

Being a connection between the freshman and a new community, guidance counselors of an academic group use active and interactive activities in the classroom and outside. These include role-playing games, psychological training, group discussions, museum tours, excursions to companies, arranging master classes by invited experts, meetings with representatives of companies, government, public organizations, etc.

Table 12.2 The major smart components used in the "role reversal" method

Smart technology components		Theoretical disciplines	Professional disciplines	Professional training in English	Professional communities
Hardware/ equipment	Ceiling-mounted projectors	Yes	Yes	Yes	Yes
	Smart boards	Yes	No	No	No
	Interconnected desktop computers	Yes	Yes	Yes	Yes
Smart curricula	Adaptive courses	Yes	No	No	No
	e-learning/b-learning modules	Yes	Yes	No	No
	Face-to-face modules	Yes	Yes	Yes	Yes
Smart pedagogy	Collaborative learning	No	Yes	Yes	Yes
	Student-generated learning content	No	Yes	No	No
	Gamification	Yes	No	Yes	Yes
	Project-based learning	No	Yes	Yes	Yes

Initial adaptation to smart education. All first-year students in the Institute of Power Engineering at Tomsk Polytechnic University (TPU) must take the Adaptation Course, which runs over two semesters. Classes meet for 2 h per week in the fall semester and 1 h per week in the spring semester, over 36 weeks giving 48 h of tuition. After taking the course, students are able to:

- prepare themselves for facing new academic challenges;
- improve their time management;
- freely navigate through the informational environment of the university and quickly receive all necessary information for study and leisure;
- present information clearly and effectively;
- make a speech confidently;
- work on their weak points after learning their traits and characteristics; and
- set and achieve goals.

The course is provided by proficient guidance counselors. At present, 17 guidance counselors work with first- and second-year students in the Institute of Power Engineering. All of them combine faculty positions (assistant professor or associate professor) with guidance counseling. One guidance counselor supports each student group.

The Adaptation Course contains 3 modules: "Informational," "Academic," and "Social and Psychological." The "Informational Module" includes introductory sessions on university life and the students learn to use the "New Student Guide", or guidelines for new students on how to start their learning process successfully. They also learn how to utilize the website to organize their student life including schedule service, teachers' contact hours, campus map, etc. The "Academic Module" consists of workshops such as "E-services and the Educational Process in TPU," "Techniques for Working with Information," "Public Speech," "Cloud Technologies," etc. The "Social and Psychological Module" consists of psychological training such as "How to Set and Achieve Goals," "Time Management," "Know Thyself," "Stress Management," "Conflict and Reconciliation Behavior," "How to Pass Exams Perfectly Well," etc.

The aim of the academic and social adaptation project is to shorten the period of student adaptation, increase stress resistance, and retain students. The use of smart technologies, such as e-learning, b-learning, personalization, interactive tutorials, learning through video games, etc., allows young people to adapt to rapidly changing conditions and ensure transition from passive to active content. Over recent decades, the Internet has proven to be the most successful global project, changing the economy and society with new forms of communication and collaboration, implementation of innovations, and new modes of working with information and knowledge. Currently, electronic educational resources are an essential part of life for any person associated with education.

Firstly, all students receive access to all the information resources of the university, which will help them in solving various troubles, problems, issues, and challenges. After registration (receiving their username and password) as users of TPU's corporate network, every student is able to work with programs and services in a personalized enclosed space, or an individual student online service. In the individual student online service, a student and a guidance counselor have equal opportunities for communication together. Additionally, there is the possibility that students are given online counseling. Moreover, there is a series of tutorials, which helps students adapt to university life. For instance:

- "Virtual TPU"—students familiarize themselves with university information and the educational environment, namely, the resources and the services available for students; and
- "The basics of work with information and library resources"—students make themselves familiar with library resources, as well as rules and techniques for information processing. They obtain information about rare and unique resources, electronic catalogues, and options for more efficient work in the reading rooms.

Smart technology for smart education. Use of internet technologies allows students not only to help in their adaptation to university life, but is also conducive to their mastering of subjects through expanding the course content, teaching, and learning methods. In addition, it increases students' motivation to learn and provides them with opportunities to study independently and acquire radically new knowledge that would be of practical importance in their future workplace. Guidance counselors are given unlimited opportunities for self-development, which, in its turn, shifts teaching process to a much higher level. It is well known [29] that group methods, which include training sessions, discussions, etc., are key to effective group work. This activity aims to develop such skills as teamwork, readiness as the leader of the group to formulate team goals, to demonstrate the importance of their future profession, etc. The cognitive process in the training sessions relies on active work by the participants and their own experience. The trainer can assist in acquainting participants with each other, creating an atmosphere of cooperative work, partnership, and mutual understanding. The interactive educational technologies allow students to assimilate new knowledge more effectively and to obtain more information owing to the opportunity to ask questions, express their opinions, practice new skills, etc. Let us consider some forms of training and their purposes:

- "Student Team Building"—communicative skills training, which aims to create a cohesive team in a student group and includes three sessions: "Making introductions," "Group logo," and "Development of group rules and regulations";
- "Time Management"—students learn techniques enabling them to use their time in accordance with their personal goals and values;
- "Stress Management"—students learn the causes of stress; physiological, emotional, cognitive and behavioral symptoms of stress; methods of coping with stress; and examine how to avoid negative impacts of stressors;
- "Conflict and Reconciliation Behavior"—students learn conflict management techniques and develop negotiation skills necessary for effective conflict management; and
- "SMART goals"—students examine the principles of goal setting and forming a personal life program.

According to a survey among students, 32% of them frequently use skills acquired during training and 46% use the skills occasionally. In addition, 8% of students plan to, in the future, use knowledge and experience gained through their training. Webinars are a modern and accessible learning method via the Internet, which are now used as an integral part of the educational process at various levels. Without leaving the office or home, students are able to attend lectures, workshops, or seminars as well as participate in the discussion of topical issues. As mentioned above, the Adaptation Course runs only over two semesters; however, continuity and regularity are vital in student guidance counseling as second-year students are also still in need of support. Therefore, seminars, lectures, and training sessions can be offered using webinar technology. Webinar platforms provide both web-streaming audio and phone-bridge options, yet a survey conducted in 2015 among the students of the Institute of Power Engineering found that 76% of students prefer personal participation in training sessions. Also, 23% noted that they were interested in online training sessions only if it was impossible to attend the classroom.

Resources for smart education. At TPU, modern electronic educational resources have been designed and introduced including software, interactive electronic documents, media resources, and educational complexes. All of these resources can be uploaded to an online Learning Management System platform to provide centralized management of the educational services. At TPU's Institute of Power Engineering, an online educational resource for guidance counselors has been designed, which includes both interactive electronic documents and an educational complex. The resource comprises the weblog "Guidance Counseling" and the electronic course book "Communication training for the Adaptation Course classes." The personal weblog "Guidance Counseling" was designed by the senior guidance counselor and consists of brief records, which are regularly added to in reverse chronological order and include educational materials, media resources, and presentations. The resources in the blog are available for all TPU guidance counselors. The structure of the blog records resembles the familiar sequential structure of a log and comprises the following pages:

- "Main page"—information on upcoming events for the guidance counselors concerned;
- "Guidance and Counseling History"—history of guidance counseling at TPU;
- "Documents"—resources for performance: the TPU Statement on Guidance and Counseling, schedule, guidelines and work plan templates for "Adaptation Program" courses;
- "Presentations"—an extensive list of presentations, from which a guidance counselor can choose the relevant one for his/her class;
- "Video"—videos for in-class activities and different video training sessions for faculty;
- "Tests and Questionnaires"—tests and questionnaires designed for the counselors to consolidate their expertise or to evaluate their level of knowledge for further professional development;
- "Wall"—a feature making it possible for counselors to exchange their opinions on different topics; and
- "Electronic manual"—where the electronic course book "Communication training: in-class application" is posted.

A blog as a means of network communication has a number of advantages in comparison to web forums, chats, and email. A blog reader receives information and feels that he or she is a part of the counseling community. The latter is a result of the guidance counselor being informed about the events held at the university and knowing colleagues' attitudes to the activities performed by other guidance counselors. Additionally, a blog guest can make comments or participate in a discussion with the blogger if the blogger agrees to such a mode of communication.

A poll conducted in 2014 among the guidance counselors of the Institute of Power Engineering found that approximately 30% of them considered the Adaptation Course classes difficult to teach. This was the reason for the development of training materials as well as an electronic course book. The goal of the electronic course book is to develop skills and knowledge for guidance counseling. After studying the e-course book, a teacher is supposed to: know what a training session is; be capable of using training plan templates and scenarios; gain experience in designing training session scenarios; and be able to hold a training session within a class. In terms of the structure, the electronic course book is based on three modules: introduction, basic information, and workshop and testing.

The advantages of using an electronic course book are having an advanced, user-friendly and sufficiently simple navigation mechanism within the tutorial; optimization of user interface adaptation of educational material to the level of students' knowledge; inclusion of multimedia fragments (graphics, audio, and video); and adaptive user interaction with the elements of the course book.

Peer guidance. Within many Russian higher education institutions, there are student organizations charged with helping freshmen to adapt to university life, so-called "peer guidance." Such a project was launched at TPU in 2013 with a clear structure, strategic objectives, and feasible techniques. It is successfully developing thanks to students' active support.

The main objective of peer guidance is the educational and social adaptation of first-year students to the university environment, as well as to student life, by contributing to the personal development of the freshmen. To ensure successful implementation of the "Peer Guidance" project, students have designed an information and communication website on the social network "Vkontakte". The website is maintained by a student-guide under the supervision of the Senior Guidance Counselor of the Institute and a psychologist in charge of the University Centre of Personal Development. They post records, photos, and video films on the varied events organized and held by students. It allows students and guidance counselors to exchange information on their experiences, compete in their achievements, and prepare an annual report, which is presented at the Guidance and Counseling Service.

The main advantage of peer guidance is that it was created for voluntary participation, based on the personal motivation of each student-counselor. Thus, the student-counselor becomes a mentor for first-year students, whom they trust, who guides them, and whose opinion is important for them. A student-counselor not only plays an important role in first-year students' lives, but also helps and contributes to the development of their competencies. Thus, guidance counseling acquires a new form and extends the capability of its activities, becoming part of the smart environment.

12.5 Testing of Research Outcomes

This educational model was tested at the Department of Electrical Power Systems at TPU in cooperation with the Department of Higher Mathematics. Students are taught disciplines supported by e-courses in LMS Moodle. In addition, first- and second-year students are offered uncomplicated mathematical problems related to power engineering. Third-year students study the professional discipline "Power Plants" in Russian in the fall semester and in English in the spring semester. Masters' students are taught in collaboration with experts from local power industry companies.

12.5.1 Issues in e-Learning

The web-based portions of the courses aim to develop students' independent learning and assess the outcomes. The effectiveness of web-courses is monitored annually. The online surveys were conducted at the beginning and end of each semester and involved 546 students. The following aspects were analyzed: (a) course site design and attractiveness; (b) student perception of online learning; (c) benefit from the course; (d) prompt and timely feedback. The majority of students consider teacher online feedback "important" and "very important" (Fig. 12.3).

The solid line represents courses in general, whereas the dashed line represents theoretical courses only [22]. Figure 12.3 shows that face-to-face communication is of extreme importance for theoretical disciplines such as mathematics.

Online consultation in particular does not seem to be an appropriate way for explaining some difficulties faced in problem solving. In our opinion, mathematics requires not only verbal contact for better explanation, but also face-to-face communication. Better results would be achieved by the personal involvement of students in problem solving. Mathematics becomes clear and understandable through watching the progress of solving problems on a board or on a piece of paper. Therefore, some massive open online courses (MOOC) teach mathematics (for example, at the University of Ohio) using simulation of a "sheet of paper and pencil." E-learning and b-learning allow not only student-teacher communication, but also student-student communication. Figure 12.4 also shows students' opinions on the ability to communicate online with their fellow students. The prevailing character of social networks could explain the low percentage.

Additionally, students noted some benefits of web-courses. Figure 12.5 shows the students' opinion of the possibility of increasing their rankings and constant access to teaching aids (dashed line—Higher Mathematics, solid line—Power Plants).

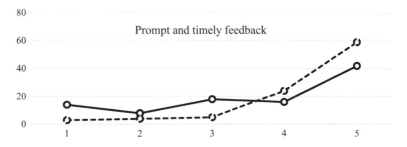

Fig. 12.3 Percentage distribution of responses of students from 1—"not important" to 5—"very important"

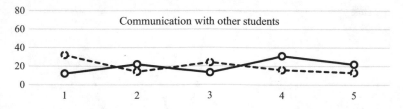

Fig. 12.4 Percentage distribution of responses of students from 1—"not important" to 5—"very important"

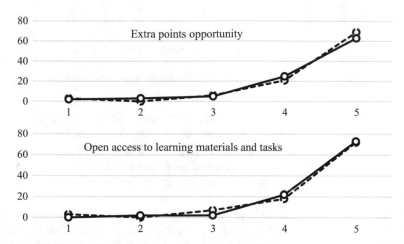

Fig. 12.5 Percentage distribution of responses of students from 1—"not important" to 5—"very important"

12.5.2   Student Perceptions on the Relevance of Course Components

A comparative analysis of the relevance of web-course components was carried out. The survey findings indicate that intensity of content requests was greater when students were involved in creating content. Figure 12.6 shows the statistics of visits to web-courses (dashed line—Higher Mathematics, solid line—Power Plants). As seen in Fig. 12.6, the majority of freshmen start to use the electronic part of the math course intensively just before the exam (dashed line). Moreover, the most popular items during this period are different tests.

However, senior students use the web-course more evenly throughout the semester (solid line). This can be explained by the fact that the professional disciplines involve innovative smart technology more frequently. For instance, the discipline "Power Plants" was supported by two web-courses in Russian and English. The Russian version consists of tests, forums, videos, teaching aids, etc. It

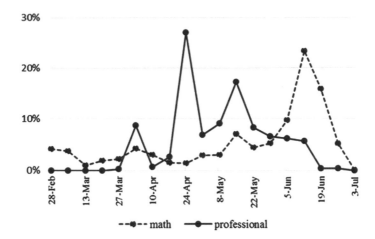

Fig. 12.6 Distribution of the intensity of use of web-courses during the semester

should be noted that students, while under instructor supervision, created the theoretical portion of the content. After discussion in the classroom, the best presentations were downloaded. It was observed that co-created content is required by students of other groups as well. At the end of the semester, an anonymous survey was conducted in order to analyze the effectiveness of this approach. The results are shown in Table 12.3.

It was observed that the "role reversal" method forces students to use different resources (books, the Internet, scientific papers, etc.) in order to find relevant information. They even visited the websites of manufacturers of the required equipment. The English version's content aims to enhance intrinsic motivation for studying professional disciplines in a foreign language. The web-course consists of scientific papers, video materials, tests, a glossary, etc.

Table 12.3 Students' opinions on the "role reversal" method of education

Questions	Yes, %	No, %
Was it your first experience of presenting a new topic for your fellow students?	100	0
Was the topic of your presentation interesting for you?	90	10
Was it challenging to choose the exact piece of information concerning your topic?	75	25
Was it challenging to organize and adapt the material for your presentation?	80	20
Were your fellow students' presentations clear and understandable?	89	11
Do you find this way of new knowledge acquisition useful and promising?	94	6
Would you like this methodology to spread to other disciplines?	90	10

Fig. 12.7 Bachelors and
masters' students on studying
the "Power Plants" discipline
in English

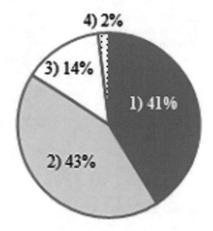

Only authentic materials are used in the web-course. Teachers and students both emphasized the significant role of the course in professional English training. Figure 12.7 shows the survey results on how bachelors and masters' students assess the necessity of using English in their professional activities.

Figure 12.7 shows that the main driving force for studying the subjects in English is:

1   the teacher demonstrates his/her proficiency in English (41%);
2   general knowledge of English (43%);
3   discipline and interest in the subject (14%); and
4   other variables (2%).

12.5.3   Providing Support for Students

The Institute of Power Engineering carried out a survey to learn the opinions of students about the guidance and counseling service. The survey involved 243 students (174 first-year and 69 second-year students). The students were asked to evaluate various aspects of the guidance counselors' performance. Overall, about 90% of respondents assessed the work as "good" and "excellent." Figure 12.8 shows the assessments on a 5-point scale: 5—"excellent"; 4—"good"; 3—"improvement needed"; 2—"bad"; 1—"very bad."

The statistics show that not all the students were satisfied with the performance of the guidance counselors. This could be explained by the fact that an academic group usually consists of 20–25 students and a guidance counselor is not able to focus on such a large number of individuals. We suggest that one guidance counselor support only 10–15 students. Nevertheless, the experience of involving senior students in guidance counseling showed good results.

Fig. 12.8 Assessment of guidance counselor support in adaptation to university life

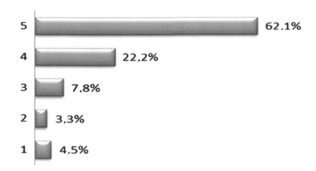

12.6 Discussion

As mentioned in the previous sections, smart education is a new evolutionary paradigm in education with huge capacity for the development and creativity of all participants of the educational process. Focusing only on two main aspects of smart education—independent and lifelong learning—we became convinced of the effectiveness of this approach in comparison to traditional education. In addition, we saw an immediate positive student reaction to such smart components as web-courses, smart boards, web labs, training sessions, etc.

Figure 12.9 shows what kind of training students find more useful and interesting. As can be seen from the diagram, at the beginning just over half of the students were in favor of a blended form. However, after graduation, about 70% of students prefer a blended approach. This agrees well with the results in [18].

The effectiveness of the proposed methods for introducing smart technologies in students' independent work was evaluated. The evaluation has shown a positive tendency in the development of independent learning skills. However, the analysis has shown that intrinsic motivation is too low to achieve 100% effectiveness of the proposed methods for the organization of independent and lifelong learning. This

Fig. 12.9 The answers to the question: which form of study would you prefer?

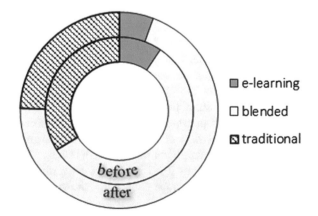

Fig. 12.10 Model A: traditional scheme

can be explained by the fact that we are taking the first steps in this direction as well as by the relative overload of students.

At present, it is not always easy to involve professional communities in the educational process. This approach has been implemented now only in one group of undergraduates due to the heavy workload of professionals. Initially, the university made use only of a traditional model, where the main components (theoretical disciplines, professional disciplines, English language and the professional community) did not overlap in the educational environment, i.e., they influenced each other, but without collaborative educational content (Model A) (Fig. 12.10).

Adding "Power Stations (in English)" to the curriculum, the area of overlapping is formed naturally and depicted as a common content of disciplines "Power Stations" and "General English" (Model B) (Fig. 12.11).

At present, the example of involving professional communities in undergraduate education led to a scheme in which the content of professional disciplines and professional communities overlaps (Model C) (Fig. 12.12).

In the future, the development of smart education and smart technologies ought to lead to the overlapping of theoretical disciplines, general English, professional disciplines, and the professional community in a smart educational environment (Model D) (Fig. 12.13).

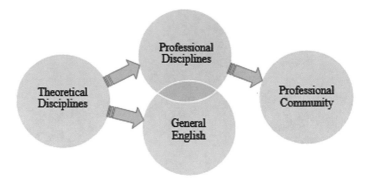

Fig. 12.11 Model B: traditional scheme including professional English

Fig. 12.12 Model C:
traditional scheme involving
the professional community

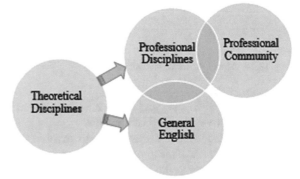

Fig. 12.13 Model D: smart
educational scheme

12.7   Conclusion

This paper examines some techniques aimed at enhancing student motivation in terms of independent work and in relation to independent and lifelong learning. The following means were considered:

- A "role reversal" method of education;
- Professional training in English;
- Involvement of the professional community in the educational process;
- e-Learning and b-Learning.

The above techniques were first designed independently by the faculty of different departments involved in teaching power engineering at Tomsk Polytechnic University. A powerful trend toward a united effort to combine forces, techniques, and experience in teaching the disciplines "Higher Mathematics," "Power Stations," and "Power Stations (in English)" enabled us to discover common issues and approaches to their solution.

The review of existing literature on the introduction of smart technologies in higher education demonstrates its current relevance. The introduction of smart technology makes the educational process much more interesting and helps to develop the latent potential of students. It is expected that it will lead to improved results in the training of professionals, which in turn requires teachers to have a high degree of competency in using smart technology [30]. Analysis of the proposed methods showed that the application of smart technologies to the independent work of students (LMS Moodle) leads to higher results in the learning of both theoretical and professional disciplines; however, professional and theoretical disciplines do not presently have common content in the smart environment. This study marks the beginning of the formation of a new curriculum aimed at a fusion of disciplines in all cycles of academic and vocational education—the creation of a system of smart education.

We began with the creation of overlapping smart content for the subjects "Higher Mathematics" and "Power Plants". Further analysis needs to be conducted to evaluate the results of the introduction of combined smart content to the educational process. Naturally, the first results may be obtained after at least four years when the students complete studying both the theoretical and professional disciplines. In the future, it is expected that the introduction of such an approach will lead to increased motivation in relation to independent learning and lifelong learning. This is achievable owing to students forming the essential skills of independent work while searching for new information and finding solutions to new problems. In addition, these graduates will be capable of making use of international practices and ideas.

References

1. Belskaya, E., Moldovanova, E., Rozhkova, S., Tsvetkova, O., Chervach, M.: University smart guidance counselling. Smart Innov. Syst. Technol. **59**, 39–49 (2016). doi:10.1007/978-3-319-39690-3_4
2. Mitchell, P.J., Mitchell, L.A.: Implementation of the bologna process and language education in Russia. Proc. Soc. Behav. Sci. **154**, 170–174 (2014). doi:10.1016/j.sbspro.2014.10.130
3. Uskov, V.L., Bakken, J.P., Pandey, A., Singh, U., Yalamanchili, M., Penumatsa, A.: Smart University Taxonomy: features, components, systems. In: Uskov, V.L., Howlett, R.J., Jain, L. C. (eds.) Smart Education and e-Learning, pp. 3–14 (2016)
4. Bertoux, P.: Line crise de croissance. Education et Devenir Press, France (1984)
5. Singer, S.R., Nielsen, N.R., Schweingruber, H.A. (eds.): Discipline-Based Education Research: Understanding and Improving Learning in Undergraduate Science and Engineering. The National Academies Press, Washington (2012)
6. Graaff, E., Kolmos, A.: Management of Change. AW Rotterdam, The Netherlands (2007)
7. Mertins, K., Ivanova, V., Natalinova, N., Alexandrova, M.: Aerospace engineering training: universities experience. In: MATEC Web of Conferences, vol. 48, Article number 06002 (2016)
8. Vassileva, J.: Toward social learning environments. IEEE Trans. Learn. Technol. **1**(4), 199–214 (2008)
9. O'Neill, T.A., Deacon, A., Larson, N.L., Hoffart, G.C., Brennan, R.W., Eggermont, M., Rosehart, W.: Lifelong learning, conscientious disposition, and longitudinal measures of academic engagement in engineering design teamwork. Learn. Individ. Differ. **39**, 124–131 (2015)
10. Hwang, G.J.: definition, framework and research issues of smart learning environments—a context-aware ubiquitous learning perspective. Smart Learn. Environ. **1**(4) (2014). http://www.slejournal.com/content/1/1/4
11. Kegan, R.: The Evolving Self: Problem and Process in Human Development. Harvard University Press, Cambridge, Mass (2012)
12. Maslow, A.H.: Motivation and Personality. Harper and Row, N.Y. (1970)
13. Neves-Silva, R., Tshirintzis, G., Uskov, V., Howlett, R., Lakhmi, J.: (2014) Smart digital futures. In: Proceedings of the 2014 International Conference on Smart Digital Futures. IOS Press, Amsterdam, The Netherlands (2014)
14. Uskov, V., Howlet, R. Jain, L. (eds.): Smart education and smart e-Learning. In: Proceedings of the 2nd International Conference on Smart Education and e-Learning SEEL-2016, June 17–19, 2015, Sorrento, Italy. Springer, Berlin-Heidelberg, Germany (2015)
15. Tikhomirov, V., Dneprovskaya, N., Yankovskaya, E.: Three dimensions of smart education. In: Chapter Smart Education and Smart e-Learning. Smart Innovation, Systems and Technologies, vol. 41, pp. 47–56 (2015)
16. Rebrim, O., Sholina, I., Syskov, A.: Smeshanoe Obuchenie. Vischie obrazovanie v Rossii **8**, 68–72 (2005). (In Russian)
17. Ellis, R.A., Calvo, R.A.: Minimum indicators to assure quality of LMS-supported blended learning. Educ. Technol. Soc. **10**(2), 60–70 (2007)
18. Ginns, P., Ellis, R.: Quality in blended learning: exploring the relationships between on-line and face-to-face teaching and learning. Internet High. Educ. **10**, 53–64 (2007)
19. Monteiro, A., Leite, C., Lima, L.: Quality of blended learning within the scope of the bologna process. Turk. Online J. Educ. Technol. **12**(1), 108–118 (2013)
20. Kyei-Blankson, L., Ntuli, E.: Practical Applications and Experiences in K-20 Blended Learning Environments. IGI Global (2013)
21. Lobregat-Gómez, N., Mínguez, F., Roselló, M.-D., Sánchez Ruiz, L.M.: Blended learning activities development. In: Proceedings of 2015 International Conference on Interactive Collaborative Learning, ICL, pp. 79–81 (2015)

22. Imas, O., Kaminskaya, V., Sherstneva, A.: Teaching math through blended learning. In: Proceedings of 2015 International Conference on Interactive Collaborative Learning, ICL, pp. 511–514 (2015)
23. Lai, C.L., Hwang, G.J.: A comparison on mobile learning preferences of high school teachers with different academic backgrounds. In: 2015 IIAI 4th International Congress on Advanced Applied Informatics, pp. 259–262 (2015)
24. Garfield, J.B.: An investigation of junior high school students' attitudes toward components of Mathematics. Unpublished master thesis. University of Minnesota (1977)
25. Tseng, W.S., Kano, T., Hsu, C.H.: Effect of integrating blended teaching into mathematics learning for junior high school students. J. Comput. Appl. Sci. Educ. **1**(2), 39–57 (2014)
26. Elmselmi, A.; Boeuf, G.; Elmarjou, A.; Azouani, R.: Active pedagogy project to increase bio-industrial process skills. In: 19th International Conference on Interactive Collaborative Learning, 21–23 September 2016, Clayton Hotel, Belfast, UK pp. 970–980 (2016)
27. Rutkauskiene, D., Gudoniene, D., Maskeliunas, R., Blazauskas, T.: The gamification model for e-learning participants engagement. In: Uskov, V.L. et al. (eds.) Smart Education and e-Learning, pp. 291–301 (2016). doi:10.1007/978-3-319-39690-3_26
28. Belskaya, E.Y.: The role of the guidance counseling in first-year student adaptation to university life in the institute of power engineering. In: Linguistic and Cultural Traditions and Innovations: XIV International Science and Practice Conference, pp. 20–26. Tomsk (2014)
29. Burke, A.: Group work: how to use groups effectively. J. Eff. Teach. **11**(2), 87–95 (2011)
30. Sysoyev, P.V., Evstigneev, M.N.: Foreign language teachers' competency in using information and communication technologies. Proc. Soc. Behav. Sci. **200**, 157–161 (2015). doi:10.1016/j.sbspro.2015.08.037

Chapter 13
Knowledge Building Conceptualisation within Smart Constructivist Learning Systems

Farshad Badie

Abstract This chapter focuses on the meeting of Constructivism (as a learning theory) and Smart Learning and, thus, theorises Smart Constructivist Learning. The main field of research is Smart Learning Environments. Relying on the phenomena of 'meaning construction' and 'meaningful understanding production' in the framework of smart constructivism, we will focus on analysing Smart Constructivist Knowledge Building. Accordingly, we will analyse Learning-and-Constructing-Together as a smart constructivist model of learning. The outcomes of this chapter could support the developments of smart learning strategies.

Keywords Smart learning · Constructivism · Meaning · Understanding · Knowledge building · Collaborative learning · Philosophy of education

13.1 Introduction and Motivation

The process of knowledge building leads to changes in humans' minds. In the context of cognitive developmental psychology, conceptual change is a type of process that focuses on the conversion of a human's conceptions and the relationships between her/his old and new conceptions [1, 2]. Thus, the most salient effect of knowledge building could be recognised to be on conceptual change of the learners'/mentors' conceptions over the course of time. We begin this chapter with our special focus on the fact that knowledge acquisition (that is the most determinative process within knowledge building processes) is a reflective activity that enables learners and mentors to draw upon their previous (and accumulated) experiences and reflect on their background as well as existing knowledge [3]. The reflective activity of knowledge acquisition supports learners and mentors in reflecting on themselves, on their society, and on their environment. Knowledge

F. Badie (✉)
Center for Linguistics, Department of Communication, Aalborg University, Aalborg, Denmark
e-mail: badie@id.aau.dk; badie@hum.aau.dk

© Springer International Publishing AG 2018
V.L. Uskov et al. (eds.), *Smart Universities*, Smart Innovation,
Systems and Technologies 70, DOI 10.1007/978-3-319-59454-5_13

acquisition enables learners and mentors to conceptualise and understand. Subsequently, it enables learners to evaluate both their present and past, so as to build up and shape their future actions (i.e., operations, practices, proceedings, movements, contributions and manners) as well as to construct and develop the construction of their latest pre-structured and pre-constructed knowledge. As described, 'understanding' has been recognised as the consequence of 'conceptualisation'. Our research [4, 5] has concluded that "an understanding could be realised to be a local manifestation of a global conceptualisation".

It is important to account for the fact that human beings become concerned with various construction processes over their pre-formed knowledge in order to obtain the opportunities necessary to develop their constructed knowledge and to produce their deeper understandings (i.e., meaningful comprehensions). Constructivist Learning (based on constructivist epistemology and constructivist models of knowing) has become the central framework of this research. Relying on this framework, our supposedly theoretical model of learning deals with how knowledge can assumedly be built by a learner/mentor. Through the lens of cognitive psychology, Piaget's developmental theory of learning says that constructivist knowledge acquisition, as well as knowledge building, is concerned with how an individual goes about the construction of knowledge in her/his own mental apparatus [6, 7]. Accordingly, for any learner or mentor, knowledge acquisition could be recognised as seeking knowledge regarding different objects, processes, events, and phenomena with regard to her/his background knowledge. As for the structural and existential characteristics of constructivism, the construction of knowledge is conceived of as a type of dynamic process. It can be informally described in terms of personal understanding in multiple actions. Consequently, constructivist learning is highly concerned with the active generation of personal understanding [5, 8].

This chapter focuses on 'Smart Constructivist Learning Systems', which are a specific sub-class of constructivist learning systems where 'constructivism' meets 'smart learning'. In accordance with the subject of this book, though, we look at the area of 'Smart Learning Environments'. According to [9], "smart education represents an integration of smart objects and systems, smart technologies, smart environments, smart features (smartness levels), smart pedagogy, smart learning and teaching analytics systems". Relying on the framework of smart education, we will focus on the development of a conceptual framework for analysing knowledge building in the framework of smart constructivism and over the flow of the learners' understandings. Correspondingly, we will characterise the main components of a smart constructivist pedagogy (and a smart constructivist model of learning). It may justifiably be assumed that the outcomes of this chapter will support designing and developing innovative learning and mentoring strategies as the products of smartness. We will conceptualise and prove that there are strong interrelationships between 'smart constructivist model of learning' and 'collaborative learning strategy'.

According to [9, 10], research in the area of smart learning systems should not only focus on software/hardware/technology features, but also on smart 'features' and the 'functionality' of smart systems. Furthermore, in order for smart learning systems to be effective and efficient for different learners, or mentors, there are certain smartness levels (smart distinctive features). The most significant feature in this research is analysing the phenomenon of 'smart learning' in the framework of constructivism. It focuses on acquiring new knowledge and building on existing knowledge. Also, this research aims to identify and recognise the concept of 'understanding' toward the awareness of learners. Therefore, this chapter will be highly concerned with the 'learning/self learning' feature. Additionally, our approach will rely on logical descriptions using, e.g., assumptions, implications, and different logical rules over conceptual analysis of the phenomenon of 'smart learning' and the concept of 'understanding'. Due to this, this research is effectively structured over logical reasoning processes and could support researchers' thoughts for the development of inferential and logical reasoning processes within smart learning systems.

Note that this research has been designed based on our own approaches to the analysis of 'meaning construction' and 'understanding production' processes [8, 11–13]. Our ideas have been based on a new scheme for interpretation based on semantics and interaction. Interaction consists of (i) interactions between learner and her/his self, (ii) interactions between learner and other agents (e.g., mentors, other learners, and smart programs), and (iii) interactions between learner and her/his environment.

13.2   Background of Thought

In this research, our conceptualised scheme for interpretation (based on semantics and interaction) will be analysed in the framework of smart constructivism. In our opinion, learning in the framework of constructivism is highly concerned with the active generation of personal meaningful understandings. This is based on personal constructed meanings and over personal mental objects. More specifically, we believe that the phenomenon of 'understanding' could be valid (and meaningful) based on learners' constructed meanings. In fact, this belief is the main building block of this research. This means that this chapter is specially concerned with 'meaning construction', 'meaningful understanding production', and 'knowledge construction' in the framework of smart constructivism. We strongly believe that there is a bi-conditional relationship between 'understanding production' and 'meaning construction' in the framework of smart constructivism. Accordingly, it shall be claimed that the phenomenon of 'understanding' could be valid and meaningful based on learners' and mentors' constructed meanings in the framework of constructivism and, in the context of smart learning environments.

13.3   Constructivist Learning Systems: Literature Review

This section conceptualises the most significant and supportive characteristics of constructivist learning. In our opinion, the following items are the most fundamental. These can be shared by Constructivism and SmartNess:

a. Understanding the learner's understanding;
b. Respecting the learner's background knowledge;
c. Paying attention to the learner's understanding of personal learning; and
d. Focusing on the learners' and mentors' reliable universal knowledge in the context of their interactions.

In fact, these items could both conceptually and epistemologically relate the concept of 'Constructivism' to the phenomenon of 'SmartNess'. This section totally focuses on the concept of 'Constructivism'. The next section, subsequently, will focus on the actuality of the junction between constructivist learning and smart learning systems.

13.3.1   Understanding the Learner's Understanding

The most fundamental point is the concept of 'understanding'. This concept is very complicated and sensitive in psychology, neuroscience, cognitive science, cybernetics, philosophy, and epistemology. There has not been any absolute, decisive, or independent description and specification of 'understanding'. It is important to note that (i) we can potentially describe our grasp (and our conceptions) of the concept of 'understanding', e.g., [14]. This is relying on the fact that it is possible to support the realization of understanding within various specific areas. Furthermore, (ii) we could describe 'understanding' in order to support its representation (e.g., [15, 16]). Finally, (iii) some descriptions could focus on specifying the components and constituents of understanding (i.e., from the perspectives of cognition and affects) [17–23]. We believe that the first item is the most crucial one. In addition, it shall be claimed that (ii) and (iii) could logically be subsumed under (i).

Let us now focus on our own realisation of the concept of 'understanding'. Assessing from the epistemological point of view, it could be concluded that there has always been a very strong bi-conditional relationship between 'understanding something' and 'explaining something'. The dependency between understanding and explanation is considerable in analytic sciences (e.g., mathematics, physics, chemistry, biology, computer science) as well as in the humanities and social sciences. The explanation or the actual explaining of a phenomenon (an object, event, or process) can shed light on the produced personal understanding of that thing. The relationship between understanding and explanation is bi-directional. Therefore, there is also a path from understanding to explanation. In fact, the well-understood phenomena could be explained more properly in order to be interpreted and realised by other agents (mentors and other learners). It is worth mentioning that there have been some descriptive models that focus on the concepts of explanatory proofs and explanatory systems along with their interrelationships with the concept of 'understanding' [24].

In our opinion, "a human being who tackles to understand something—directly or indirectly—becomes concerned with the taxonomy of various concepts relevant for that thing, and thus, she/he needs to move toward the chain of various related concepts in order to approach to the more specified concepts" [12]. Additionally, she/he must be able to propose strong explanations of those related concepts. We shall, therefore, say that 'concept' and 'generality' could be interpreted as the most significant ideas that could support the structuralist account of understanding and could support understanding the concept of 'understanding'. Consequently, constructivist learning (based on a constructivist epistemology and constructivist models of knowing) is highly concerned with an individual's knowledge building processes based on her/his own produced understandings. The constructivist learning systems make the learners and the mentors concerned with the understanding of more specific concepts with regard to the special focuses on their understanding of more general concepts. In fact, the constructivist learning systems focus on developing the concept of 'understanding of more specific concepts'.

13.3.2   The Importance of the Learner's Background Knowledge

Any background knowledge, by activation, becomes actualised and directed to the more-developed construction of knowledge. Living and experiencing different things are the first metaphorical teachers of all human beings. Additionally, in the context of learning environments, background knowledge could be defined as knowledge that learners have. This may come either from their previous learning environments and learning materials or from their own life experiences [25, 26]. Constructivist learning systems focus on knowledge building over learners' background knowledge. In fact, through the lens of constructivism, the concept of 'learning' is seen as the 'process of construction' over personal background knowledge. Furthermore, constructivism focuses on the individual learners' comprehensions of their own objectives with regard to insights based on their background knowledge. The theory of constructivism could also focus on the individual mentors' comprehensions of their own objectives with regard to insights based on their background knowledge and on knowledge of what will be taught.

13.3.3   The Learner's Understanding of Personal Learning

Here, the focus is on learners' conceptualisations and realisations of the phenomenon of 'learning' (e.g., [27–29]). More clearly, learners are concerned with (i) their own conceptions of the phenomenon of 'learning', as well as their conceptions of their personal learnings, and with (ii) the reflection of their personal learning on themselves and society. It shall be stressed that the most significant matter in constructivist learning is transforming the phenomenon of 'learning' into the constructions of knowledge. In fact:

- Constructivism focuses on transformation of the phenomenon of 'learning' into the learners' comprehensions of their personal constructed meanings.
- Constructivism focuses on transformation of the mentors' comprehensions of their personal constructed meanings into the phenomenon of 'mentoring'.

13.3.4   The Learner's and Mentor's Reliable Universal Knowledge in the Context of Their Interactions

Constructivist learning could work as an explanatory, heuristic, and developmental framework. It must be considered that there exists a kind of reliable global and universal knowledge between constructivist learners and constructivist mentors. It is constructed and developed by both groups. For example, this knowledge evolves in learners' and mentors' action-grounded conversational exchanges [30, 31]. According to our research in [8], the produced meanings by learners and mentors support the constructions of their own worlds. Subsequently, regarding Laurillard's conversational learning framework [32, 33], the learners' and the mentors' constructed worlds become interacted and the learner-mentor interactions manifest themselves between their constructed worlds. The outcomes of these interactions become reflected in the learners' and the mentors' conceptual knowledge that support their reliable universal knowledge. These processes express how the constructed meanings could be reflected in their constructed reliable universal knowledge.

13.4   Smart Constructivism: Research Project Objectives

This section, based on the identified concepts in the last section, investigates some conceptual and epistemological linkages between constructivist learning and smart learning systems. The conclusions could potentially express how educationalists and educators in smart learning environments could benefit from constructivist learning systems.

13.4.1   Smart Constructivism: Understanding the Learner's Understanding

As mentioned, we believe that comprehending the learner's understanding is the most crucial conception relevant to the concept of 'understanding'. According to [34], 'learning behaviour and learning pattern analysis' could be one of the most significant research issues of smart learning. It shall be taken into consideration that these outcomes are applicable to understanding learners' behaviours and learning patterns in the integrated real-world and virtual-world environments. Comprehension of learners' understandings, as the consequences, could support educationalists and educators in designing and developing more effective learning strategies. In fact, this issue is a very good example of grasping the idea of learners' understanding within smart learning environments.

In addition, we interpreted the concepts of 'concept' and 'generality' as the most significant concepts that could support the structuralist account of understanding and comprehending the concept of 'understanding'. Taking into consideration the concept of 'generality', the smart learning approaches must motivate deeper and more complicated levels of learners'/mentors' understandings. Accordingly,

- supporting any individual learner in producing her/his own deeper understanding of the world, and
- supporting any mentor in producing her/his own deeper understanding of the learners' understandings and of the problems of the learners

could be considered the most important objectives of smart learning systems. The most salient characteristic of smart constructivist learning systems is their special attention to the learners' understandings and, respectively, to the mentors' understandings, with respect to their own produced meanings and with regard to their own generated meaningful understandings. An individual's understanding of more specific concepts could be achieved with regard to her/his understanding of more general concepts. For example, a learner's understanding of the concept of 'InductiveLogicProgramming' is absolutely dependent on and supported by her/his understanding of the concept of 'LogicProgramming'. Therefore:

i. Smart constructivist learning systems must focus on explaining more general concepts, and, inductively, move toward explaining more specific concepts.

Also, similar to what [34] suggests for recording the details of the students' learning behaviours, we can conclude that:

ii. Since smart constructivist learning systems respect the learners' and mentors' produced understandings of the world, these learning systems can record the individuals' understandings of the world. This can provide good opportunities for educationalists to achieve valuable understandings of the learners' understandings. Note that the educationalists, educational psychologists, and learning theorists could also achieve valuable understandings of the learners' and the mentors' understandings. Furthermore, the long-term analysis of multiple levels of the learners'/mentors' understandings can definitely support researchers in knowing more about the efficiencies and productivities of any smart educational system.

13.4.2   Smart Constructivism: The Learner's Background Knowledge

In the framework of smart learning, any learner must be informed about the learning program's objectives. Subsequently, she/he could be able to identify her/his personal objectives. Accordingly, she/he

- activates her/his background knowledge,
- compares her/his own objectives with the program's objectives,
- focuses on processing different kinds of information, and
- works on self regulating and organising her/his self.

We shall claim that activating background knowledge is the most crucial process within these processes. Furthermore, referring to [35] and relying on constructivist theory of learning, one of the most important characteristics of an effective, efficient, and engaging smart learning environment is one that can adapt to the learner/mentor and can personalise instruction and learning support. This characteristic is highly relevant to (i) the wide variety of learners with different levels of prior knowledge, different psychological backgrounds, and different interests, and (ii) the attitudes and policies of mentors with their background knowledge of any learning environment, their background knowledge of any learner, and their knowledge of what they are going to teach/train. It shall be concluded that:

- Smart constructivist learning systems must respect the learners' and the mentors' background knowledge and attempt to construct, as well as develop, knowledge over their existing background knowledge. These systems do not destruct or destroy the pre-constructed knowledge of learners. Rather, they only focus on repairing, mending, and developing.
- Smart constructivism must produce and develop a kind of self-organisation process for any learner with respect to her/his own insights. This can be based upon her/his life experiences, her/his previous learning experiences, and her/his identified personal objectives.
- Smart constructivist learning systems can adapt to any learner in order to support her/his learning process by suggesting her/him the right learning strategies with regard to her/his background knowledge. The outcomes could, to a very high degree, support and advance the learners' lifelong learning.

- Smart constructivist learning systems can be adapted to the mentors, the adaptive teachers, and smart programs, as well as personalise their own instruction and teaching strategies with regard to the personalised learning environments and the conceptualised learners.

13.4.3 Smart Constructivism: The Learner's Understanding of Personal Learning

In smart learning environments, any learner transforms the phenomenon of 'learning' into 'demonstrations of understanding of what she/he is learning'. Accordingly, the learner reflects on her/his own learning strategy and promotes it over time.

Smart constructivism must consider the transformation of the phenomena of 'learning' and 'mentoring' into knowledge constructions. Smart constructivist learning systems must support learners/mentors in reflecting their own conceptions of 'what they assume they have to do as learners/mentors' on their learning/mentoring processes, respectively, on their knowledge constructions, and, consequently, on themselves and on their society.

13.4.4 Smart Constructivism: The Learner's and Mentor's Reliable Universal Knowledge in the Context of Their Interactions

Smart constructivist learning systems must aim at supporting learners and mentors in developing their universal conceptual knowledge. By taking into consideration

  i. the learners' constructed worlds,
 ii. the mentors' constructed worlds,
iii. the learners' conceptual knowledge, and
 iv. the mentors' conceptual knowledge,

smart constructivist learning systems must support the development of their reliable universal knowledge. It shall be stressed that any learner and any mentor can try to adapt the universal conceptual knowledge to her/his own constructed world. This means there is always a bi-directional relationship between 'own constructed worlds' and 'the universal conceptual knowledge' in the form of 'reflections' and 'adaptations', respectively.

13.5   Smart Constructivism: Methods Used in Research Project and Their Outcomes

13.5.1   Learners' Developing Conceptions of Learning in Smart Constructivism

This section focuses on learners' conceptions of the phenomenon of 'learning' and, subsequently, on their conceptions of the phenomenon of 'smart learning'. Note that any learner's conception(s) of the phenomenon of 'learning' play(s) a fundamental role in her/his study behaviour [29, 36]. Regarding behavioural and cognitive analysis of human beings' qualitative interpretations of the phenomenon of 'learning', any learner observes, interprets, and evaluates the world through the lenses of her/his own conceptions. In fact, the amalgamation of her/his mental images of the concept of 'smart learning' and her/his mental representations of the words 'smart' and 'learning' in 'smart learning' are manifested in the form of her/his conceptions. Accordingly, they are expressed in her/his actualisations and interpretations that all support her/his own understandings of smart learning. [37] provides information about the amalgamations of 'mental images' and 'linguistic expressions' regarding the philosophy of mind and language.

Note that the design and development of any smart constructivist learning system must be learner-centered [38]. Considering the significant importance of learner-centered analysis of the concept of 'smart learning' and in order to propose more analytic descriptions of smart learning, we need to put ourselves into the learners' shoes and observe the phenomenon of 'learning' from their perspective. Regarding this requirement, we take into account the significant products of [27–29]. The model sketches on Säljö's seminal studies on learners' conceptions of 'learning'. In more proper words, this model—qualitatively—focuses on adult learners' experiences of (and thoughts about) the phenomenon of 'learning'. This model could be interpreted as a layered model (Fig. 13.1). Any of its inner/deeper layers are supported by its outer/shallower ones.

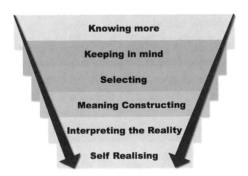

Fig. 13.1 Learners' developing conceptions of the phenomenon of 'Learning'

Let us describe and analyse them:

1. On the shallowest layer, the learner recognises that the phenomenon of 'learning' is equivalent to 'knowing more and knowing new things'. Such a learner is strongly dependent on her/his learning environment, learning materials, and teachers such as trainers, instructors, tutors, and mentors. This learner heavily needs ideas to be expressed, be explained, and be imparted explicitly. Furthermore, she/he needs her/his teacher/mentor to isolate and classify the flow of well-structured information into separated and individual facts. Such a learner needs the teacher/mentor to break down the procedures into isolated facts. In this layer, the learner needs to know more isolated and realisable facts. This learner needs to attain the abilities of 'naming' and 'identifying'. Identifying multiple facts prepare the learner for 'describing' facts and primary procedures.

2. The second layer could be identified by the concept of 'keeping in mind'. The concept of 'keeping' indirectly relates the learner to 'reusability' and 'reproduction'. In fact, she/he aims to memorize an acquired and known fact in order to apply and activate it regarding her/his own requirements and tasks. The one who attempts to keep something in mind is still trying to know more. Reusing and reproducing prepare the learner for 'describing' and 'combining' various facts, and, respectively, procedures.

3. The third layer is identified by the concept of 'selecting'. The ability of selection and refinement prepares the learner for pragmatism and for practical approaches. The learner expects her/his teacher to motivate her/him through selection processes. Selection and refinements connect the learner with 'comparing', 'contrasting', 'relating', and 'explaining'. Additionally, it indirectly makes a connection with 'justifying' and 'analysing'.

4. The fourth layer is identified by 'meaning construction'. She/he has become concerned with interpretation, analysis, justification, primary reasoning, and primary criticising. In our opinion, this layer is the most crucial one due to the fact that it makes a linkage between learners' fundamental and their advanced understanding of the concept of 'learning'. We shall focus emphasis on identifying this level by 'meaning construction'. It is not equivalent to ignoring the fact that meaning construction is an infinite process of any learner. The fourth layer is identified by 'meaning construction' because the process of meaning construction reaches its highest point and finds its most extreme significance in this layer. This layer provides a crucial interval for signifying the phenomenon of an individual learner's 'learning' within her/his learning processes.

5. The fifth layer makes the learner concerned with 'interpreting the reality'. Learning as an interpretative, explanatory, and expository process must be capable of supporting the learner in 'interpreting', 'explaining', and 'understanding' the reality of the world. This means she/he has become concerned with explaining the causes and reasons, criticising, formulating, and theorising. In this layer, many learners characterise 'learning' as the process of self-development.

6. The sixth layer is identified by 'self realisation'. The learner has become concerned with 'creation', 'generation', and 'reflection/mirroring' when it comes to her/his self and society. It's very important to know that learning, as the transcendental process of self-realisation and self-organisation, is continual, successive, and concatenated.

Regarding the described layers of learners' conceptions, we can realise that any outer/shallower conception, whether logically, conceptually, or cognitively, supports its inner/deeper layer. Assessed by logics, the conjunction of the outer layers is subsumed under their inner ones. For instance, the conjunction of the concepts of 'knowing more', 'keeping in mind', and 'selecting' are subsumed under 'meaning construction'. Then, through the lens of formal semantics, the provided logical model of 'meaning constructing' satisfy 'knowing', 'keeping in mind', and 'selecting'. Informally, those who are concerned with meaning construction have previously been concerned with 'knowing', 'keeping in mind', and 'selecting'. Accordingly, the succession of the layers' contents from 'knowing more' to 'self realising' could represent the flow of the concept of 'understanding' in learners' perspectives. In fact, there is a succession that could be considered as a flow of understanding regarding the expressed model. The succession could be described as: (1) knowing new isolated facts … (2) identifying them … (3) keeping them in mind … (4) describing them … (5) reusing them … (6) combining them … (7) selecting them … (8) comparing them with each other … (9) relating them to each other … (10) explaining them and explaining by applying them … (11) interpreting them and interpreting by using them … (12) analysing them and analysing other things using them … (13) justifying for their existences and justifying by employing them … (14) reasoning for [and based on] them … (15) criticising for [and based on] them … (16) theorising for [and based on] them … (17) developing them and developing other things based upon them … (18) reflecting on selves and on society (with regard to them).

An important question is "How could we establish a connection between a flow of understanding with regard to the learners' developing conceptions of the phenomenon of 'learning' and the phenomenon of 'smart learning'?" In other words, how could we characterise the concept of 'understanding' with regard to the learners' conceptions of learning within smart learning environments? To answer these questions, we shall stress that any smart learning environment should be filled with available and well-organised learning materials and should also be aesthetically pleasing. Any smart learning environment must be 'effective' [35, 39, 40]. What is likely to make a learning environment effective, efficient, and engaging for a wide variety of learners with different levels of background knowledge, psychological backgrounds, and interests is one that can adapt to the learner, personalise instruction, and support learning. This suggests that appropriate adaptation is a hallmark of smart behaviour. The concept of 'smart learning environments' has been presented as one "… that makes adaptations and provide appropriate support (e.g., guidance, feedback, hints or tools) in the right places and at the right time based on 'individual learners' needs, which might be determined via analysing their learning behaviours, performance and the online and real-world contexts in which they are situated. ..." [34]. Furthermore, [34] states that a smart learning environment is able to offer adaptive support to learners through immediate analyses of the "needs of individual learners from different perspectives". It shall be taken into consideration that any smart learning environment meets the personal factors (e.g., learning styles and preferences) and learning status (e.g., learning performance) of individual learners. In fact, all individual learners and their needs are the most central components and incorporators of smart learning environments. It is worth mentioning that IBM has also recognised smart educations as student-centric education systems [41].

Taking all the characteristics of smart learning mentioned here into consideration, any smart learning system utilized as a student-centric system must prepare a background for the learners' flow of understanding and support them within different aspects of their understandings. Also, as mentioned earlier, the most central focus of constructivist smart learning systems is on learners' understandings with regard to their own produced meanings and their generated meaningful comprehensions. At this point we shall state that smart constructivist learning systems must be developed over the individual learners' conceptions and requirements. These developments must be supported by the special focus on the flow of understanding of learners.

Let us take into consideration some significant results of our discussions with undergraduate students. A number of students wanted to know which facts would be required and helpful for them. We can transform this requirement into (i) 'How could a learner know the required and helpful facts?' Also, a few students told us that they know that they need to select facts in order to conceptualise them and to have a better understanding of them, but they don't know which facts must be selected. Again, we can transform this requirement of learners into (ii) 'How could a learner find the ability to select the rightful and beneficiary facts in order to construct meaning over them?' Also, a student wanted to know how she could let her mentor know about her constructed meanings. This question could be translated into (iii) "How could a learner announce her/his constructed meanings to their mentor or other learners?". Questions such as (i), (ii), and (iii) are prevalent to any learner.

Smart constructivist learning systems must be able to provide a kind of requirement analysis and to suggest rightful choices to individual learners. In the beginning, the learning system, the learner, and the mentor should not look at each other, but should actually look at the same point and discover the appropriate facts together. Accordingly, the conceptions of the mentor could influence the learner and vice versa. Furthermore, the learner's and the mentor's conceptions could be influenced and modified with regard to what the system has suggested to them. Smart constructivism must be capable of locating the learner in her/his best position to go toward her/his production of meaningful comprehension. Respectively, the mentor must be guided to find her/his most appropriate position in relation to the learners' positions.

In order to express and analyse the concepts of 'meaning', 'meaning construction', and 'meaningful comprehension', our theoretical model needs to be supported by a proper educational and pedagogical model. This can provide an organized framework for representing different levels of learners' understandings. We need to employ a model of learning concerned with various complexities of understanding at its different levels/layers in order to support the conceptualised idea of 'understanding', to analyse the flow of understanding in experts'/educationalists' points of view, and to model it in smart constructivist learning systems.

13.5.2   Smart Constructivism and the Structure of Observed Learning Outcomes

The Structure of Observed Learning Outcomes (SOLO) taxonomy is a proper model that represents multiple layers of learners' understandings within learning and knowledge acquisition processes [42]. SOLO provides an organised framework for representing different levels of learners' comprehensions. It is concerned with various complexities of understanding at its different layers. In the framework of SOLO, learners are concerned with five levels of understanding (Fig. 13.2).

As an analytic example, we focus on a learner, Martin, who is learning Java Programming:

- Pre-structured knowledge: Martin does not really have any knowledge about Java. This kind of knowledge about Java has been constructed over his mental backgrounds and from his previous experiences, e.g., experiencing different products that are developed in Java, meeting Java's official and related websites, discussing with Java programmers, etc. The most important fact is that Martin does not have any special constructed knowledge about Java.
- Uni-structured knowledge: Martin has limited knowledge about Java and may know few isolated facts. Thus, he mainly focuses on identifying those isolated facts. For example, he knows that Java works based on classes of objects and

Fig. 13.2 SOLO taxonomy: levels of constructed knowledge and levels of produced understanding

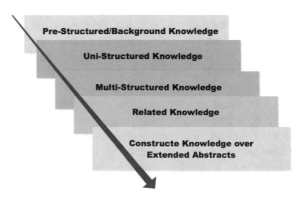

that Java is an object-oriented language. He may know that Java derives its syntax from C. Based on this, Martin has a very shallow understanding of Java. The known facts are isolated and he is not able to either relate them together or apply them.

- Multi-structured knowledge: Progressing from the previous level to this level simply means that Martin knows a few facts about Java, but he is still unable to find logical and conceptual linkages between them. Martin (i) has extended the domains of his factual knowledge about the isolated facts, (ii) has become concerned with combinations of various isolated facts, and (iii) has become concerned with descriptions of the results of those combinations. For example, he knows about object-oriented languages, he knows that object-oriented programming is a paradigm based on the concept of 'objects' and 'things', and he knows that object-programming languages focus on 'objects' rather than 'subjects' and 'actions'. Martin has produced some mental combinations of these facts. He is preparing himself for producing logical and relational models based on his produced combined facts.
- Related Knowledge: Martin has started to move towards higher levels of conception about Java. He has also begun moving towards deeper levels of understanding of Java. At this level, he is able to link different facts together and explain several conceptions of Java. The important fact is that he has become concerned with analysis, argumentation, explanation, justification, comparison, and applications relevant to Java. Now, Martin can explain and analyse the elements of his factual knowledge and can relate them together. He can now relate the characteristics of object-oriented systems and Java programming. He knows why object-oriented paradigms are in favour of 'objects' and not in favour of other phenomena. He is able to explain and analyse the characteristics of Java as well as apply different methods to them.
- Extended Abstract: This layer is the deepest and the most complicated level of Martin's understanding. Here, Martin is not only able to link a huge number of related conceptions together, but he can also link them to other specified and complicated conceptions. Now, he is able to link multiple explanations and justifications in order to produce more complicated extensions relevant to Java. Martin has become concerned with theorising, hypothesising, creating, and criticising.

According to Fig. 13.2, the extended abstracts are the products of deeper realisations and understandings of relational structures and constructed related knowledge. Relational structures are the products of deeper comprehensions of multi-structures and constructed multi-structured knowledge. In a similar manner, the multi-structures are the products of deeper comprehensions of uni-structures and constructed uni-structured knowledge. Finally, the uni-structures are the products of deeper pre-structures and pre-structured background knowledge.

At this point, we need to focus on the HowNess of satisfaction of the flow of understanding from 'pre-structured and background knowledge' to 'constructed knowledge over extended abstracts' by smart learning development and design.

Smart constructivist learning systems must be able to support the development of knowledge constructions over any learner's background and pre-structured knowledge. The central idea is that smart constructivism must generate a kind of self-updating process for any learner with respect to her/his own insights based on her/his background knowledge in order to prepare her/him for her/his individual processes of semantic interpretation, meaning construction, and understanding production. Let us be more specific on the concepts of 'semantic interpretation' and 'meaning construction'.

As characterised earlier, one of the most significant features of smart constructivist learning systems is their special focus on the learners' understandings with regard to their own produced meanings and generated meaningful comprehensions. In addition, we have mentioned that there is a bi-conditional relationship between 'understanding production' and 'meaning construction'. Therefore, we shall stress that the following items have a logical bi-conditional relationship:

- The process of knowledge construction as "pre-structured knowledge → uni-structured knowledge → multi-structured knowledge → related knowledge → knowledge over extended abstracts"; and
- The learners' meaning construction.

At this point, we employ a linguistic approach to explain and analyse this bi-conditional relationship. This approach, in dynamic semantics, has considered meaning as a context-update function [43, 44]. You can also find one of its particular applications in [45]. Considering meaning as a context-update function, the input of the Meaning function is a context and the output is its updated form. Any context comprises different types and different numbers of conceptions. Terminologically, we can consider conceptions as the sub-class of contexts. Therefore, we describe any 'meaning' as a conception-update function like *Meaning: Conception → Conception '*. This function iteratively organises itself in multiple loops and repetitions. It shall be claimed that the constructed meanings of any learner, based on her/his constructed knowledge over extended abstracts, are the updated forms of her/his constructed meanings within relational structures. Also, the constructed meanings in the ground of her/his related knowledge on mental relational structures are the products of her/his constructed meanings based on her/his multi-structured knowledge on mental multi-structures. In a similar manner, the constructed meanings, based on multi-structured knowledge and mental multi-structures, are the updated products of the constructed meanings based on uni-structured knowledge. Finally, the constructed meanings on uni-structured knowledge are the updated constructed meanings over mental pre-structures and pre-conceptions.

When it comes to semantic interpretations, our approach recognises the learner's semantic interpretation as the connector of her/his various levels of constructed meanings [46]. In other words, the interpretations semantically support the succession of the updated meanings. Relying on this conception, an interpretation could be known as the continually adjusted relationship between two things. It is quite important to consider the following when it comes to smart constructivism:

1. The learner's intention behind her/his conceptions, and
2. The learner's actual mental universe of her/his conceptions, which are based on her/his accumulated experiences.

As concluded earlier, smart constructivism must consider the fact that any individual learner transforms 'what she/he is learning' into "uni-structures of knowledge, multi-structures of knowledge, related structures of knowledge, and constructed knowledge over extended abstracts". In fact, any learner, based on her/his tasks and roles as a learner, increases the complexities of her/his constructed meanings in order to be closer to her/his own deepest understanding. Smart constructivist learning systems must be capable of supporting learners in reflecting their own multiple conceptions of a phenomenon. This occurs when it comes to mirroring the concatenation of the produced conceptions on their own learning as well as on different levels of their constructed knowledge.

13.6 Knowledge in Smart Constructivist Learning Systems: Analysis of Methods' Outcomes

Relying on the framework of constructivism, the current theoretical analysis of smart learning is not focusing on ontologies or the existence of knowledge. The central focus, though, is on the tenets of humans' knowledge construction and development. This involves the creation of mental models when encountering new, unusual, or otherwise, unexplained experiences [35]. We have taken into account that learners create their own mental representations in order to make sense of their experiences and learning tasks. By interpreting the phenomenon of 'learning' as the process of knowledge construction, we need to put any individual learner at the center of the proceeding of knowledge construction. The personal characteristics of any learner, the mental backgrounds, personal experiences, and the pre-structured and uni-structured knowledge all support the foundations of knowledge construction. This section deals with how multiple categories of knowledge can assumedly be constructed in the framework of smart constructivism.

13.6.1 Categories of Knowledge in Smart Constructivism

We adopt Bloom's taxonomy in order to clarify what we mean by 'categories of knowledge'. Bloom's taxonomy is a framework for classifying educational and pedagogical objectives. These could be interpreted as the statements of what educators and educationalists expect the learners to have dealt with [47, 48].

Considering Bloom's taxonomy and taking into account the constructivist theory of learning, we could express the view that the concept of 'knowledge' has a strong relationship with 'recognition' of multiple phenomena. In fact, knowledge construction is supported by any individual's insights, based on her/his own recognition of various materials, methods, procedures, processes, structures, and settings in the form of her/his conceptions. According to Cambridge dictionary [49], having knowledge about something or about some phenomenon could be realised as being related to the following items: (i) Having a piece of knowledge about that thing/phenomenon and (ii) judging about that thing/phenomenon based on personal experiences and information.

We shall claim that we are allowed to divide knowledge into separated classes (for example, into $Class_1$, $Class_2$, ..., $Class_n$) if and only if we have aimed at clarifying and specifying the humans' conceptions of any of them (e.g., $Class_i$) and, respectively, of all of those separated classes (i.e., $Class_1$, $Class_2$, ..., $Class_n$). In the end, we must consider the union of all classes as the phenomenon of 'knowledge'. Let us focus on analysing how Bloom has dealt with the phenomenon of 'knowledge'. Bloom's taxonomy categorises knowledge into multiple classes, e.g., distinct classes for knowledge of terminologies, knowledge of ways and means, knowledge of trends and sequences, knowledge of classifications and categorisations, knowledge of criteria, knowledge of methodologies, knowledge of quantifications, knowledge of principles, knowledge of generalisations and specifications, and knowledge of theories and structures. Since then, [48] has proposed a knowledge dimension in the revised version of Bloom's taxonomy. The revised taxonomy consists of (i) factual knowledge (e.g., terminological knowledge), (ii) conceptual knowledge (e.g., knowledge of theories, models and structures), (iii) procedural knowledge (e.g., knowledge of methods and algorithms) and (iv) meta-cognitive knowledge (e.g., contextual knowledge, conditional knowledge).

We strongly believe that these four classes could support us in clarifying and analysing the interconnections between the phenomena of 'learning (and knowledge acquisition)' and 'knowledge building'. We shall, therefore, claim that the phenomenon of 'learning' consists of a sort of transformations from constructed knowledge in the world (e.g., by experts, by theoreticians, etc.) into the sets of 'facts', 'procedures'. and 'concepts' in different 'contexts'. We believe that procedures are constructed over the chain of separated, connected, and related facts. Then, in our opinion, any procedure is just the concatenation of a number of facts. Therefore, learning provides multiple functions from constructed knowledge into 'facts' and 'concepts'. Learners need to deal with those facts and concepts while they need to construct their own knowledge with their insights based on what they construct over those facts and concepts. In [50], we have argued as following: "... In our opinion, there is a concept behind every fact. Then any factual knowledge can be supported by a conceptual knowledge. For instance, according to a fundamental characteristic of terminological knowledge (as a type of factual knowledge), we can represent terminologies by means of taxonomies. A taxonomy could be constructed based upon concepts. Then a terminological knowledge has been supported by a conceptual knowledge. Also, as another instance, we can define a body of the related elements and interpret it as a set of constructors for denoting various concepts and their interrelationships. That's how the concept languages and descriptive languages appear. Then, we could be able to represent knowledge over concepts, their instances and their relationships ...". Thus, we shall claim that everything is translatable into, and mentally representable in the form of, a concept. Accordingly, concepts are manifested in the learners' conceptions and they could be declared in the learners' hypotheses. A concept might be interpreted to be a linkage or interconnection between the mental representations of linguistic expressions and the other mental images (e.g., representations of the world, representations of inner experiences) that a learner has in her/his mind [37].

It shall be concluded that the phenomenon of 'smart learning' must provide multiple transformations from 'knowledge', either 'received from outside' or 'experienced within inside', into concepts. Learners represent those concepts in their minds and propose their own conceptions of those concepts. Consequently, learners construct their own knowledge with insights based on their produced conceptions. It is a fact that learners' conceptions could elucidate others and could be shared with them through Internet, social networks, virtual classes, and media. Learners can propose/announce their own conceptions of what they have constructed in the form of texts, voices, videos, etc. The collection of these processes could be identified by 'construction of own packages of knowledge by learners' in smart constructivist learning systems.

13.6.2   A Conceptual Framework for Knowledge Building in Smart Constructivism

The main objective of this section is to propose a conceptual framework for representing the stream of understanding within knowledge construction processes in smart constructivist learning systems. First, we shall refer the readers to our research in [51], which focused on formal semantic analysis of interrelationships between multiple categories in learners' developing conceptions of the phenomenon of 'learning' [27–29]. We need to employ the results of that research. More particularly, that research has focused on the conceptualisation of the phenomenon of 'learning' within the top-ontology of adult learners' developing conceptions of learning. Self-realisation (and self-awareness) is the most excellent conception of learners. It can conclude all other conceptions within its lower categories. Assessed by logics, all conceptions of learners within lower categories of conceptions are subsumed under 'self realisation'. Relying on Description Logics, [51] has focused on discovering the main constructive concepts and their interrelationships under 'self awareness' as well as a semantic representation of adult learners' developing conceptions has been sketched out. Figure 13.3 represents a network that has been developed over an important piece of the proposed semantic representation in [51].

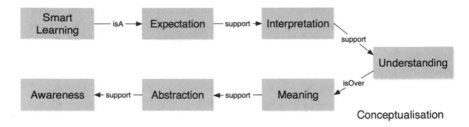

Fig. 13.3 A semantic representation of concept of 'Understanding' in smart learning environments

Figure 13.3 represents a structural analysis of 'smart learning' on the highest conceptual level and from the perspective of the most excellent learning conceptions; this semantic representation is meaningful in the context of 'conceptualisations'.

This network shows that the concept of 'smart learning' is a kind of expectation. In some cases, it is an 'outlook'. Smart learning, as an expectation, supports learners' interpretations and understandings of the world. In fact, by relying on individuals' interpretations, this expectation produces a strong belief that the phenomenon of 'smart learning' will be valid and meaningful. Furthermore, humans' interpretations support their personal understandings, making it is possible to say that any personal understanding is a kind of limited interpretation in the context of conceptualisations. Learners, through relying on their conceptualisations and by engaging their interpretations, explicate what they mean by classifying a thing, process, event, or phenomenon as an instance of a concept. The interpretations prepare learners for producing their personal meaningful descriptions over their own conceptions, and, in fact, over their constructed concepts. Therefore, an 'understanding' could be realised to be the sub-process of an 'interpretation'.

On the other hand, though, all interpretations are not necessarily understandings. In fact, all the interpreted concepts may not be understood, but all the understood concepts certainly have been interpreted, see our research in [5]. Then, understanding, in the framework of smart constructivism, is produced over 'interpretations' of things, processes, events, and phenomena as well as within smart learning environments. Additionally, as analysed, understanding could be considered as constructed over individuals' constructed meanings. Meanings on the deepest layers of understanding, as well as on highest floor of the constructed knowledge, support 'abstractions' and 'production of knowledge over the extended abstracts' by learners. These abstractions support individual meaningful comprehensions over individual constructed meanings. Figure 13.3 structurally and conceptually supports Fig. 13.4. Figure 13.4. represents a conceptual framework for 'knowledge creation' over the stream of learners' understandings within smart constructivist learning systems. It represents a conceptual description of 'knowledge building' toward 'deepest understanding levels of learners' within smart learning environments.

Fig. 13.4 A conceptual framework for knowledge building within smart constructivist learning systems

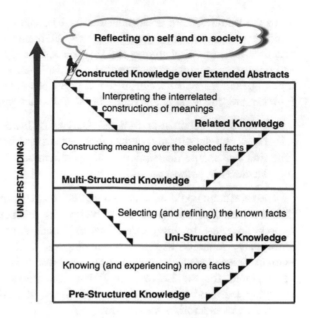

13.7   Conceptualising a Smart Constructivist Pedagogy: Testing of Research Outcomes

This section employs the outcomes of [52–58] in order to conceptualise a smart constructivist pedagogy based on the proposed model of knowledge building. According to Fig. 13.3, the phenomenon of 'smart learning' in the framework of constructivism is an expectation that is supported by any individual's interpretations and meaningful understandings. Consequently, both learners and mentors are interpreters, organisers, and constructors within the process of smart constructivist learning and in the context of smart learning environments. In fact, they are the developers of a collaborative process of constructing. Therefore, it shall be emphasised that smart constructivism doesn't assess the phenomenon of 'learning' as an outcome of a development. It does, however, recognise it as a development. Here, learners are inventors. They must be allowed to generate their hypotheses based on their own conceptions of the world. The main characteristics of these conceptions are as follows:

1. Conceptions are learner-centered (individual-centered).
2. Conceptions are central-organizing.
3. Conceptions are generalised across experiences and direct observations.
4. Conceptions require reorganisable pre-conceptions.
5. Conceptions make sense to communities by becoming shared.

In the framework of smart constructivist pedagogy, learners must have opportunities to announce their pre-conceptions, their presuppositions, their hypotheses based on their presuppositions, and their possible suggestions over them. Learners, as constructors of meanings, need to organise their experiences and, correspondingly, generalise and specialise the experiences into their personal hypotheses. Furthermore, mentors, adaptive teachers, and smart programs must be able to:

  i. work on conceptual and logical analysis of learners' hypotheses,
 ii. check the validity and definability of learners' hypotheses,
iii. find reasonable descriptions and specifications for denying and refusing the learners' hypotheses.

The third item could be done deductively based on rules or inductively based on different cases of study. In other words, in order to be disclaimed, learners' hypotheses must be illuminated and explored. Any kind of error, mistake, or inaccuracy would be assessed as an outcome of learners' misconceptions. The learners' misconceptions could be found and organised. Thus, their mistakes would be explored for themselves. Note that counterexamples are quite efficient in resolving learners' misconceptions and errors. It shall be concluded that smart constructivist mentoring focuses on:

  i. discovering conceptions/misconceptions of any individual learner,
 ii. discovering the common conceptions/misconceptions among a group of learners,
iii. conceptualising learners' conceptions/misconceptions,
 iv. conceptualising and attempting to understand learners' understandings over their conceptions/misconceptions, and
  v. motivating proper conceptions and resolving misconceptions.

It shall be stressed that smart constructivism could consider 'improvable and re-organisable conceptions of learners' as the main building blocks of its knowledge building pedagogy. [59] is in line with the conceptualised theory and has had a special focus on the learners' productive use of the principle of improving their conceptions within their relationships with their 'constructed knowledge'. At this point, we shall conclude that the presented conceptualisation of knowledge building has had a special attention to 're-organisable conceptions of learners within their connections with their collaborative constructed knowledge'. Table 13.1 is presented in order to itemise the most important components of Smart Constructivist Pedagogy and its significant characteristics. Later on, Fig. 13.5 schemes the conceptual interrelationships between those components and the phenomena of 'knowledge' and 'conception'.

Table 13.1 Main components of smart constructivist pedagogy

Components	Characteristics
Smart constructivist learning	• The phenomenon of 'learning' in the framework of smart constructivism is interpreted as a process of knowledge construction. The constructed knowledge is idiosyncratic • Smart constructivist learning is strongly concerned with self-regulation, auto-organisation, self-development and, finally, self-learning • Learning in the framework of smart constructivism is an active and dynamic (not passive) process • In the framework of smart constructivism, the constructed knowledge by any individual learner is not innate, passively absorbed, or invented, but it is 'constructed' and developable • In the framework of smart constructivism, learners interpret their world and, correspondingly, construct their own versions of the world based on their personal conceptions • The most significant objectives of smart constructivist learning are 'meaning construction' and 'meaningful understanding production' • Smart learning in the framework of constructivism proceeds toward developing constructed structures. • Experiences and prior understandings of learners play fundamental roles in smart constructivist learning • Smart constructivist learning encourages and motivates any individual learner to explore and discover the world by her/him self • Smart constructivist learning encourages any individual to make her/his own sense of the world • In the framework of smart constructivism, the phenomenon of 'learning' is situated in the context in which it occurs • Smart constructivist learning is strongly supported by social interactions and conversational exchanges
Smart constructivist mentoring (by human beings, adaptive mentors, smart programs)	• The phenomenon of 'mentoring' in the framework of smart constructivism is a process of knowledge construction. • Mentoring in the framework of smart constructivism is an active and dynamic (not passive) process • Smart constructivist mentoring conceptualises learners' conceptions of the world • In the framework of smart constructivism, the constructed knowledge by mentors is not innate, passively absorbed, or invented, but it is 'constructed' and developed by the mentor with regard to the learners' opinions, actions, transactions, questions, and answers • In smart constructivist learning systems, the mentor is an expert and advanced learner and has a special respect for learners' choices • In smart constructivist learning systems the mentor is an organiser around significant conceptions that could motivate learners • In smart constructivist learning systems, the mentor must get to know about any individual learner and her/his backgrounds • In smart constructivist learning systems, the mentor assists learners and links them with their background knowledge

(continued)

Table 13.1 (continued)

Components	Characteristics
	• In smart constructivist learning systems, the mentor mainly focuses on (i) constructing meanings for her/him self, (ii) giving feedbacks to learners with regard to their constructed meanings, and (iii) developing meaningful understandings for her/him self
	• In the framework of constructivism, smart mentoring conceptualises learners' understandings based on their conceptions of the world
	• In the framework of constructivism, smart mentoring builds a world of developed understandings
	• In the framework of constructivism, smart mentoring proceeds toward developing constructed knowledge structures
	• In the framework of smart constructivism, any learner must be driven by her/his mentor to understand the world and to change her/his understanding with regard to her/his misconceptions. In fact, smart mentoring discovers/recognises learners' misconceptions, mistakes, and errors
	• In the framework of constructivism, smart mentoring focuses on making senses. It's highly affected by the learners' senses of the world
	• In the framework of constructivism, smart mentoring is situated in the context in which the phenomenon of 'smart learning' occurs
	• In the framework of constructivism, smart mentoring is strongly supported by social interactions and conversational exchanges
	• In the framework of constructivism, an effective smart mentoring aims at presenting open-ended identifiable, describable, specifiable, justifiable, and analysable problems to learners

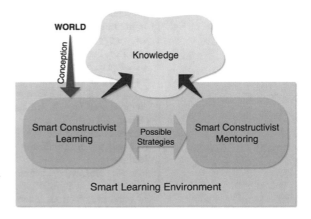

Fig. 13.5 The conceptual relationship between main components of smart constructivist pedagogy

13.8 Linking Smart Constructivism and Collaborative Learning Strategy: Verification of Research Outcomes

This section picks up the collaborative learning strategy that is highly relevant to smart education in order to focus on explaining its possible connections with 'the smart constructivist model of learning'. This section describes why 'collaborative learning strategy' could cope with and could be furnished by the presented and conceptualised approach.

The central focus of this research has been on knowledge building. This means we need to take into consideration the phenomenon of 'knowledge building' in order to check the validity and reliability of the constructivist model of learning in conjunction with 'collaborative learning strategy' and smart learning environments. First, it seems useful to take a look at Popperian epistemology [60] in order to work on conceptual analysis of knowledge building in smart learning. More specifically, the concept of 'knowledge building' could be derived from an epistemology that treats conceptions of human beings as entities in their own right that can have properties, connections, and potentialities. Consequently, it's quite important to be concerned with the concepts of 'pervasive knowledge building' and 'knowledge of community'. In fact, we need to focus on the fact that in collaborative learning, or 'Learning-and-Constructing-Together', the constructed knowledge must be capable of becoming spread widely throughout a group of learners. In the context of collaborative learning, any individual learner constructs her/his own knowledge. This means she/he attempts to construct the universal knowledge and also develop the construction of the knowledge of her/his community.

This section (i) relies on [61] and its conceptual analysis of the phenomenon of 'togetherness' in learning environments, (ii) follows the analysed policies of [62–64], and (iii) uses the methodological notions of [65], to focus on conceptualisation of 'Learning-and-Constructing-Together' while it's concerned with knowledge building within junctions of 'smart constructivism' and 'collaborative learning'.

13.8.1 Essential Value 1: The State of Knowledge

Creative knowledge work could be interpreted as a work that advances the state of knowledge of a community. The 'state of knowledge' is an emergent collective phenomenon and might be interpreted for a group of people. According to the concept of 'state of knowledge', knowledge building pedagogy is supported by the premise that authentic creative knowledge work can take place in any learning environment or in any smart learning environment. The state of knowledge of a group of learners within a smart learning environment only indirectly reflects the knowledge of individual learners. This conclusion could be implicated by smart constructivism. In fact, relying on smart constructivism, the state of knowledge of an individual learner, based on her/his constructed meanings, could highly reflect the knowledge of the community and, inversely, the state of knowledge of the community could only indirectly reflect the knowledge of the individual learners. Then, learners could re-organise and update their constructed meanings. Therefore, it is reasonable to expect that individuals' achievements go along with developments and advancements of community knowledge. This conclusion seems to be in parallel with the proposed approach of Zhang and colleagues in [66]. This characteristic, based on the state of knowledge, could highly affect course-by-course, program-by-program, and semester-by-semester changes in plans and strategies of any smart learning environment. Note that the mentor, the adaptive mentor, or the smart program, is another member of any learning community and, therefore, her/his/its constructed meanings reflect the knowledge of the community. In addition, it shall be considered that the mentor's knowledge is, regarding the feedbacks and transactions of learners, developable.

13.8.2 Essential Value 2: The Phenomenon of 'Discourse'

According to [64], discourse could come from sharing knowledge as well as subjecting conceptions to criticism. For example, in online meetings, web conferences, webinars, and Massive Open Online Courses (MOOC), any individual learner could become concerned with a kind of discourse which could be interpreted as 'a filter that determines what could be accepted into the canon of justified beliefs' [67]. However, it could be argued that modern learning strategies must support any individual learner and, also, any individual mentor, in playing her/his own creative roles in order:

i. to improve her/his own conceptions, and
ii. to judge and to make decisions more rationally beside her/his manners of criticism.

We shall claim that this kind of discourse-based judgement and decision-making is the consequence of any individual's, and, consequently, of a community's construction of factual and conceptual knowledge. It can be labelled as 'Social Constructivism in the Framework of Smart Constructivism'. Relying on practical and empirical approaches, this kind of social constructivism would be more concerned with shared goals of advancing understanding beyond what is currently interpreted and understood. In fact, the practices could support the processes of meaning construction. Consequently, the produced social meanings, in the context of interactions and conversational exchanges between individuals within a smart learning environment, could be updated and be more organised.

13.8.3   Essential Value 3: Authoritative Information and Their Reliability

Smart constructivism in collaborative learning supports learners in:

i. using their own authoritative information that is achieved based on their own experiences, explorations, studies, etc. and
ii. bringing other authoritative information (e.g., from other individuals, from e-books and e-references, from learning applications) as evidences of their own authoritative information.

The latter supports the development and reorganisation of all individuals' constructions based on received authoritative information from others within their social interactions. It shall be claimed that the interconnections between (i) and (ii) elaborate the 'state of knowledge of community' in the long term. Accordingly, the interrelationships between (a) and (b) increase the state of knowledge of the community:

a. A learner's constructions based on her/his own authoritative information.
b. A learner's development of her/his constructions with regard to others' authoritative information.

13.8.4   Essential Value 4: Explanation and Understanding

The Organisation for Economic Co-operation and Development (OECD) has emphasised the importance of conceptual understanding as a basis for creative knowledge work of all kinds: "Educated workers need a conceptual understanding of complex concepts, and the ability to work with them creatively to generate new ideas, new theories, new products, and new knowledge" [68]. It might be assumed that any individual learner has to understand appropriately in order to develop her/his own knowledge constructions. Similarly, as discussed earlier, learners' understandings are strongly supported by explanations. Accordingly, it must be stressed that the development of knowledge building in smart learning societies is highly related to the phenomena of 'explanation' and 'understanding'.

Smart constructivism, as a theory of learning, must support the conceptual understanding of learners in different communities and organisations. Special attention must be given to guiding, instructing, and mentoring any individual learner. Any learner in such a framework must be guided in order to construct her/his own meanings and to support her/his society with her/his constructed meanings. In addition, the smart constructivist theory of learning within collaborative strategies focuses on developing the communities' understandings. In our opinion, a proper strategy must follow the conceptual framework presented in Fig. 13.4. In addition to this, smart constructivism must focus on developing 'knowing HowNess combined with knowing WhyNess' as 'explanatorily coherent practical knowledge'. A similar principle for practical knowledge has been analysed in [69].

13.9 Smart Constructivist Learning Communities: Validation of Research Outcomes

According to [70], smart learning communities must be sensible, connectable, accessible, ubiquitous, sociable, sharable, and visible/augmented. We shall claim that our research has interconnections with the features of 'being connectable', 'accessibility', 'being sharable', and 'visibility'.

At this point, we shall draw your attention to Vygotsky's theory of social constructivism [71–73]. In our opinion, Vygotsky's ideas are quite helpful in conceptualising smart constructivist learning communities. Vygotsky's theory, based on his ideas in human cultural and biosocial development, has supported the development of social constructivism. Vygotsky believed that 'social interaction' plays a fundamental role in the process of humans' cognitive development. In his opinion, an individual with stronger understandings and higher abilities in particular domains could be a so-called 'teacher'. He specified the concept of 'teacher' by defining the notion as an MKO (i.e., More Knowledgable Other). Additionally, Vygotsky defined ZPD (i.e., the Zone of Proximal Development) in order to express the concept of 'learning' by an individual learner under MKO's supervisions and/or in her/his collaborations with other individuals. Vygotsky believed that learners could learn in this zone. It shall, therefore, be concluded that we can have a similar conception of smart learning communities. In fact:

i. A mentor, an adaptive mentor, or a smart program is considered more knowledgable as an individual due to their stronger understandings and higher abilities in particular domains. They supervise learners.

ii. Learners have interactions and conversational exchanges with each other and develop their personal constructions of knowledge.

iii. The phenomenon of 'smart learning' occurs over actions, transactions, questions, and answers between any learner and mentor as well as between any learner and other learners.

13.9.1 Conceptualising Smart Constructivist Learning Communities

The fundamental characteristics of smart constructivist learning communities are as follows:

- Smart learning communities are communities using a discourse engaged in activity, reflection, interaction, and conversation.
- The main goals of any smart learning community are (i) Learning-and-Constructing-Together and (ii) producing the Collaborative Understanding.
- The main belief of any smart learning community is that the phenomena of 'smart learning' and 'development' are integrally tied to any individual's communicative and social interactions with other individuals.
- The second important belief of smart learning communities is that the use of information technologies (IT) and information communication technologies (ICT) is more likely to create a constructivist perspective towards the phenomenon of 'smart learning'.
- Smart learning communities must be given senses by (i) learners' made senses of the world based on their own experiences, explorations, and discovered key concepts and by (ii) their shared conceptions of the world.
- In the context of smart learning communities, any individual learner must be permitted to express, explain, defend, prove, and justify her/his conceptions of the world. Subsequently, all learners must be allowed to communicate their conceptions to each other as well as to their smart learning community.
- Smart learning communities must involve instructed interactions that guide any individual learner to recognise and resolve her/his conceptual inconsistencies and to modify conceptions through her/his interactions and conversational exchanges.
- In the context of smart learning communities, both interactions and conversational exchanges between two agents support bi-directional meaning constructions and collaborative understanding developments.
- In the context of smart learning communities, any constructed knowledge by an individual learner supports collaborative knowledge construction.

13.9.2 Smart Constructivist Learning Communities and Knowledge Building Technologies

In the context of smart constructivist learning communities, any conception is a building block of a knowledge construction. Any conception of an individual learner must be connected to and related to all others' conceptions. For example, any conception of a learner could be expressed in the form of her/his notes, paintings, sound clips, video clips, etc. Accordingly, the conceptions can be recorded and archived in the digital library of the relevant smart learning environment. Therefore, the smart learning environment must record a huge collection of conceptions. These could be represented by, e.g., data models, conceptual models, graphical models, statistical models, and concept maps. This can be seen in Fig. 13.6.

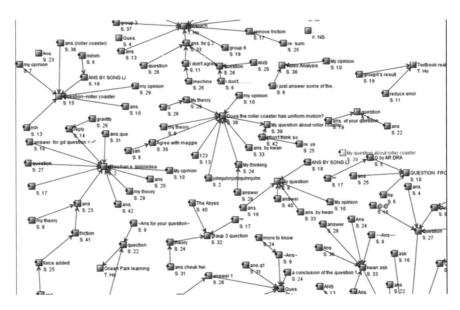

Fig. 13.6 Knowledge building view in knowledge building communities

Consequently, any conception would be viewable in multiple views as well as from different perspectives. For example, John's conception could be viewed from the perspective of Bob's and Mary's conceptions or from the perspective of their mentor's conception. In addition, there could be different possible interpretations for any linkage between two conceptions. These all could be recorded in the digital library. For example, Elizabeth may have an interpretation of John's conception, but she has observed and conceptualised John's conception from the perspective of Mary's conception. Accordingly, Elizabeth's interpretation, over the arc/line between John's and Mary's conceptions, could produce a new conception that could be recorded in the digital library.

The 'Knowledge Forum' [59, 62, 64, 74] is a proper knowledge-building environment. This multimedia knowledge-building environment could be recognised as a kind of smart learning environment. Such a smart learning environment focuses mainly on knowledge building. Knowledge Forum becomes organised by all of its users. All users are the constructors and developers of a huge collaborative knowledge construction. It might be assumed that such an environment can be an appropriate developable environment for 'knowledge building within smart constructivist learning communities'. Such a smart system can represent the advancing knowledge of any individual and of any community.

It's undeniable that smart learning communities are dependent on discourses engaged in activity, reflection, and interaction. We cannot deny that the most important objective of a modern learning community like Knowledge Forum is Learning-and-Constructing-Together. It must be taken into consideration that a smart constructivist learning community believes that 'smart constructivist learning' and 'knowledge development' are both integrally dependent on any individual's interactions and collaborations with other agents. Furthermore, we cannot ignore the importance of Collaborative Meaning Construction and Understanding Production in smart constructivist learning communities.

13.10 Discussion and Concluding Remarks

This research is focused on the area of Smart Learning Environments. Our theory has been presented (i) based on traditional constructivist theory of learning and (ii) by considering new requirements of learners in the digital age. It has—with special focus on 'constructivist epistemology' and 'constructivist models of knowing'—conceptualised Smart Constructivist Learning Systems.

In this research, knowledge acquisition has been recognised as the process of seeking knowledge [by human beings] about different phenomena, objects, processes and events with regard to their personal background knowledge. The concepts of 'knowledge building' and 'understanding production' have been the most sensitive terms in this article. More clearly, our theoretical model deals with (i) how knowledge may reasonably be assumed to be built by an individual, and with (ii) how her/his meaningful understandings could be assumed to be produced. The constructivist theory of smart learning, and, respectively, the smart constructivist theory of learning is a modern learning theory that is conceptualised over the phenomenon of 'smartness'. What we have offered has been a 'conceptual, logical and epistemological description' which has justified the importance of Smart Constructivist Knowledge Building Strategies. More specifically, this research has presented a specification of conceptualisation of:

a. smart constructivism,
b. smart constructivist learning,
c. meaning construction and understanding production in the framework of smart constructivism,
d. knowledge building in the framework of smart constructivism,
e. smart constructivist collaborative learning,
f. smart constructivist learning communities,
g. smart knowledge building environments, and
h. collaborative meaning construction and understanding production in the framework of smart constructivism.

As for the structural characteristics of smart constructivism, knowledge construction is conceived of as a type of active process, and it can be informally described in terms of personal understanding in multiple actions. Furthermore, it has been theorised that the phenomenon of 'understanding' could be valid and meaningful based on learners' [and mentors'] constructed meanings in the framework of constructivism and in the context of smart learning environments.

Accordingly, the concept of 'knowledge building' is interpreted as the consequence of 'meaning construction', 'understanding production' and 'sense making' by any individual learner.

Subsequently, this chapter has worked on designing a conceptual (and logical) framework for analysing knowledge building in the framework of smart constructivism and over the flow of learners' understandings. Considering that framework, we have identified the most significant characteristics of a smart constructivist pedagogy. It has been assumed that the conceptualised theory must be able to support other learning/mentoring strategies as the products of the phenomenon of 'smartness'. Accordingly, we have—relying on the characterised concept of 'smart learning communities'—picked up the 'collaborative learning strategy' and worked on checking the validity of Learning-and-Constructing-Together (as a model of learning) within smart learning communities. Subsequently, the most fundamental characteristics of knowledge building within smart learning communities are conceptualised. We shall claim that smart constructivism—besides Learning-and-Constructing-Together—could support some strategies like, e.g., Learning-and-Constructing-by-Doing and Learner-based programs of study with variable structures adaptable to types of learners.

We strongly believe that the theory of smart constructivism, and, subsequently, the constructivist model of learning within smart learning environments can support subsequent developments of smart learning strategies. This theory could support renewed qualitative developments of knowledge building and understanding production within smart learning environments.

References

1. Chi, M.T.H.: Conceptual Change within and across ontological categories: examples from learning and discovery in science. In: Giere, R.N. (ed.) Cognitive Models of Science, vol. 15, Minnesota studies in the Philosophy of Science, Minneapolis, MN: University of Minnesota Press, pp. 129–186 (1992)
2. Limon, M.: Conceptual Change in history. In: Limon, M., Mason, L. (eds.) Reconsidering Conceptual Change: Issues in Theory and Practice, pp. 259–289. Kluwer, Dordrecht (2002)
3. Watkins, C., Carnell, E., Lodge, C., Wagner, P., Whalley, C.: Effective Learning, National School Improvement Network (2002)
4. Badie, F.: Concept representation analysis in the context of human-machine interactions. In Proceedings of the 14th International Conference on e-Society (pp. 55–62), International Association for Development of the Information Society (IADIS), Algarve, Portugal (2016a)
5. Badie, F.: Towards concept understanding relying on conceptualisation in constructivist learning. In Proceedings of the 13th International Conference on Cognition and Exploratory Learning in Digital Age, pp. 292–296, International Association for Development of the Information Society (IADIS), Mannheim, Germany (2016b)
6. Piaget, J.: Origins of Intelligence in the Child. Routledge & Kegan Paul, London (1936)
7. Piaget, J., Cook, M.T.: The Origins of Intelligence in Children. International University Press, New York, NY (1952)
8. Badie, F.: A conceptual framework for knowledge creation based on constructed meanings within mentor-learner conversations. In: Smart Education and e-Learning 2016, Springer

International Publishing. Volume 59 of the series Smart Innovation, Systems and Technologies, pp. 167–177 (2016c)

9. Uskov, L.V., Howlett, J.R., Jain, C.L. (ed.) Smart Education and e-Learning 2016. Springer International Publishing (2016)

10. Uskov, V.L., Bakken, J.P., Pandey, A., Singh, U., Yalamanchili, M., Penumatsa, A.: Smart University Taxonomy: Features, Components, Systems, Smart Education and e-Learning 2016. Springer International Publishing (2016)

11. Badie, F.: A semantic basis for meaning construction in constructivist interactions. In: Proceedings of the 12th International Conference on Cognition and Exploratory Learning in Digital Age, pp. 369–376. International Association for Development of the Information Society (IADIS), Greater Dublin, Ireland (2015a)

12. Badie, F.: Towards a semantics-based framework for meaning construction in constructivist interactions. In: Proceedings of the 8th International Conference of Education, Research and Innovation, pp. 7995–8002. International Association of Technology, Education and Development (IATED), Seville, Spain (2015b)

13. Badie, F.: Towards semantic analysis of mentoring-learning relationships within constructivist interactions. In: Emerging Technologies for Education. Springer International Publishing. Springer Lecture Notes in Computer Science. Proceedings of International Symposium on Emerging Technologies for Education, Rome, Italy (2017a)

14. von Foerster, H.: Understanding Understanding. Essays on Cybernetics and Cognition. Springer, New York (2003)

15. Peschl, M.F., Riegler, A.: Does Representation Need Reality? Rethinking Epistemological Issues in the Light of Recent Developments and Concepts in Cognitive Science, Understanding Representation in the Cognitive Sciences. Springer US, pp. 9–17 (1999)

16. Webb, J.: Understanding Representation. SAGE Publications (2009)

17. Chaitin, G.J.: Algorithmic Information Theory. Cambridge University Press (1987)

18. Kintsch, W., Welsch, D., Schmalhofer, F., Zimny, S.: Sentence memory: a theoretical analysis. J. Memory Lang. Elsevier (1990)

19. Di Pellegrino, G., Fadiga, L., Fogassi, L., Gallese, V., Rizzolatti, G.: Understanding motor events. a neurophysiological study. Exp Brain Res **91**, 176–180 (1992)

20. MacKay, D.: Information Theory. Inference and Learning Algorithms. Cambridge University Press, Cambridge (2003)

21. Zwaan, R.A., Taylor, L.J.: Seeing, language comprehension. J. Exp. Psychol. Gen. **135**(1), 1–11 (2006)

22. Uithol, Sebo, van Rooij, Iris, Bekkering, Harold, Haselager, Pim: Understanding motor resonance. Social Neurosci. Routledge **6**(4), 388–397 (2011)

23. Uithol, Sebo, Paulus, Markus: What do infants understand of others' action? A theoretical account of early social cognition. Psychol. Res. **78**(5), 609–622 (2014)

24. Grosholz, Emily: Herbert Breger. The Growth of Mathematical Knowledge, Springer Science and Business Media (2013)

25. Marzano, R.J.: Building Background Knowledge for Academic Achievement: Research on What Works in Schools, Alexandria, VA 22311-1714 (2004)

26. Fisher, D., Frey, N.: Background Knowledge: The Missing Piece of the Comprehension Puzzle, Portsmouth. Heinemann, NH (2009)

27. Säljö, R.: Learning in the Learner's Perspective: Some Commonplace Misconceptions. Reports from the Institute of Education, University of Gothenburg (1979)

28. Van Rossum, E.J., Schenk, S. M.: The relationship between learning conception, study strategy and learning outcome. Brit. J. Educ. Psychol. (1984)

29. Van Rossum, E.J., Rebecca, H.: The Meaning of Learning and Knowing, Sense Publishers, The Netherlands (2010)

30. Boyd, G.M.: Conversation Theory. In: Jonassen, D.H. (ed.) Handbook of Research on Educational Communications and Technology, 2nd edn, pp. 179–197. Lawrence Erlbaum, Mahwah, NJ (2004)

31. Pask, G.: Developments in Conversation Theory (part 1), Int. J. Man-Mach. Stud. Elsevier Publishers (1980)
32. Laurillard, D.M.: Rethinking University Teaching: A Framework for the Effective Use of Educational Technology. Routledge, London (1993)
33. Laurillard, D.: Rethinking University Teaching, A Conversational Framework for the Effective Use of Learning Technologies. Routledge, London (2002)
34. Hwang, G.J.: Definition, Framework and Research issues of Smart Learning Environments— A Context-Aware Ubiquitous Learning Perspective. Smart Learn. Environ. Springer Open J. 1, 4 (2014)
35. Spector, J.M.: Smart Learn. Environ. Conceptualizing the Emerging Field of Smart Learning Environments, Springer, Berlin Heidelberg (2014)
36. Pratt, D.D.: Conceptions of Teaching. Adult Educ. Q. **42**(4), 203–220 (1992)
37. Hans, Götzsche: Deviational Syntactic Structures. Bloomsbury Academic, London/New Delhi/ New York/ Sydney (2013)
38. Coccoli, M., Guercio, A., Maresca, P., Stanganelli, L.: Smarter Universities: a vision for the fast changing digital era. J. Vis. Lang. Comput. **25**, 1003–1011 (2014)
39. Bates, T., Spector, M., David Merrill, M. (eds.): Special issue: Effective, efficient and engaging (E3) learning in the digital age (2008)
40. Merrill, M.D.: First Principles of Instruction: Identifying and Designing Effective. Efficient and Engaging Instruction. Wiley, San Francisco, CA (2013)
41. IBM: Smart Education, http://www.ibm.com/smarterplanet/us/en/
42. Biggs, John B., Collis, Kevin F.: Evaluating the Quality of Learning: The SOLO Taxonomy. Structure of the Observed Learning Outcome). Academic Press, New York (2014)
43. Chierchia, G.: Dynamics of Meaning: Anaphora, Presupposition, and the Theory of Grammar. University of Chicago Press (2009)
44. Gabbay, D.M., Guenthner, F.: Handbook of Philosophical Logic, vol. 15. Springer Science & Business Media (2010)
45. Larsson, S.: Formal Semantics for Perception. Workshop on Language, Action and Perception (APL), Center for Language Technology, Gothenburg, Link: http://clt.gu.se/dialogue-technology-lab/sltc2012-apl (2012)
46. Simpson, J.A., Weiner, E.S.C.: The Oxford English Dictionary. Oxford University Press (1989)
47. Bloom, B.S., Engelhart, M.D., Furst, E.J., Hill, W.H., Krathwohl, D.R.: Taxonomy of Educational Objectives: The Classification of Educational Goals, Handbook I: Cognitive Domain. David McKay Company, New York (1956)
48. Krathwohl David, R.: A Revision of Bloom's Taxonomy: An Overview, Theory into Practice. Routledge Publishers (2002)
49. Cambridge Dictionary: http://dictionary.cambridge.org/dictionary/english (2017)
50. Badie, F.: A conceptual mirror: towards a reflectional symmetrical relation between mentor and learner. Int. J. Inf. Educ. Technol.: IJIET 2017 **7**(3), 199–203. ISSN: 2010-3689 (Proceedings of the 3rd International Conference on Education and Psychological Sciences (in 2016), Florence, Italy (2017b))
51. Badie, F.: A semantic representation of adult learners' developing conceptions of self realisation through learning process. In: Proceedings of the 10th Annual International Technology, Education and Development Conference, pp. 5348–5353. International Association of Technology, Education and Development (IATED), Valencia, Spain (2016b)
52. von Glasersfeld, E.: A constructivist approach to teaching, in constructivism in education. In: Steffe, L.P., Gale, J., Hillsdale, N.J. (eds.) Lawrence Erlbaum Associates, pp. 3–15 (1995)
53. Wasson, B.: Instructional Planning and contemporary theories of learning: is this a self-contradiction? In: Brna, P., Paiva, A, Self, J. (eds.) Proceedings of the European Conference on Artificial Intelligence in Education, pp. 23–30. Colibri, Lisbon (1996)
54. Fosnot, C.T.: Constructivism: A Psychological Theory of Learning. In: Fosnot, C.T. (ed.) Constructivism: Theory, Perspectives and Practice, pp. 8–33. Teachers College Press, New York (1996)

55. Boethel, M., Dimock, K.V.: Constructing Knowledge with Technology, Austin. Southwest Edu ca tional Development Laboratory, Texas (2000)
56. Fox, R.: Constructivism examined. Oxford Rev. Educ. **27**(1), 23–35 (2001)
57. Maclellan, E., Soden, R.: The importance of epistemic cognition in student-centered learning. Instr. Sci. **32**, 253–268 (2004)
58. Yilmez, K.: Constructivism: its theoretical underpinnings, variations, and implications for classroom instruction. Educ. Horizons **83**(3), 161–172 (2008)
59. Caswell, B., Bielaczyc, K.: Knowledge forum: altering the relationship between students and scientific knowledge. Educ. Commun. Inf. **1**, 281–305 (2001)
60. Popper, K.R.: Objective Knowledge: An Evolutionary Approach. Clarendon Press, Oxford (1972)
61. Baker, M., Andriessen, J., Järvelä, S. (eds.): Affective Learning Together: Social and Emotional Dimensions of Collaborative Learning. Routledge, London (2013)
62. Scardamalia, M., Bereiter, C., Lamon, M.: The CSILE project: Trying to bring the classroom into World 3. In: McGilley, K. (ed.) Classroom lessons: Integrating cognitive theory and classroom practice, pp. 201–228. MIT Press, Cambridge, MA (1994)
63. Bereiter, C., Scardamalia, M.: Theory building and the pursuit of understanding in history, social studies, and literature. In: Kirby, J.R., Lawson, M.J. (eds.) Enhancing the Quality of Learning: Dispositions, Instruction, and Learning Processes, pp. 160–177. Cambridge University Press, New York (2012)
64. Scardamalia, M., Bereiter, C.: Knowledge building: theory, pedagogy, and technology. In: Sawyer, K. (ed.) Cambridge Handbook of the Learning Sciences, pp. 397–417. Cambridge University Press, New York (2014)
65. Sorensen, E.K.: Networked e-Learning and Collaborative Knowledge Building: Design and Facilitation. Contemp. Issues Technol. Teach. Educ. **4**(4), 446–455 (2005)
66. Zhang, J., Scardamalia, M., Reeve, R., Messina, R.: Designs for collective cognitive responsibility in knowledge building communities. J. Learn. Sci. **18**(1), 7–44 (2009)
67. Latour, B., Woolgar, S.: Laboratory life: The Social Construction of Scientific Facts. Sage Publications, Beverly Hills, CA (1979)
68. Organization for Economic Co-operation and Development (OECD): 21st Century Learning: Research, Innovation and Policy. OECD, Paris (2008)
69. Bereiter, C.: Principled practical knowledge: not a bridge but a ladder. J. Learn. Sci. **23**(1), 4–17 (2014)
70. Adamko, A., Kadek, T., Kosa, M.: Intelligent and adaptive services for a smart campus visions, concepts and applications. In: Proceedings of the 5th IEEE International Conference on Cognitive Infocommunications, 5–7 Nov 2014, Vietri sul Mare, Italy. IEEE (2014)
71. Vygotsky, Lev S.: Interaction between learning and development. Read. Dev. Child. **23**(3), 34–41 (1978)
72. Vygotsky, L.S.: Mind in Society: Development of Higher Psychological Processes (1978)
73. Vygotsky, L.S.: Collected Works of L. S. Vygotsky, Vvol. 1: Problems of General Psychology (trans: Minick, N.). Plenum, New York (1987)
74. Scardamalia, M.: CSILE/Knowledge Forum®. In: Kovalchick, A., Dawson, K. (eds.) Education and technology: An Encyclopedia, pp. 183–192. Santa Barbara, ABC-CLIO (2004)

Author Index

Printed in the United States
By Bookmasters